## ACLS History E-Book Project

Reprint Series

The ACLS History E-Book Project (www.historyebook.org) collaborates with constituent societies of the American Council of Learned Societies, publishers, librarians and historians to create an electronic collection of works of high quality in the field of history. This volume is produced from digital images created for the Project by the Scholarly Publishing Office and the Digital Library Production Service at the University of Michigan, Ann Arbor. The digital reformatting process results in an electronic version of the text that can be both accessed online and used to create new print copies. This book and hundreds of others are available online in the History E-Book Project through subscription.

Many of the works in the History E-Book Project are available in print and can be ordered either directly from their publishers or as part of this series. For information refer to the online Title Record page for each book. Inquiries regarding this series can be directed to info@hebook.org.

ACLS
HISTORY E-BOOK

http://www.historyebook.org

LIVERPOOL STUDIES IN GEOGRAPHY

GENERAL EDITOR: H. C. DARBY

# MAPS AND POLITICS

A REVIEW OF THE ETHNOGRAPHIC CARTOGRAPHY OF MACEDONIA

Musée Ethnographique de Beograd

MACEDONIAN PEASANTS FROM KUMANOVO, *c.* 1870

# MAPS AND POLITICS

## A REVIEW OF THE ETHNOGRAPHIC CARTOGRAPHY OF MACEDONIA

BY

H. R. WILKINSON, M.A.

*Lecturer in Geography*
*in the University of Liverpool*

LIVERPOOL
AT THE UNIVERSITY PRESS
1951

*Printed in Monotype Aldine Bembo 11 on 13 point by C Tinling & Co Ltd of Liverpool London and Prescot for the University Press of Liverpool and published 1951 by the University Press of Liverpool*

# PREFACE

A DEFINITION of Macedonia in William Smith's *Dictionary of Greek and Roman Geography* contains the observation that "it was a small country *with a peculiar population*." The reference is to the Macedonia of antiquity the population of which, according to Pliny, was made up of "no less than one hundred and fifty nations." Pomponius Mela was also impressed by the number of nations seated within Macedonia, "as many as there were cities."

The modern Macedonians have this much in common with those of antiquity—they too have affinities with many, if different, nations. But it was not until the beginning of the nineteenth century that the modern ethnographic diversity of the Macedonian population aroused any widespread interest. When that interest did mature, fostered by political considerations, a conflict of opinion on the ethnographic affinities of the Macedonians developed which became the classic controversy of its kind in the world. Political geographers have much to learn from that remarkable controversy. The many ideas on the ethnographic structure of Macedonia which sprang from it are recorded in a variety of ethnographic maps. These maps are valuable sources for the study of the formative phase in the modern political geography of Macedonia and its adjoining territories. They are historical documents which lend themselves readily to geographical analysis, and they incorporate vital evidence about the origins and growth of the ethnographic dispute which has always been at the heart of the Macedonian problem.

In this book, the author has investigated the evidence recorded in these maps with two aims in view. The first has been to demonstrate the origins and development of the Macedonian problem by arranging and presenting the evidence to the best possible advantage with the aid of cartographic devices. The second has been to use the Macedonian maps to exemplify the characteristics of ethnographic maps in general, the merits and the limitations of which do not appear to be generally appreciated.

The author has been well aware, because of the delicate political issues involved, of the necessity to observe at all times a strict impartiality of outlook. He has never entertained any idea of justifying any one of the variety of claims put forward by the many rival parties which have contested this region in the past, or which may have interests there at present. Nor has there been in the author's mind any question of urging boundary revisions in the Macedonian region.

The illustrations in the text are based on original ethnographic maps. They are, however, interpretations rather than facsimile reproductions of such maps. In order to facilitate comparison, it has been necessary to reduce all maps to a similar scale, and to standardize methods of representation.

For the purpose of easy identification and reference, the name of each author, his professed nationality, and the date of publication are given in the case of each ethnographic map. Where the author is unknown some simplified reference has been used instead, e.g., League of Nations' Map, 1926. For the sake of clarity, and because it is not difficult to fix places by reference to coastlines, lakes and towns, graticules have been omitted from the maps. Unless otherwise indicated, north points are understood to be at the top of each map in the conventional manner.

The sources of all maps are fully acknowledged in the bibliography. This takes the form of a list of the original titles of maps to which reference is made in the text, or which the author has at some time consulted. The list is arranged in chronological order according to date of publication. Sizes of maps and scales are given where available.

The place-names in this book are based as closely as possible on the spellings in use in the country where the places are situated. A note on the source of these spellings has been incorporated elsewhere. Place-name policy is, however, only as rigid as common sense dictates.

It would be manifestly absurd, for example, to dispense with the familiar anglicized form of the place-name, Macedonia, and to replace it by Makedonia (Serbian), or Makedhonía (Greek).

In the spelling of ethnographic names, anglicized forms are preferred wherever possible, e.g., Albanian rather than Shquiptar, Greek rather than Hellene. The name ' Bulgar ' is not used to describe the modern Bulgarian because of the historically, non-Slav connotation attached to it. The original Bulgars were not Slavs whereas the modern Bulgarians are. The name ' Serb ' is retained but those Serbs living within the political boundaries of Serbia are referred to as ' Serbians.' The name *Macedo-Slav* is italicized throughout because of its association with J. Cvijić. It is a better descriptive term than ' Macedonian ' which is often used to refer to all nationals of the Jugoslav Republic of Macedonia, including Slav, Albanian and Turkish elements.

## ACKNOWLEDGEMENTS

I WARMLY acknowledge all the help I have received from many sides in the writing and production of this book. The idea of writing a monograph on the ethnographic maps of Macedonia was the outcome of some discussion with Professor H. C. Darby in the autumn of 1946. For Professor Darby's part in the inception of the work, and for his subsequent help and encouragement at all stages in its preparation, I am deeply indebted. Professor W. G. East was kind enough to read the original manuscript and to offer valuable suggestions and amendments for which I am very grateful. My thanks are due also to my colleagues in the Department of Geography at Liverpool who have always been ready to proffer advice and useful criticism and to help in the reading of proofs. My wife has shared in the tasks of compilation, proof-reading and indexing and has given me much helpful criticism in the arrangement of material.

Gifts of maps, and pertinent information bearing on Greek Macedonia have been made to me from time to time by Mr A. A. Pallis, Director of the Greek Department of Information in London. Mr Pallis has always been willing to place at my disposal his own great knowledge of Macedonian affairs. I am under an obligation to many Jugoslav geographers of the Jugoslav

Institutes of Geography at Belgrade, Zagreb, Skoplje and Ljubljana, for contributions of maps and for general information on Jugoslav Macedonia. I should like to express my thanks to the map curators of the British Museum Map Room and of the Bodleian Library for their kindness in helping me to locate certain maps.

The maps were redrawn for publication by Mr A. G. Hodgkiss with his customary skill and patience. Only after much experiment was a formula devised to interpret coloured maps in black and white. Mr D. H. Birch drew some of the supplementary maps which depict changes of boundary.

The University of Liverpool generously assisted me with grants both in aid of the photographic copying of maps, and of the carrying out of field-work in Macedonia. I received much hospitality in Macedonia and my stay there was extended at the kind invitation of the Ministry of Science and Culture at Belgrade.

Mr R. A. Downie, Secretary of the University Press of Liverpool, has throughout given help and advice on the technical matters of production and I am pleased to conclude these acknowledgements on a note of thanks to him.

H.R.W.

*Liverpool,*
1951

## CONTENTS

# LIST OF MAPS AND DIAGRAMS

# A NOTE ON PLACE-NAMES

*Jugoslavia*

The recognition by the Government of Jugoslavia of an increased number of official languages in 1945 has correspondingly increased the variety of current forms of place-names in southern Jugoslavia. Macedonian is now the official language of the Jugoslav Republic of Macedonia, and Albanian of the Autonomous Region of Kosovo-Metohja, but so far no revised topographical maps have been issued incorporating Macedonian and Albanian place-names. The difference between Serbo-Croat and Macedonian forms is not always great, e.g. in the case of Skoplje (Skopje), but it is often of sufficient magnitude to render inconsistent the employment of Serbo-Croat forms. In the absence of data referring to the modification of place-names as a result of political re-organization, however, the principle laid down in the P.C.G.N. policy based on the pre-war official Jugoslav languages, has had to be accepted for the purpose of this book. In practice this means the spelling of place-names in the Serbo-Croat form as they appear on the G.S.G.S. 1 : 100,000 map-series covering Jugoslav Macedonia.

Exceptions, however, have been made in this usage :

(*a*) No attempt has been made to modify spellings of names already familiar to English readers, e.g., Old Serbia, Macedonia, etc.

(*b*) In the case of lakes and rivers, and other land features geographical terms have been translated, e.g., Lake Ohrid (Ohridsko Jezero).

(*c*) In historical contexts it has been necessary occasionally to employ historical forms, e.g., the Vilayet of Uskub (Skoplje).

*Greece*

The principle adopted by the P.C.G.N. in 1941 for Greek nomenclature has been followed. Nearly all the place-names employed are to be found in *A Gazetteer of Greece* (London, 1942). Unfortunately the use of latinized forms, widespread in western Europe, renders many important places scarcely recognizable in their Greek forms and consequently concessions have been made to convention in certain cases, e.g., Macedonia itself (Makedhonía), and Salonika (Thessaloníki).

It has been necessary also to resort to convention in two other instances—for the sake of historical accuracy, and in order to avoid a plurality of forms for places common to two or more countries. In the latter case one form has been retained throughout, e.g., Bulgarian, Rodopi mountains (Greek form, Rodhópi), Bulgarian, Mesta river (Greek form, Néstos).

As a guide to the pronunciation of Greek names, in words of more than one syllable primary stress has been shown by an acute accent on the last vowel-letter of the stress syllable, unless it happens to be a capital letter.

*Other countries*

As far as possible P.C.G.N. policy has been employed in the case of Turkey, Albania and Bulgaria, with the same reservations as those stated above for Jugoslavia and Greece. It has been thought advisable to keep to convention in the case of such well-known general place-names as the Balkans, Thrace, Anatolia, the Danube and the Aegean sea.

CHAPTER I

# INTRODUCTORY

MACEDONIA defies definition for a number of reasons. Hardly two authorities can be found to agree on its exact delineation, although many agree on its general location. The name itself is the Latin form of an ancient Greek place-name, one of many which has persisted right down to our own day, for describing this part of Europe. This persistence has largely been due to the effects of the Turkish conquest. Ignorance of Balkan languages, difficulties of transliteration, lack of topographical survey, all combined to restrict the use of contemporary place-names so that the opening of the nineteenth century still found western European scholars thinking of Balkan geography in terms of Ptolemy and Strabo. Macedonia to them meant the Roman province, framed by a natural boundary of mountains marching with geometrical precision on all sides—the *Pindhus*, the *Scardus*, the *Rhodope*.[1] These somewhat oversimplified interpretations may be the root-cause of the modern tendency to define Macedonia as a 'natural region,' a tendency apparent in the works not only of Bulgarian, but of British, American, French, Italian and German geographers.[2]

[1] For good examples of the persistence of this concept see (1) H. Dufour, *Atlas Universel* (Paris, 1860), (2) Philip's *Classical Atlas* (*n.d.*).

[2] See such views expressed in (1) " The Balkan States," *Peace Handbook*, Vol. IV, Historical section of the Foreign Office (1920), (2) D. M. Brancoff, *La Macédoine et sa population chrétienne* (Paris, 1905), (3) D. Jaranoff, *Carte de la Macédoine dans ses limites géographiques, à l'échelle de* 1:3,000,000 (Sofia, 1933), (4) Vaughan Cornish, *Borderlands of Language in Europe* (London, 1936), (5) H. N. Brailsford, *Macedonia: its races and their future* (London, 1906), (6) Leon Dominian, *The Frontiers of Language and Nationality in Europe* (New York, 1917), (7) E. Driault, *Histoire Diplomatique de la Grèce* (Paris, 1926), (8) L. Schultze-Jena, *Makedonien, Landschafts-und Kulturbilder* (Jena, 1927).

The use of the terms 'natural' boundaries and 'natural' region in connection with these definitions is highly misleading in view of the diversity of structure, relief and climate found within the prescribed region, as well as the debatable nature of the so-called limits. On examination of a detailed orographical map the natural framework is found to consist of complicated systems of mountain ranges which achieve only here and there a lineal form, and within which prevails a diversity of land-forms and a complex drainage system. Of

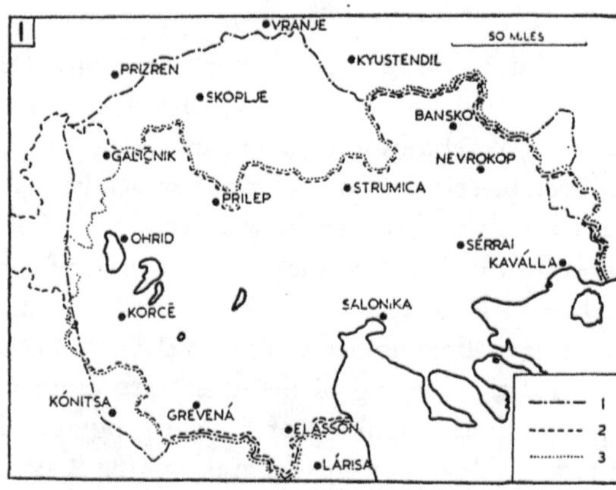


FIG. 1. SOME DEFINITIONS OF MACEDONIA

The references in the key are as follows: **1.** Greek, C. Nicolaides, 1899; **2.** Anonymous, *Carte des Ecoles Chrétiennes de la Macédoine*, 1905; **3.** A Greek definition given by D. M. Brancoff, 1905.

all the attempts to define Macedonia, that which makes its appeal to physical geography is the least profitable, and also the easiest to refute.

Perhaps equally facile are the attempts to define Macedonia, historically, by invoking its past political boundaries. V. Colocotronis, the Greek historian, for example, devoted four hundred pages and a score of facsimile maps to demonstrate, inconclusively, that the northern boundary of classical Macedonia roughly approximated to that of modern Greece.[1] He is not the least offender. But historians less concerned with the propagandist value of their researches have revealed

[1] *La Macédoine et L'Hellenismé: Etude historique et ethnographique* (Paris, 1919).

the traditional, ephemeral character of Macedonian political boundaries. From being reduced to the confines of Salonika at certain periods of its classical history, Macedonia has at other times reached the Adriatic, the *Haemus* and the Danube. No useful purpose therefore is achieved by insisting on the hypothetical stability of the historical boundaries of Macedonia. It is noteworthy that the Osmanli Turks, who inherited so much from their Byzantine precursors, never recognized any Macedonian administrative unit. It is not unreasonable to conclude that history no more sets its seal upon the boundaries of Macedonia than does physical geography.

Some of the conflicting views on the extent of Macedonia have been summarized in Figs. 1 & 2. There is a lack of agreement manifest in the definitions given in these sketch-maps and therefore to avoid confusion it is important to intimate the definition of Macedonia that has been adopted in this monograph. The name is used throughout the text as a convenient means of referring to the region which lies between the Sar mountains in the north, the Aegean sea in the south, the lower Mesta river and the Rodopi mountains in the east, and the Albanian highlands in the west (Fig. 3). This region is distinctive not on account of any physical unity or common political experiences but rather on account of the complexity of the ethnic structure of its population. It is a zone where the Albanian, Greek, Bulgarian and Serbian linguistic provinces meet and overlap, and where in addition exclaves of Romanian and Turkish speech are found ; it is a region where the concept of national sentiment, associated with language, exists side by side with the perhaps older concept of community based on religious affinity ; it is a region where many influences, economic, cultural and political, emanating from different parts of Europe, Asia and Africa, meet and mingle but where the process of fusion has not always taken place. The traditional incapacity of this region to absorb and to transform might be correlated with its physical diversity and with its function as a through-route between the new world north of the Alps and the old world of the eastern Mediterranean and the Near East.

Its role as a routeway is the second characteristic feature of Macedonia. This role partly explains its ethnic complexity but it has had other consequences also in that it has made Macedonia a name synonymous with strategy. It was from here that Alexander the Great led his armies into Persia and as far afield as the plains of India. It was by clinging to Macedonia that the Byzantine Emperors contrived to maintain their influence in the Balkan peninsula. It was on Macedonia that the mediaeval empires of Bulgaria, Wallachia and Serbia were successively based. It was the conquest of Macedonia that paved the way for the expansion of the power of the Osmanli Turks to eastern Europe on the one side, and to Africa and the Near East on the other. Thus the strategic function of Macedonia is derived

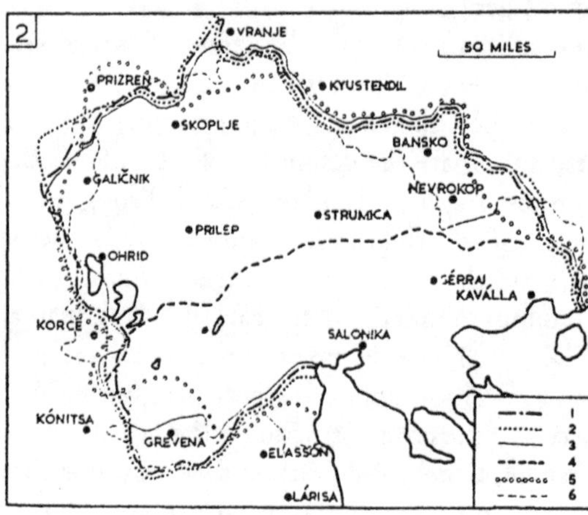


FIG. 2. SOME FURTHER DEFINITIONS OF MACEDONIA

The references in the key are as follows: 1. Bulgarian, V. Kănčev, 1900; 2. Austrian, F. Meinhard, 1899; 3. Bulgarian, D. M. Brancoff, 1905; 4. Serbian, S. Gopčević, 1889 (northern limit); 5. Italian, *Enciclopedia Italiana*, 1930; 6. German, L. Schultze-Jena, 1927 (*cf.* Figs. 58 & 77).

from its command of the great corridor-route which leads from central Europe to the Mediterranean along the Vardar and Morava valleys, a route which has witnessed the passage of countless armies, Greek, Roman, Slav and Turkish.

In the nineteenth century Macedonia's strategic function was no less important than it had been in previous years but interest in it mounted higher because of the patent decline of Turkey, the armies of which had for so long dominated its routeways. The governments of the four great empires of Austria-Hungary, Russia, Turkey and

Great Britain were alike conscious of maintaining or acquiring, as the case might be, control of the whole or part of Macedonia, which they rightly regarded to be the key to the Near East. Furthermore, after the rise of the national states during the course of the century, the possession of Macedonia was hotly contested by Greece, Bulgaria, Serbia, Romania, and, later, by Albania. Italy, too, was interested in its possibilities. In addition to the great strategic benefits accruing from its possession Macedonia also controlled important economic

FIG. 3. DIAGRAM OF THE ETHNOGRAPHIC DIVERSITY OF MACEDONIA AND ITS ADJOINING TERRITORIES

An hierarchy of complexity of local ethnographic structure may be distinguished in the south-western Balkan peninsula. The arrows indicate affinities of local populations with neighbouring politico-ethnographic groups. Macedonia has relations with six, Thrace with three, and Northern Albania, Northern Ipiros (Epirus) and the Niš region, each with two such groups, respectively.

outlets. The Drin valley routes leading to the Adriatic were best approached from Macedonia and its Aegean ports possessed a potential hinterland which transcended the Sar and the Rodopi mountains.

The two distinguishing characteristics of Macedonia which have been emphasized, the complexity of its ethnic structure and the importance of its routeways, are related because accessibility often gives rise to a heterogeneous population. But they are related in another way, and this latter relationship provides the key to the understanding of the political geography of Macedonia. During the course of the nineteenth century the ethnic structure of local populations became increasingly influential in deciding their political affinities. Where that structure was obscure, as in the case of Macedonia, the onus of establishing the affinities of its population fell upon parties interested in its strategic and economic possibilities rather than in its people. All powers, both great and small, imperialistic and nationalistic, discerned the importance of putting exactly that interpretation on the ethnography of Macedonia which might best extend their influence in the area and so prepare the way for the establishment of local hegemony, or Near Eastern ascendency, as the case might be. It is for these reasons that a detailed study of ethnographic maps of Macedonia forms a fruitful approach to its political geography.

At the same time, the exploration of this cartographic source-material serves another useful purpose. It throws light on the nature of ethnographic maps themselves, on the development of a special type of cartography, on methods of representation, compilation and production. The validity of criteria may be examined, palpable misrepresentations and inaccuracies determined, and something of the uses and misuses, the merits and the limitations of such maps, be demonstrated.

This monograph, therefore, is concerned with a survey and analysis of some two hundred ethnographic maps of Macedonia. They include some of the first and some of the latest of such maps to be compiled. Most of them are products of the nationalist fervour of the nineteenth and twentieth centuries. Few were published before 1840. The maps selected incorporate all possible points of view. Some are more detailed than others. Scales vary but often a small sketch-map is as illuminating as a large plan. Many of them deal specifically with Macedonia, or part of Macedonia and its adjoining territories, others refer to Macedonia only as part of the Balkan peninsula, or as part of Turkey-in-Europe, or even as part of Europe as a whole.

The maps are treated in chronological order because so many of them are derivative, and each may be properly studied only in its historical context. A comparative method of approach has also been attempted which has necessitated a policy of standardization and simplification of maps, and the adoption of a common formula of analysis for the contents of each. A survey of the ethnic classifications used by various authorities since 1821 reveals more than one hundred and fifty diverse groups (see Summary). Whilst it has been necessary to describe this variety, which is in itself a significant pointer to the difficulties and limitations of ethnographic mapping, it has been found practicable to do so under the following selected headings : 1. Turks ; 2. Greeks ; 3. Slavs ; 4. Vlachs ; 5. Albanians. Only occasionally has it been necessary to modify this plan of description. In the accompanying maps, where possible, a standard pattern of representation for the distribution of each of these main groups has been applied, each map being reduced to a comparable scale. Except for the illustrations in the summary chapter it was not found practicable to use a common base-map.

## Chapter II

# TURKS, GREEKS AND ILLYRIANS, 1730–1843

### Europa Polyglotta, 1730

One of the first attempts to show the distribution of the vernacular languages of modern Europe was made in a small map, published in Nuremberg in 1730 (Fig. 4). The anonymous author distinguished Scottish, Irish, Welsh and Anglo-Saxon in the British Isles, Moorish and Catalan in Spain, and showed other indications of being relatively well informed on the distribution of the major linguistic groups of western, central and northern Europe. It is reasonable to assume, therefore, that the author's interpretation of the distribution of spoken languages in south-eastern Europe would accord with the best informed opinion of his time. In the Balkans, south of the Danube, he recognized only three linguistic groups—Turkish, Greek and another which he described as *Illiri-co-Sclavonica.*

*Distribution of the Turkish Language*

The author mark .d on his map the whole of the eastern half of the peninsula, from the lower Danube to the borders of Thessalía, including the greater part of Macedonia, as part of the Turkish linguistic province. Evidently the belief that the Turkish language and culture were well established over the greater part of the Balkans was widely held in Europe in 1730. It was unlikely that a devout Christian such as the author (witness the sentiment expressed on the map) would willingly have over-estimated the strength of the Turks.

*Distribution of the Greek Language*

The author limited the district, in which Greek was spoken, to the Greek archipelago proper, and he did not show any extensions of the Greek tongue into either Macedonia or Thrace. A lengthy time

FIG. 4. EUROPA POLYGLOTTA, 1730

Only a part of the original map is portrayed above. The distribution of languages is indicated by the opening phrases of the Lord's Prayer written in the vernacular over the territory where each language is spoken.

was to elapse before the idea matured which credited the Greeks with any considerable territory on the mainland.

*The Illiri-co-Sclavonica*

The third linguistic group distinguished by the author which may be termed Illyro-Slav was the largest, since it was not limited to the

Balkans but stretched from the Adriatic, in a north-easterly direction, as far as Poland and Russia. It was divided into two by a wedge of Hungarian. The southern part extended from the Sar mountains to Carinthia and Styria and it is to be inferred from the map that here both Illyrian and Slav were spoken. But neither the Albanians, as such, nor the Serbs, nor the Bulgarians, nor yet the Romanians, were specifically referred to on the map.

*Conclusion*

So in 1730, when the major linguistic affinities of the rest of Europe were already fairly well established, the principal groups of the Balkans, as they are known to-day, had not even been envisaged by European scholars. The existence in European Turkey of ethnic groups, the origins of which often pre-dated the Osmanli conquests, remained almost entirely unsuspected. No one fully appreciated the fact that the Turkish possessions constituted one great preserve, wherein diverse cultures, languages, religions and traditional economies had maintained their respective individualities for over three hundred years. Nearly seventy years more were to elapse before travellers began to bring back tales of the Christian subjects of the Sultan, and to report that they not only outnumbered the Turks in Macedonia, Thrace and Bulgaria, but that they even took an active part in the commerce and administration of the Turkish Empire.

In the opening decades of the nineteenth century, a host of itinerant scholars and adventurers began to inform an incredulous western public that the Turks had not succeeded in colonizing even a tenth of their European domain, nor yet had they had much success in imposing their language and religion upon the peoples whom they had conquered. It transpired that the fanatical fury with which the Osmanli Turks had swept through the Balkans to the very gates of Vienna had dwindled, after the first flush of conquest, to a mild toleration which even if it derived from a policy of *dividere et imperare* was none the less remarkable. F. C. L. H. Pouqueville, D. Urquhart, H. Ubicini and other early nineteenth-century travellers confirmed in their reports the overwhelming numerical superiority in the Balkans of the Christian subjects of the Sultan, and the ethnographic maps of the second decade of the century recorded the revolution in ideas which had taken place consequent upon these revelations.

## F. A. O'Etzel's Map of 1821

F. A. O'Etzel was a cavalry captain and a member of the General Staff of the Prussian Army. He was a geographer of some repute, with a particular interest in ethnographic cartography. Using language as his criterion, he compiled an ethnographic map of Europe which was produced by the Technical Bureau of Berlin in 1821. This map clearly summarized the ideas he held concerning ethnographic groupings and distributions in the Balkans at the beginning of the century. Altogether O'Etzel distinguished six groups (Fig. 5).

### *The Turks*

He indicated the Turks only as a minority within the peninsula, giving no details of their distribution because his method of showing minorities was very inadequate : it consisted of inscribing the name of the minority concerned, in small Roman letters over their habitat. Capital letters were used to depict majority populations and some attempt was made to show their extent by means of a dotted outline. But even using this crude method, O'Etzel managed to convey the fact that he did not believe that the Turks formed the bulk of the population of the Balkans. It is of interest to note that he marked separately the *Yüruks* in Macedonia. They were nomadic pastoralists of Turkoman extraction, who later were usually classed with, or as, Turks.

### *The Greeks*

Instead of the Turks, O'Etzel depicted the Greeks as the predominant ethnic group in the Balkans. He marked them in southern Macedonia as far north as Lake Ohrid, in Thrace, and along the coast of the Black sea. O'Etzel evidently believed that if the inhabitants of most of southern Macedonia and Thrace were not Turks, they must be Greeks. The reasons for this changed view-point will be examined later.

*Pure Illyrians*

The Illyrians were, according to O'Etzel, the chief inhabitants of the territory to the north of the Greeks but his intelligence concerning them was but little in advance of that of the Nuremberg map of 1730. His general thesis was that the Illyrians formed the indigenous population of the Balkans. They were still to be found in a pure state amongst the *Clementini* of the Albanian Alps and the Montenegrins of Montenegro. Most of the Illyrians, however, he believed had been romanized or slavized.

*Romanized Illyrians*

O'Etzel was of the opinion that during the prolonged period of the Roman occupation of the Balkans, a group of the Illyrians had come under the influence of Roman culture and language. This group, he referred to as Wallachian (Vlach), found both north and south of the Danube. A peculiar feature of O'Etzel's map was the extraordinary extent of territory he ascribed to those Vlachs south of the Danube. It included northern Macedonia, the upper Struma valley, the Rodopi mountains and the whole of the upper valley of the Maritsa, including Edirne (Adrianople). It is quite true that the Vlachs, during the middle ages, were much more numerous than they are to-day. The knowledge of this fact, coupled with travellers' vague reports of the presence of Vlachs in these areas, probably influenced O'Etzel and gave rise to this piece of cartographical fantasy.

*Slavized Illyrians*

O'Etzel maintained that another group of the Illyrians had resisted, to some extent, the Roman culture, only to succumb to that of the Slavs. The slavized Illyrians were, as one might expect, an extremely varied group. They included the following branches :

(i) the Albanians who occupied the Adriatic coast between the Drin river and the Gulf of Arta and extended as far inland as Lake Ohrid (O'Etzel evidently believed that they had become Slavs) ;
(ii) the Bulgarians, who were confined on the map to a small area between the Danube and the Balkan crest-line—a significant underestimation of their distribution which helped to give rise

FIG. 5. F. A. O'ETZEL'S MAP, 1821
Only a part of the original, which is in colour, is portrayed above.
Ethnographic groups which have a relative majority are indicated in capitals, and minorities in lower case.

to the misconception that the Bulgarians formed only a minority of the Balkan peoples ;

(iii) the Morlachs and Uskoks of Dalmatia and Croatia who were later to be regarded as slavized Romanians.

*The Slavs*

Finally, O'Etzel distinguished the 'pure Slavs' showing them as occupying the land between the coastal ranges of the Adriatic and the Danube above Vidin. He included in this category the Serbs, the Bosnians and the Croats.

*The Significance of O'Etzel's Map*

O'Etzel's map was widely accepted in Europe in 1821. M. A. Denaix, the French geographer, based the ethnographic map of Europe in his *Atlas Physique, Politique et Historique de l'Europe*, 1829, on that of O'Etzel. The second edition of this atlas, in 1855, retained the map with only slight modifications. Unlike O'Etzel, however, Denaix believed that the process of slavization amongst the Illyrians had proceeded to the stage when all Illyrians might well be regarded as Slavs even although they were still known by their old names as, for example, the Morlachs.

The chief difference between the 1730 map and O'Etzel's map was that on the former the Turks had been regarded as all important, whilst on the latter the Greeks assumed the role of the major group of the southern Balkans. The change in ideas had immense significance for the future political geography of the peninsula. There was a number of reasons why, towards the end of the eighteenth century, the belief in the Greek character of the Balkans became widespread, and why, at the beginning of the nineteenth century this belief reached its zenith. These reasons may be grouped under four separate headings.

(1) *The Greek National Movement.* Towards the end of the eighteenth century, the Greek national movement, in the form of a literary renascence, was already under way, led by such men as Rhegas (1753–98), author of Greek revolutionary songs and the organizer of fiercely patriotic secret societies, and Adhamándios Koraïs (1748–1833),

who edited the Greek classics and helped to reform and purify the Greek literary language.[1] As the movement grew, it aimed at the liberation of *Hellas*, the homeland of the ancient Greeks, based on the archipelago. There, throughout the period of Turkish rule, the Greeks had managed to retain a fair amount of administrative freedom, particularly in Pelopónnisos (Morea), which had remained virtually an autonomous state, although still under Turkish suzerainty. Turkish control of the difficult country in the archipelago had never been effective. Osmanli influence there had been countered, not only by the Greeks themselves, but also by the power of the Italian republics, particularly by Venice. In the archipelago furthermore, the Greek fighting spirit had been maintained by the *armatoli* and the *klephts*, whose stirring adventures form the subject of many Greek folksongs. The *armatoli* were Greek militiamen enrolled for the purpose of policing the autonomous Greek districts. The *klephts*, on the other hand, were brigands loth to acknowledge Turkish suzerainty, patriots, outlaws and adventurers.

The Greeks in the islands of the Aegean and Ionian seas were also ripe for emancipation. The Turks had never been a seafaring people ; the Greeks naturally manned the Turkish navy, and, what was more important, assumed almost complete control of the commerce of the Empire. Under the Turkish flag (and after 1783, under the Russian flag, in the Black sea), the Greek mercantile marine had grown extraordinarily rich and powerful. It was able to exploit to the full the advantages arising out of Turkish control of the Straits of Dardanelles and the Bosporus, and of the coasts of the eastern Mediterranean. The Greek mariners had, by 1821, become a highly privileged and wealthy body of men. Their voyages often took them out of the Mediterranean, and their presence abroad brought to the notice of the rest of the world the importance of the Greek element within the Turkish possessions.

In 1821, the year in which O'Etzel's map appeared, the national

[1] The 'purification' (katharévouas) of the language ultimately restricted its political influence and had unfavourable repercussions on the development of the 'Grand Idea' (see p. 118). Of incidental interest is the fact that Rhegas was a Vlach.

movement in the archipelago passed from the preparatory to the active stage, and the insurrection in Pelopónnisos against the Turks marked the first phase in the political emancipation of the modern Greeks.

(2) *The Revival of Byzantine Greece.* Greek political activity within the Turkish Empire was not, however, confined to the archipelago, which had been acknowledged as Greek even on the map of 1730. There was yet another centre of Hellenism in the Balkans, in Constantinople itself, for the Greeks were not only heirs to Hellas, but also to Byzantium. The other aspect of the Greek revival was due to the survival of Byzantine influences. The Byzantine Empire was not destroyed by the Osmanli conquest. A. A. Pallis pointed out in a recent publication that the Turks after their victory modified but little the machinery of the Byzantine state : " The Sultans took the place of the Greek Emperors on the throne of the Caesars, the ceremonial of the old Seraglio continued that of the Sacred Palace, the Turkish beys or pashas stepped into the shoes of the Byzantine or Frankish nobles or owners of the great landed estates."[1] But although the Turks had inherited the administrative machinery of the Byzantine Empire, they had to depend upon the Greeks to keep it functioning. At different times Greeks had held various key posts in the Sultan's administration and they were, for example, invariably elected to the 'Hospodarships' of the Danubian principalities. By the end of the eighteenth century the Turkish administrative system, from the lowest to the highest grades, had become permeated with Greek officials.

Even more important than the administrative power wielded by the Greeks, was the fact that the spiritual welfare of nearly all the non-Moslem inhabitants of the Balkans was in the hands of the Greek Patriarch and his Greek prelates. The Turks, from the very beginning of their rule in Europe, had allowed the Orthodox Church to occupy a privileged position in their Empire, partly in order to offset the influence of the Roman Catholic Church in their dominions. So whilst the Moslem faith took the place of Christianity as the state religion, the Greek Orthodox Church continued to flourish ; even its

[1] *Greece's Anatolian Venture—and After* (London, 1937).

estates were left untouched, and the Greek ecclesiastical courts continued to function side by side with the Moslem religious tribunals. The Turks had realized from the beginning that the secret of successful control of the whole of the Near East, including the Balkans, lay in indirect rule through the media of the various Churches within their domain. In the whole of this vast area, languages were not nearly such powerful bonds as religions. There were a great many languages but only a few religions, if one ignored sectarianism; religious belief formed a convenient criterion for distinguishing the various nationalities under the power of the Sultan. Hence the official recognition by the Turks of *milleti*, or national-religious communities, which were represented at the Porte by the heads of the Churches concerned (see p. 42). Thus the Turkish Sultans assembled in Constantinople an imposing array of ecclesiastical rulers of whom by far the most important, and most influential, was the Greek Patriarch. In the provinces the Greek metropolitans were as well established as the Moslem *cadis*, and the old ecclesiastical dioceses, with their Byzantine names, were still maintained. It is true that for a time the Christian populations of the Balkans were represented at the Porte by both Slav and Greek Patriarchs, but the latter had gradually assumed control of the Orthodox Slavs as well as of the Greeks. In the eighteenth century both the Serbian Patriarchate of Peć and the Bulgarian Patriarchate of Ohrid had been abolished. With them had vanished the vestiges of Slav cultural and spiritual independence in this part of Europe. Later, all Orthodox Christian subjects of the Sultan were regarded for all official purposes as Greek, insomuch as they formed part of the Greek or *Rumi millet*. Official Turkish statistics at this time, for example, recognized only two 'nationalities' in the Balkans—Turkish and Greek.

Control of the Church also gave to the Greeks a monopoly of educational and cultural institutions. Greek became the language, not only of the Church but also of the schools, and therefore of the educated classes, no matter what the local vernacular happened to be. Greek was the only written language of the Balkans until well into the nineteenth century. The following lines written in Greek by a Vlach priest in 1802 and translated by A. J. B. Wace and M. S. Thompson in 1914, summarize the sentiment prevailing at the beginning of the nineteenth century before the Romantic Movement developed (see p. 28)[1]:

Albanians, Bulgars, Vlachs and all who now do speak
An alien tongue rejoice, prepare to make you Greek,
Change your barbaric tongue, your rude ways forgo,
So that as bygone myths your children may them know.

In the opinion of another well-known historian: "South-eastern Europe was ruled by the Turks; but [ . . . ] its religion, education, commerce and finance were in the hands of the Greeks." These words of Sir Charles Eliot sum up admirably the role of the Greeks in the Turkish Empire in the late eighteenth century.[2]

The fight for Greek emancipation thus assumed a dual aspect. It took the form, on the one side, of a narrow national movement largely centred in the Greek archipelago, which looked to ancient Greece for its inspiration, to Athens for its capital, and to Attic Greek for its literary language. On the other side, it took the form of a wider movement towards the revival of the Byzantine Empire, a movement which had its headquarters in the Phanar quarter of Constantinople, a movement which looked to the glory of Byzantium for its inspiration, and to Constantinople for its capital. This divergence of view was to become a characteristic feature of Greek history. It split the Greeks ultimately into two hostile camps and had a tragic aftermath.

(3) *Russia and the Greeks.* The first of the Great Powers to recognize the spiritual and commercial importance of the Greeks in the Balkans, and to work for Greek independence in that theatre, was Russia. The eighteenth century had witnessed the enormous expansion of Muscovy, to the Baltic sea in the north and to the Black sea in the south. Russian foreign policy ever envisaged the ultimate control of Constantinople: the common bond of the Greek Orthodox faith between the Russians and the Greeks of Constantinople was the means whereby successive Russian foreign ministers worked towards the

[1] A. J. B. Wace and M. S. Thompson, *The Nomads of the Balkans* (London, 1914).

[2] *Turkey in Europe* (London, 1900).

achievement of this end. As early as 1782, the Tsarina Catherine II of Russia had proposed a complete reconstruction of the political map of the Balkan peninsula. With the help of the Habsburg Emperor she had hoped to expel the Turks from Europe, and resurrect the Greek Empire. Her whole scheme turned upon the revival of Byzantine Greece. The Slav nationalities did not even figure in her plans. Catherine's second grandson, Constantine, was to have been the new Greek Emperor. He received instruction in Greek, the tongue of his future subjects. Catherine's scheme forcibly illustrated the blossoming of the idea that the Balkans constituted a Greek province and that Greek nationals formed its chief inhabitants.

On many subsequent occasions Russia found a means of championing the Greek Church within the Turkish domain and thereby furthering the cause of 'Pan-orthodoxy.' It was, for example, on Russian soil that the most famous of the Greek revolutionary secret societies was founded—the *Philike Hetaireia*—by the Greek merchants of Odessa in 1814. The Russians, of course, did proffer help to the Serbs and Montenegrins in the late eighteenth and early nineteenth centuries but the fact remains that bonds of religion were more important and useful than linguistic affinities in 1782, and even as late as 1821 the future of the Balkans seemed to lie with the Greeks rather than with the Slavs.

(4) *Philhellenism.* In central and western Europe the tendency to regard the population of the southern Balkans as Greek was partly due also to the new enthusiasm for the achievements of the ancient Greeks, which was particularly strong in the British Isles in the second decade of the century. The philhellenists included not only classical scholars and romanticists, but liberals also. The intensity of their pro-Greek feeling reached a peak on the occasion of the Greek insurrection of 1821. "The mere name of the Hellenes, heard once more upon the lips of men after centuries of complete oblivion, thrilled the hearts of those who owed to Greek philosophy, Greek art and Greek literature a debt larger than they could acknowledge or repay."[1] The neo-classicists assumed, rightly or wrongly, a continuity of culture and race between the ancient and the modern Greeks. Among the ardent

[1] J. A. R. Marriott, *The Eastern Question* (Oxford, 1917).

supporters of their cause was Lord Byron and his friend J. C. Hobhouse, who did all in their power to foster Philhellenism.

It is not surprising, all things considered, that at the beginning of the nineteenth century pro-Greek sentiment was in the ascendency. The power and influence of the Greeks themselves, the practice of regarding religion as the chief measure of national sentiment, and the readiness of western Europe to look for a Greek revival in the Near East were all factors operative in creating the dangerous illusion that the Balkan peninsula was to all intents and purposes part of the Greek world.

## W. Muller's Map of 1842

*J. P. Fallmerayer's Thesis, 1830*

The philhellenists and neo-classicists were rudely shaken by the work of Fallmerayer which appeared in 1830.[1] He maintained that the classical Greeks had been completely wiped out during the period of barbarian invasions, and that the modern Greeks were not the descendants of the Hellenes. His ideas were not always sound and were later disproved, but they influenced European thought considerably at the time and stimulated interest in the application of racial affinity to ethnographic distributions in the Balkans. Fallmerayer's influence was to be discerned in Müller's portentous work on ethno-geography, published in both Paris and Leipzig in 1842. According to his representations there were five important ethnographic groups to be considered in the Balkans (Fig. 6).

*The Turks*

Müller did not even mark the Turks as a minority within the peninsula. From being reckoned the dominant group in 1730, they had now disappeared from the map altogether.

*The Pelasgians and the Hellenes*

The term 'Pelasgian' was resurrected by Müller to refer to the inhabitants of southern Macedonia and Thrace. The term had been

[1] *Geschichte der Halbinsel Morea während des Mittelalters* (Stuttgart und Tübingen, 1830–36).

formerly used by the ancient Greeks and by classical scholars to refer to the plebeian population of the ancient Hellenic Empire. Müller's purpose in re-introducing the Pelasgians into the ethnographic map of south-eastern Europe, was to emphasize that these peoples were not

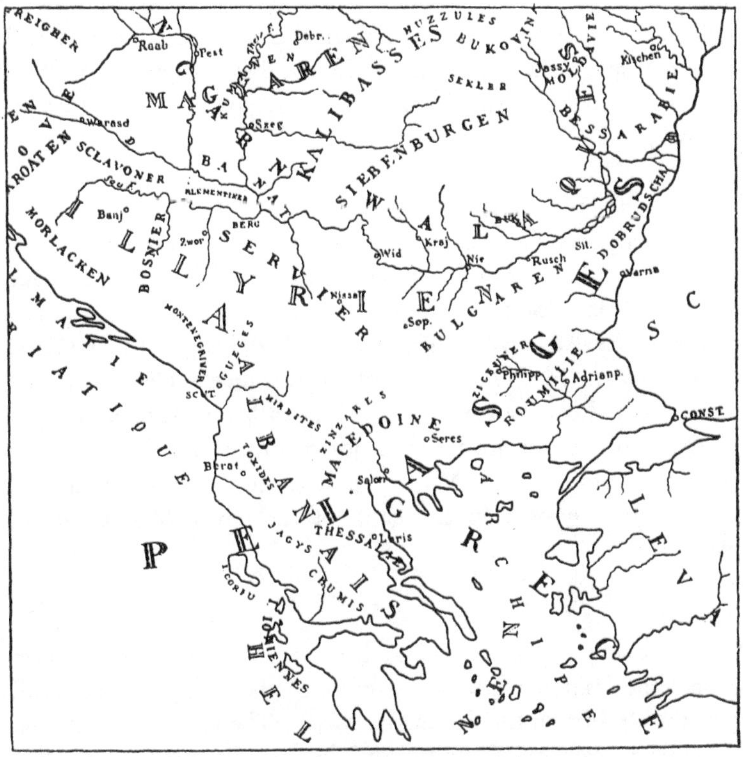


FIG. 6. W. MULLER'S MAP, 1842

Only a part of the original, which is in colour, is portrayed above.

the descendants of the ancient Greeks of classical renown and that they were to be distinguished from 'the Hellenes proper,' whom he confined to Pelopónnisos and to the adjacent islands.

*The Illyrians*

North of the Pelasgians, Müller depicted another major group, the Illyrians. But unlike O'Etzel, he used the term Illyrian to mean Slav.

Thus he included the Bosnians, the Serbs and the Bulgarians in his Illyrian classification.

*The Vlachs*

Müller referred to the Vlach as *Zinzares.* The superfluity of 'z' sounds in their speech had given rise to this somewhat derisive nickname. He gave them a much more limited distribution than O'Etzel, but he still marked them as the dominant population of northern Macedonia.

*The Albanians and the Gypsies*

On Müller's map the Albanians ranged from Lake Shkodrs (Skaḍar) in the north, to the Gulf of Corinth in the south. Inland they extended into western Macedonia. He distinguished the three most important branches of the Albanians—the *Geuges* (Gegs), the *Toxides* (Tosks) and the *Mirdites.* He believed that the *Jagys* and the *Chumis* of Ipiros (Epirus) were also Albanians. According to his interpretation therefore, the Albanians formed the majority of the inhabitants of the Greek archipelago. Finally, Müller separately distinguished on his map the *Zigeuner* or gypsies in the basin of the upper Maritsa, around Plovdiv.

*Conclusion*

It is reasonable to assume from the evidence offered by Müller's map that the fashion of depicting most of the Balkans as ethnically Greek was already on the decline by 1842. The fact that Müller had used the term 'Pelasgian,' the fact that he had distinguished so many Albanians, *Zinzares* and *Zigeuner*, where O'Etzel had marked only Greeks, was further proof of a trend in ideas away from the conception of an Hellenistic Balkans. These new ideas were fostered by the interest manifested by many scholars in the racial affinities of European peoples. Even in the eighteenth century Linnaeus and later Blumenbach had attempted to classify man by his physical appearance. Supposed genealogical connections between peoples of the classical world and the modern inhabitants of the Balkans became the subject of speculation towards the middle of the nineteenth century. About this time Darwin had already begun his researches into natural selection. J. C. Prichard's *Natural History of Man* appeared in 1843. It was only to be expected that anthropologists would apply the criterion of common descent to ethnic groupings. Hence the emphasis given in some of the ethnographic maps of this period to such pseudo racial-cultural groups as the 'Illyrians' and 'Pelasgians.' The inconsequential confusion of somatic and cultural data used to classify and delimit ethnic groupings reached its height about mid-century with the publication of R. Knox's *Races of Man*, in which was preached the infallibility of hereditary traits.

## G. Kombst's Map of 1843

Influenced by these new ideas on ethnic affinities was G. Kombst's interpretation of the ethnography of Europe (Fig. 7). He was a German anthropologist commissioned to produce a map in A. K. Johnston's atlas (1843 edition). The author of an *Ethnographical Map of Europe in the earliest times illustrative of Dr. Prichard's Natural History of Man*, etc. referred to Kombst's map in 1843 as the most authoritative and up-to-date ethnographic map of Europe. That Kombst's map should appear in Johnston's atlas was in itself an indication of his popularity.

### *The Turks*

On Kombst's map the Turks reappeared once more in southern Macedonia, in the whole of Thrace, and in eastern Bulgaria. The success of the Russians in the Near East, as exemplified by the Treaty of Unkiar Skelessi (1833), had disconcerted the Great Powers of Europe and there was a natural reaction in favour of Turkey as a stabilizing power in the eastern Mediterranean. Palmerston had declared in 1839, "All that we hear about the decay of the Turkish Empire and it being a dead body or a sapless trunk, and so forth, is pure and unadulterated nonsense." Palmerston's foreign policy was based on the survival of the Turkish Empire. Thus it is interesting to note that the twin bastions of British Mediterranean policy in the nineteenth century—the support of Hellenism and the maintenance of

Turkish control of the Straits, were both related closely to contemporary ethnographic ideas.

*The Pelasgo-Greeks*

Kombst utilized the term 'Pelasgo-Greek' to refer to the inhabitants of the Greek archipelago. He showed the Pelasgians neither in Macedonia nor in Thrace, where Müller had shown them.

*The Illyrians*

Another major group on his map comprised the 'Illyrians,' south of the Danube, a term he used to denote all the Southern Slavs. They extended south to reach the sea near Salonika and were sub-divided into the Bulgars (in a narrow zone south of the Danube), the Serbians, the Morlachs, the Croats, the Montenegrins and the Dalmatians.

*The Albanians and the Suliotes*

Kombst marked the Albanians in Ipiros, together with the *Suliotes*, who apparently were neither Albanians nor Pelasgo-Greeks. Their identity remained in doubt for some considerable time. As late as 1869, H. F. Tozer noted that the *Suliotes* were generally believed to be Greek but were in actual fact Albanians.[1]

## The Trend of Ethnographic Ideas Between 1730 and 1843

It may be noted that all these maps so far discussed were concerned with Europe as a whole, rather than with the Balkans. They were all on a small scale and little attempt had been made to delineate accurately the limits of the various ethnic groups in Turkey-in-Europe. The political significance of the ethnographic frontier had yet to become a vital issue. Both terminology and classification were confused and often misleading. 'Illyrian,' for example, was used both in contradistinction to, and synonymous with, the term 'Slav.' The word 'Pelasgian' was used in a racial, and a cultural sense. Linguistic, religious and racial criteria were hopelessly mixed. In spite of their obvious deficiencies, however, these maps do provide a

[1] *Researches in the Highlands of Turkey*, vol. 2 (London, 1869.)

FIG. 7. G. KOMBST'S MAP, 1843

Only a part of the original, which is in colour, is portrayed above.

valuable commentary on changes in ethnographic ideas taking place over the course of a hundred years. In 1730 the Turks were believed to have settled extensively in Macedonia and Thrace. By 1821 the opinion was widely entertained that there were only a very few Turks in Europe, but by 1843 opinion was once again swinging in favour of the Turks. In 1730 the Greeks had been shown only within their archipelago. By 1821, a school of thought in Europe believed that they inhabited nearly the whole of the Balkan peninsula, but twenty years afterwards this belief was already on the wane. In 1730 the idea that the Slavs and Illyrians populated parts of the northern Balkans was fashionable. As late as 1821, Slavs and Illyrians were still separately distinguished, but by 1840 the terms, ' Illyrian ' and ' Slav,' had become synonymous. Opinion on the dispositions of the Albanians varied enormously during this period. The Vlachs were as late as 1821 believed to inhabit the greater part of the Rodopi mountains as well as the Maritsa valley but they did not even appear on Kombst's map of 1843.

It had already become apparent by 1843 that ethnographic distributions in the Balkans varied according to the nature and accuracy of the criteria adopted to classify the population. Language gave one picture, ' race ' gave another, religion yet a third, whilst by using various combinations of criteria numerous interpretations might emerge. Thus even before the term ' ethnographic ' had become politically significant, there appeared to be no general agreement between anthropologists, ethnographers and geographers as to its exact definition.

## CHAPTER III

# THE SLAVS, 1842-1869

### INTRODUCTION

TOWARDS the middle of the nineteenth century there were signs throughout the whole of Europe of a new spirit of inquiry abroad ; a thousand and one scientific societies were springing up, ranging from small local institutions to great national academies. The comparatively new studies of philology, ethnology and archaeology, together with geography, commenced to assemble an encyclopaedic body of knowledge covering every aspect of human activity. This genuine interest in the pursuit of knowledge was enhanced by the rapid accumulation of material wealth which directly and indirectly provided facilities and leisure for research. Improved methods of communication and expanding markets brought home to scholars the significance of the ever-increasing interdependence and inter-relationships of the different parts of the world. The publication of *Petermann's Mittheilungen* began in 1855, and at the same time *La Société de la Géographie de la France* began to issue their famous *Bulletin*. In 1857, *Die Kaiserliche und Königliche Geographische Gesellschaft in Wien* was established ; indeed the proceedings of the learned academies of most of the cities of eastern Europe date back to this period. The members had the whole world as their field of study, but naturally, interest was focussed more particularly on the Balkans. In spite of the geographical proximity of the peninsula to the highly organized parts of Europe, much of it remained as mysterious and unknown as the interior of darkest Africa. As E. Réclus pointed out in his introduction to the geography of the Balkans (1875), as late as 1870 the erroneous belief in the existence of a single central mountain chain traversing the whole peninsula from east to west, was still widespread.[1] Knowledge of topography was poor ; details were unknown and such survey as had been carried out was based on false astronomical data. The abysmal lack of information on the physical and human geography of south-eastern Europe was gradually dispelled, consequent upon the explorations and investigations of A. Boué, G. Lejean, F. Kanitz, H. Mathieu, H. Barth, R. Cyprien and of the English ladies, Miss G. Mackenzie and Miss A. P. Irby, to mention a few of the scholars whose combined efforts had begun to throw some light on conditions there between 1840 and 1870. Animated by the 'geographic' spirit so conspicuously lacking in most studies of Turkey before 1840, their approach to the problems of the population of the Turkish Empire in Europe ultimately revealed ethnographic distributions completely different from those to be inferred from earlier maps. Their 'discoveries,' coming at a time when national sentiment was growing in the Balkans, had a profound political significance.

*The Romantic Movement*

The new interest in ethnographic distributions was not in the first place purely political. It was inspired by the Romantic Movement which was sweeping the whole of Europe and promoting the study of folklore, folksongs and vernaculars. The works of Hans Christian Anderson (1805-1875) and J. L. C. Grimm (1785-1863) had become very popular and served to emphasize the importance of the linguistic traditions, spoken or written, of the peoples of Europe. The doctrine that every language had some contribution to make to the commonweal of literature invested its speakers, however humble, with a new self-respect and with the means of asserting their individuality. Partly due to the new interest in language, ethnographers tended to dispense with religion as the chief criterion of 'community' and in its stead to apply more and more the test of spoken language.

The phenomenon of national sentiment based on linguistic affinity may be said to be peculiar in origin to western and central Europe.

[1] *Géographie Universelle,* Tom I (1875). This fallacy, according to J. Cvijić, was due to the deference paid to the map of Ptolemy !

It is a phenomenon associated with well-developed literary traditions and with high standards of universal education. It penetrated into the Balkan peninsula but slowly, gradually dispelling the power of religion as a force drawing men together into some form of political cohesion but never entirely supplanting religious allegiances. Indeed, many of the national problems of the Balkan peninsula become easier to understand when the region is conceived as the meeting-place of two fundamentally opposed ideas on political allegiance. The one, emanating from western Europe, maintained that linguistic unity was the *sine qua non* of the modern state whilst the other, the older idea still characteristic of the Middle East, regarded the state as a religious community.

The growing impotence of religion as a unifying force was amply demonstrated as early as 1821, when the Romanians, although Greek Orthodox in faith, failed to flock to the banner of Hypsilanti, their Greek Hospodar, when he raised his banner in a revolt against the Turks. From that time onwards more and more emphasis began to be put on language as a test of nationality, and less and less on religion. This tendency was already apparent in 1828, when A. Balbi published his famous ethnographic atlas of the world and put forward in his classification the idea that the Illyrians were 'Southern Slavs.'[1] Under the heading *Slavonne, Servienne, Serbe ou Illyrien* he grouped the Serbians, Montenegrins, Bosnians and Bulgarians. Of the Bulgarians, he wrote that they were the descendants of the famous Bulgars, but that they had forgotten their language; that they spoke Slav with a slight admixture of Turkish words and that their language was characterized by the use of an article after the noun. The growing accent on linguistic criteria gave credence to the idea that the Slavs were more important, numerically, than had been thought in 1821. But it was not until later in the century that this idea received general recognition.

### *The Slav Renascence in the Turkish Empire*

The Slav renascence in the Turkish Empire was led by the Serbs, who, from their stronghold on the Danube, on the very margins of the Osmanli possessions, were able to engage in national activities with

[1] *Atlas Ethnographique du Globe ou classification des Peuples Anciens et Modernes d'après leurs langues.* Par Adrian Balbi (Paris, 1828).

less fear of the consequences than the Slavs further south, in Bulgaria and Macedonia. They also could look for inspiration to the Croats and the Slovenes as well as to the westernized Czechs of central Europe. Moreover, the Serbs had a base of operations outside Turkey, in Hungary, because there were many Serb communities living north of the Danube. It was one of these 'Hungarian Serbs,' Dositije Obradović (1739-1811), who together with Vuk Karadžić (1787-1864), standardized the Serbian literary language and brought it into conformity with the vernacular. Hand in hand with the literary renascence, Serbian revolts developed against the local Turks, which the Porte found more and more difficult to quell. The insurgents could take shelter in the rocky limestone fastnesses of Montenegro or simply flee into Hungary until the Turkish forces dispersed. From 1804 to 1813 these spasmodic revolts were led by Kara George, and from 1815-17 by Milos Obrenović. Finally, in 1817, Turkey was compelled to recognize the Sumadija region as an autonomous Serbian principality. The Treaty of Adrianople, which ended the Russo-Turkish War of 1828-29, established the new Serbian principality on a firm basis. Turkey had to cede more territory to Serbia and to grant ecclesiastical independence, in the form of a Serbian Metropolitan in Belgrade. Since 1766, when the old Serbian Patriarchate of Peć had been abolished, the Greek prelates had had complete control of Serbian churches. The restoration of ecclesiastical independence emancipated the Serbs not only from Turkish but also from Greek influence and allowed a freer development of vernacular schools.

The success of the Serbs had repercussions in Bulgaria. In 1824, the first book in the Bulgarian vernacular was printed, and after 1835, the Bulgarians began to set up their own printing presses. Their tentative attempts at national self-expression found considerable stimulus in the presence of Russian troops in the Balkans during the first decades of the century.

### *Pan-slavism*

A marked feature of the period under review was the growth of the idea of Pan-slavism, first developed by the Hungarian Slovak, J. Kollar, in 1824. The spread of the idea into the Balkans, that all the

Slavs of Eurasia were racially and culturally akin, profoundly influenced the development of Slav nationalism and affected, also, the foreign policy of the Great Powers in this theatre. So little was known about Slav society at this time, that it was a common practice to invest all the various groups of Slavs with a common cultural inheritance. As the idea spread, every Slav felt himself to belong to a common brotherhood, a community split asunder by the misfortunes of history, but destined to coalesce eventually into a mighty political federation. Pan-slavism stirred Bulgarian and Serb alike to common action against the Turk and undermined the authority of the Porte. This fellow-feeling between the Slavs of the Balkans, inspired by Pan-slavism, made the ethnographers' task, between 1830 and 1870 relatively simple, insomuch as Slav did not dispute with Slav as to the validity of inter-Slav ethnographic divides. The distinction then, between Serb and Bulgarian was less important than that between Slav and Greek, or that between Greek and Turk. Hence, nearly all the ethnographers of this period reached a measure of agreement on the position of the Bulgaro-Serbian ethnographic frontier. The relations between the two groups of Slavs became so close after 1840, that the Turks initiated the policy of planting Tatar and Circassian (Cherkesi) colonies in the district of the Sar Mountains in order to seal off Serb from Bulgarian.[1]

Pan-slavism also exercised a directing influence on Muscovite policy in the Balkans. The Tsar, not unnaturally, assumed the role of champion of the Pan-slav movement, since Russia was the only independent Slav nation which was, at the same time, a Great Power. As early as 1812, when Russia had concluded the Treaty of Bucarest with Turkey, the liberties of the Serbians had been made the subject of treaty obligations between Russia and Turkey. Russian agents had then begun an active campaign of Pan-slav propaganda in the Balkans, with the object of uniting all the 'Southern Slavs' (including both Serbs and Bulgarians) against the Porte. Russianized Slavonic

[1] G. Muir Mackenzie and A. P. Irby, *Travels in the Slavonic Provinces of Turkey-in-Europe*, 2nd Edition (London, 1877). In a map compiled by E. G. Ravenstein for the *Universal Geography*, these Circassians are marked in the vicinity of Old Serbia and the Niš-Vranje area.

was introduced into many parts of the Balkans as a common literary language, and the Russians even went so far as to prevent the publication of Serbian manuscripts, so that the fallacy of a common Slavonic literary language might be preserved as long as possible. The cultivation of the Slav nationalities of the Balkans by Russia rendered the Tsar's old relationship with the Greek Patriarch difficult to maintain, and his former association with the Greek national movement rapidly waned; in fact Russia gradually drew away from Greece, until a point was reached, when the Tsar opposed any further extension of the Greek political boundaries in the Balkans. The Russian attitude was determined, less by the greater possibilities of Pan-slavism, as compared with those of 'Pan-orthodoxy,' than by the entry into eastern Mediterranean power politics of Great Britain in 1821, as the staunch upholder of Hellenism.

*The New Interest in Ethnographic Maps, 1840-1870*

With all these heterogeneous forces at work, the new interest in folklore, the revival of geographical research in the Balkans, the Slav national movements, the Pan-slav idea, Russia's lively interest in the 'Southern Slavs,' it was not surprising that towards the middle of the century a series of ethnographic maps appeared, very different from those already discussed in the last chapter. First, the maps were concerned more specifically with the Balkans and in particular with Turkey-in-Europe. Hence their scales were generally larger, and distributions were more accurately portrayed. Second, viewpoints were no longer confined to those of French and German scholars, but Russians, Serbs, Czechs and English also ventured into the field of cartographical ethnography. Third, the new maps were based on the reality of personal observation, and the ideas they incorporated were not obscured by considerations of genealogy or by preconceived notions emanating from classical history. The 'Illyrians,' 'Thracians' and 'Pelasgians' who had for so long figured prominently on ethnographic maps, were excluded and the major ethnic groups of the Balkans, as we know them today, were distinguished in their stead.

## P. G. Safarik's Map of 1842

It was significant that a Czech, P. G. Safařik, was the author of the first of these modern ethnographic maps based on linguistic affinity. The Czechs had for a time assumed the intellectual leadership of European Slavs and championed the cause of their unfortunate brethren in the Balkans, who were in the " pitiable position of hardly knowing their own origins." Safařik's map which appeared in 1842 was the result of twenty years' careful study of Slav antiquities. Although he himself did not travel as widely in the Balkans as later scholars, he was at least in close contact with Balkan affairs, having spent the greater part of his academic career at Novi-Sad, at that time in Hungary, where he came into contact with many Serbs. The town had long had close association with the Serb resistance movement, since being in Hungarian territory it was beyond the reach of the Turks.[1] Here Safařik was well placed to receive intelligence of ethnic dispositions from all types of Slavs fleeing across the Danube to seek refuge in Hungary. He had an intimate knowledge of Slav languages and early Slav history. He was thus well fitted for his role as the pioneer of Slav studies in the Balkans. He was the first ethnographer to recognize the six major groups of the Balkans—Turks, Greeks, Serbo-Croats, Bulgarians, Romanians and Albanians (Fig. 8).

### *The Turks*

Safařik attempted to plot the distribution of the Turks in Europe more accurately than hitherto. Kombst's map had favoured the idea that the Turks continuously occupied a large part of the southern Balkans. Safařik repudiated this belief and showed on his map that their settlement was scattered, confined for the most part to the official quarters of large towns and to strategic routeways.

### *The Slavs*

The most revolutionary aspect of Safařik's map was the depiction

[1] There is a saying in Serbia that Montenegro saved the Serbians from despair during the Turkish regime and the Vojvodina saved them from ignorance.

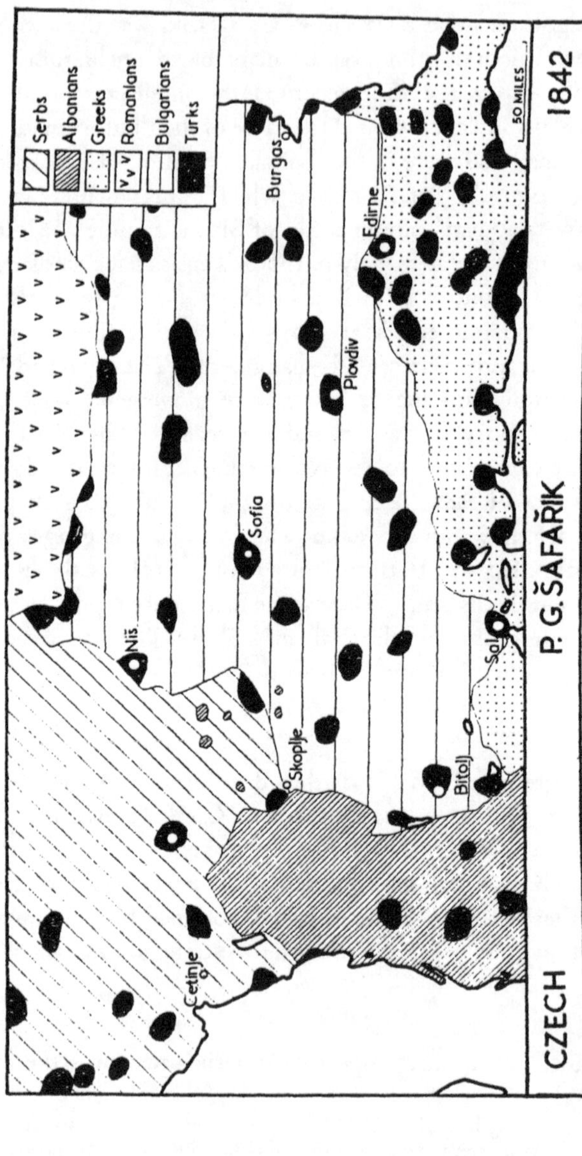
Serbs
Albanians
Greeks
Romanians
Bulgarians
Turks
Burgas
Edirne
Plovdiv
Sofia
Niš
Skoplje
Bitolj
Sal
Cetinje
50 MILES
P. G. ŠAFAŘIK
1842
CZECH

FIG. 8

of Slavs over great areas of the Balkans, regarded at the time as populated by Greeks or Turks. Furthermore, his radical reclassification of the Balkan Slavs gave coherence and form to their groupings. He recognized only two main groups of Slavs in the peninsula—the Bulgarians and the Serbo-Croats. Montenegrins, Illyrians, Bosnians, Rascians, Winds, Uskoks and Morlachs, he classified as Serbo-Croats. The Bulgarians he showed to inhabit a huge area stretching from the Dobrudja in the east to the Macedonian lakes in the west, and from the Sar mountains in the north, to a line running from Salonika to Edirne in the south.

*The Albanians*

Safařik eliminated many misconceptions regarding the host of cognate tribes known as the Albanians. He was the first scholar to attempt a clarification of the boundary between the Albanians and Slavs. With the exception of a few exclaves of Albanians in Old Serbia, he defined the Albano-Slav frontier as the Bojani river, Lake Shkodrs, the Prokletije mountains and thence as a line running southwards to Prizren and the western shores of Lake Ohrid. It is of interest to note that the present political boundary of Albania roughly coincides with this line, except that it excludes the Crni Drim valley, which Safařik regarded as Albanian.

*Conclusion*

Safařik's map suffered from many defects. It was later demonstrated, for example, that the Albanians occupied a much greater area than that which he allowed to them ; that he had greatly over-estimated the territory occupied by the Greeks in Thrace, and that he had not realized the existence of important Romanian minorities. All things considered, however, he succeeded admirably in delineating the major ethno-linguistic groups. His map was probably never meant to be a final summary. His intention was obviously to give the Slavs their proper place in the ethnographic mosaic. In so doing, he virtually revolutionized the prevailing ideas on the distribution and character of the peoples of south-eastern Europe. For the next fifty years his map set the fashion for nearly all ethnographic maps of this area.

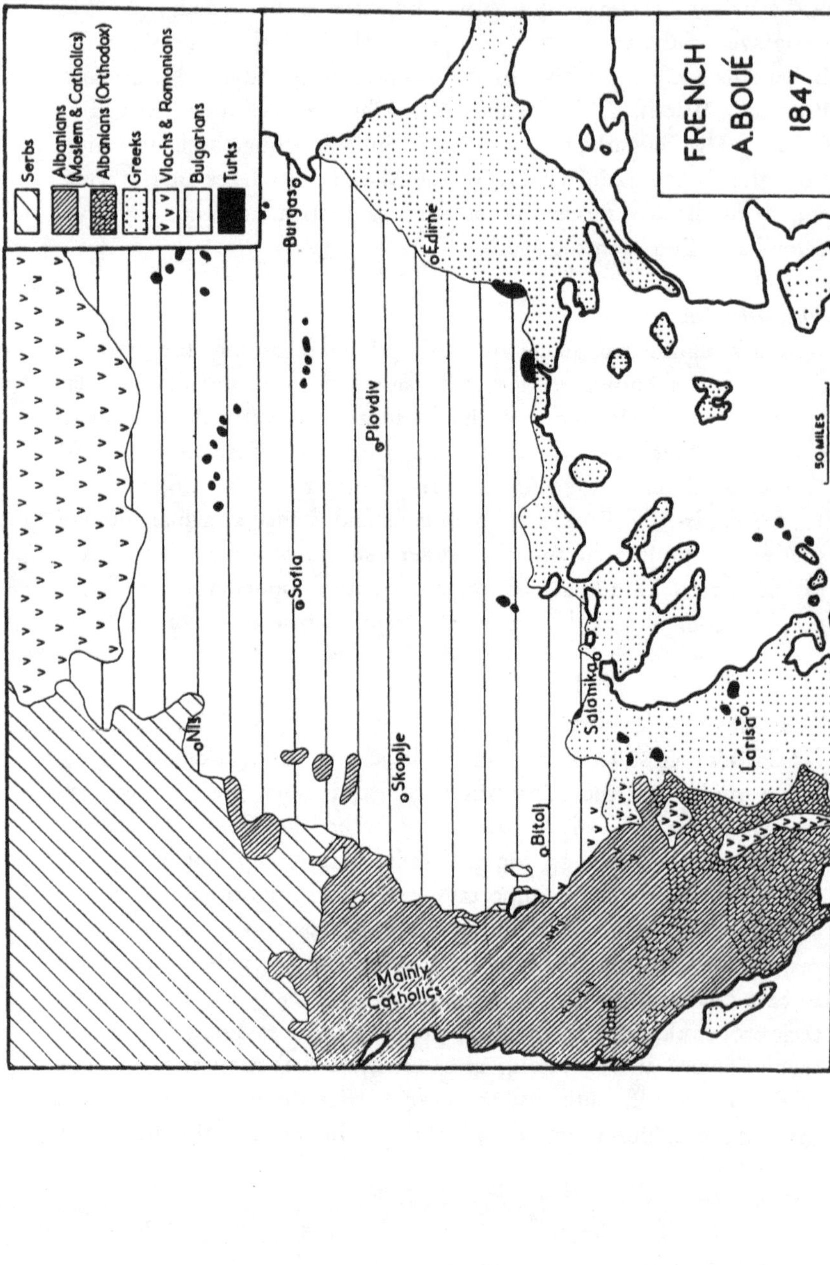
Serbs
Albanians (Moslem & Catholics)
Albanians (Orthodox)
Greeks
Vlachs & Romanians
Bulgarians
Turks
FRENCH
A.BOUÉ
1847
Burgas
Edirne
Plovdiv
Sofia
Niš
Skoplje
Bitolj
Salonika
Larisa
Mainly Catholics
Vlonë
50 MILES

FIG. 9.

## A. Boue's Map of 1847

Only five years after the appearance of Safařik's map, Ami Boué's ethnographic map of Turkey-in-Europe was published in Berghaus's atlas (Fig. 9). From a purely cartographical point of view it was considerably in advance of Safařik's work—an improvement only to be expected since Boué was a professional geographer. Boué had many qualifications which fitted him for the task of improving upon the ethnographic map of Turkey-in-Europe. In the first place, he was the foremost authority in Europe on the region. His four volume *magnum opus* on Turkey-in-Europe, which had appeared in 1840, was based on knowledge derived from extensive travel and personal observation.[1] It remains even today a primary source of information for geographer, historian and ethnographer alike. Unfortunately, copies of his work are rare, and little is known of Boué by British geographers. Although his map was not made public until 1847, we can infer from the text of Boué's writings, that he had already perfected his picture of ethnographic distributions even before Safařik's map appeared. There is no record of whether the two men exchanged information or whether they both arrived independently at the same major conclusions.

### *The Turks*

Boué was of the opinion that the Turks in Europe nowhere formed any considerable element in the population. He did emphasize, however, two aspects of their distribution. The first was the tendency of the Turks to settle in the towns, and the second was the existence of small Turkish enclaves of rural population in particularly well-favoured areas. These rural Turks Boué marked on his map. He showed the urban Turks by means of a symbol instead of in colour, so that the Turks appear to be very much under-represented on his map. He also stressed the importance of Turkish influence in Bosnia.

[1] Ami Boué, *La Turquie d'Europe, ou observations sur la géographic, la géologie, l'histoire naturelle, la statistique, les moeurs* [ . . . ] *histoire et l'état politique de cet Empire* [ . . . ] *avec une carte nouvelle de la Turquie d'Europe*, 4 tom. (Paris, 1840).

*The Greeks*

Boué did not show as much Greek territory as Šafařik had done. He perceived that a devotion to the Greek Orthodox faith did not necessarily imply a sense of Greek nationality, and that there were many Christians of the Greek Orthodox Church who regarded themselves not as Greeks, but as Albanians, Slavs or Romanians. Nevertheless, he was not ungenerous in his Greek dispositions. As early as 1840, he had recorded that the Greeks inhabited all the southern plain of Thrace and the neighbouring coastal plain of the Black sea ; that they were to be found in the Tekir Dag region of eastern Thrace, on the banks of the Maritsa, in and below Edirne, and in Plovdiv. He noted that, mixed with Bulgarians and Asiatics, they also made up the population of the area south of the Rodopi mountains and of the whole of Khalkidhikí. The Greeks were to be found also in southwestern Macedonia, where they were mixed with *Zinzares* (Vlachs) and Bulgarians. In this area, they formed the bulk of the population to the south of Kastoría and Siátista. Thessalía was eminently Greek except for a few Turks in Lárisa and in the towns ; the only other inhabitants of Thessalía were the *Zinzares*, who lived, stated Boué, on friendly terms with the Greeks and belonged to the Greek Church. Greeks were also found, mixed with Albanians, in Ipiros, in an area which stretched from the Greek boundary (of 1821) to Kónitsa. Here, whilst many of the villages were pure Greek in character, the towns were exceptionally mixed.

*The Slavs*

Boué was of the opinion that the Bulgarians occupied Bulgaria (meaning the district between the Balkans and the Danube) and 'Lower Moesia' as well as the greater part of 'Upper Moesia' (Moesia was a Roman province stretching from the Drin river to the Dobrudja). Furthermore, they formed the major part of the population of Macedonia, with the exception of the district southwest of Kastoría and the lower Néstos valley. He maintained that the mountains between the basins of Flórina and Kastoría were an effective language divide between Bulgarian and Greek, and that the mountains between Konia and Siátista, between Lake Vegorrítis and

Véroia and between Edhessa and Náousa, served a like purpose. The Bulgarians were also to be found, both dispersed and in villages, all over Thrace as far east as the Tekir Dag (*Bulgarsköri*).

Boué's distribution of Bulgarians was later put to good use by the Bulgarian 'revisionists.' His map was incorporated, for example, in P. Milioukov's pro-Bulgarian work on ethnographic maps published in 1900, and it formed part of D. Rizov's atlas produced in Berlin in 1917. Indeed, Boué's map put the Bulgarian case in such a favourable light that in 1918 A. Belić, the distinguished professor of Slavonic languages at Belgrade, was forced to make Boué's map the subject of a special article, in which he endeavoured to explain away the pro-Bulgarian views adopted by the latter. He maintained, but without conviction, that Boué's text could be reconciled with the Serbian view, that the Slavs of Macedonia, although they had Bulgarian affinities, were more nearly Serbs than Bulgarians (*cf.* Fig. 43).

Boué classified all the miscellaneous Slav peoples between the Sar mountains and Slovenia as Serbo-Croats, in the same manner as did Safařik. He, too, dispensed altogether with the term 'Illyrian.' His map contained little evidence to support the cause of the Serbs in Macedonia, although they were later to emerge as one of the principal claimants to this region. He did, it is true, recognize small Serb exclaves in Macedonia, north and east of Lake Ohrid, but he confined the bulk of the Serbs to an area well to the north of the Sar mountains and west of the Timok river. Hence he virtually excluded them from Macedonia, agreeing in this respect with Safařik. It is an interesting point that contemporary Serbian maps agreed in the main with these conclusions. For example, D. Davidović's map published in Belgrade in 1848, a year after the appearance of Boué's map, did not recognize the existence of any Serbs in Macedonia, or for that matter in Old Serbia (Fig. 10). Neither did the limits set by G. Desjardins in his map of 1853 differ appreciably from those set by Boué, in his depiction of the districts in which the Serb language was spoken (Fig. 10). Later, Serbian scholars, particularly S. Gopčević and J. Cvijić, maintained that Boué was not sufficiently conversant with the Slavonic languages to be able to distinguish Serbian from Bulgarian. Their criticisms were made, however, at a time when the political situation in the Balkans had changed and at a time when Serbian expansionist aspirations had been canalized towards the south by Austria's hardening hold on the Adriatic coastlands (see p. 146).

*The Vlachs*

The Vlachs, friends of the Greeks, as Boué described them, were plotted on his map, not as a compact, but as a scattered community. The pre-Safařik maps had shown various distributions of Vlachs in

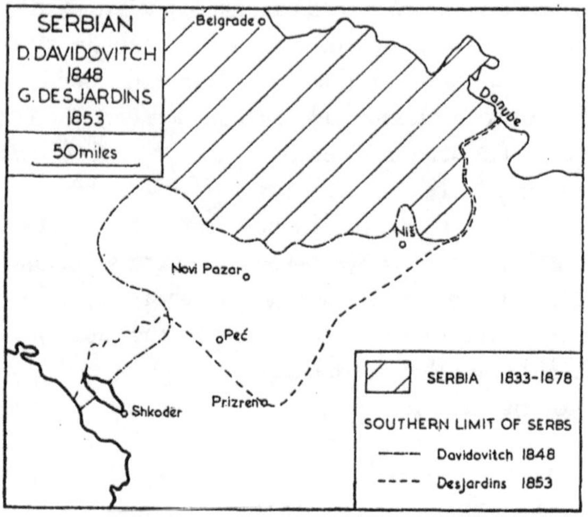


FIG. 10

'Eastern Roumelia.' Boué shattered the belief that the Vlachs were in some way still closely related to Roumelia, but perhaps he went too far in stressing their association with the Greeks. He stated that the *Zinzares* or Vlachs were sometimes difficult to distinguish from the Greeks. They had been established for a long time in Turkey, dispersed and living like the Greeks as isolated families in Turkish districts and towns, and again like the Greeks, they displayed proof of commercial aptitude. Finally, he declared that statisticians in putting the Vlachs at 600,000 had over-estimated their numbers by nearly a half.

*The Albanians*

Boué was inclined to favour a widespread distribution of Albanians and, although not quite so liberal as Müller had been in 1842, marked as Albanian all the territory between Lake Shkodrs and the Gulf of Corinth and between the Adriatic and the Drin river. He depicted Albanians *en masse*, not only in Ipiros, but also in western Thessalía. Boué suggested the estimate given by the statisticians of 1,600,000 *Skipetars* (Albanians) might err on the side of moderation when it was considered that they extended from " Epirus to the eastern part of Upper Moesia." They intermingled with the Serbs in the plain between Prizren and Peć, and between Vranje and Mitrovica. They also inhabited the country round the sources of the Lepenac, the environs of Suva Reki and the south-eastern fringe of the plain of Pristina (Kosovo). Also they were found in Kratovo, Kuršumlija, Prokuplje and Metohia. He stated that along the entire length of the Macedonian frontier east of the Crni Drim valley, the Albanians mixed very little with the Bulgarians, but to the south of Lake Ohrid they intermingled with the Vlachs and to a greater degree with the Greeks. Indeed so much fusion in the form of intermarriage had taken place in the area, that it was difficult, he maintained, to distinguish Greek from Albanian.

*The Influence of Boué's Map*

Boué's map and his works on Turkey, appearing at a time when the Balkans were assuming international importance, exerted a tremendous influence both culturally and politically. His conclusions drew attention to the possibility of Slav political hegemony in the Balkans and to the idea of a South Slavonic confederation. Being of French origin and credited with academic integrity and an impartial approach, Boué, as a prophet, created a more profound impression than did Safařik. His work did much to stimulate Bulgarian national activity and it was not long afterwards that, aided by Russia, the Bulgarians began their campaign for a national Church, independent of Greek control (see pp. 58-61). Boué's findings, incidentally, lent moral support to Russian designs in the Near East. It was perhaps an unfortunate combination of circumstances that led to the unhappy association of the Bulgarian national cause with the scheme for Russian imperialistic expansion. The important role which the Bulgarians might assume in the future history of the Balkans was not appreciated by the Governments in power in western Europe at the time (1847). An earlier appreciation of these new ethnographic ideas on the part of Great Britain, for example, might have gone a long way to avoid the tragic events which followed in later years.

## G. Lejean's Map of 1861

To lend weight to Boué's interpretation came G. Lejean's influential work on the ethnography of Turkey-in-Europe, produced as a supplement to *Petermann's Mittheilungen* in 1861 (Fig. 11). Lejean had journeyed to Turkey-in-Europe in 1857-58 at the request of the French Government, which, in common with the governments of the other Great Powers at this period, was vitally interested in assembling all intelligence possible concerning the situation in south-eastern Europe. He had at his disposal the maps of P. G. Safařik and of A. Boué, and K. Czoernig's masterly treatment of the ethnography of the Austro-Hungarian Empire provided him with a method of approach which he applied to the study of the ethnography of the Osmanli possessions in Europe.[1] Lejean's sources also included the works of F. C. L. H. Pouqueville, of W. M. Leake and of W. Jochmus. In addition, he used the lists of Moslems and *Rajas* provided by the Turkish Government; but these statistics merely consisted of an incomplete listing of Turkish and Christian households and did not form a very reliable source.[2] Above all, Lejean was an historian. He pointed out in his comments that many of the deficiencies in the earlier maps had been due to too great a dependence on language as a criterion.

[1] *Ethnographische Karte der österreichisch-Ungarischen Monarchie* [770 × 570 mm.] Von Carl Freiherrn von Czoernig, Direktor des österreichischen Statistischen Instituts (Wien, 1855).

[2] At this time, only the *Rumi* (or Greek) *Millet* and the *Islami Millet*, were officially recognized by the Turks. A *Millet* in the words of Sir John Hope Simpson was "a minority corporation recognized by the government and the law of Turkey, which enjoyed extensive ecclesiastical, scholastic and judicial autonomy." The *Bulgari* (Bulgarian) *millet* was not recognized until 1872.

He advocated an historical approach in order to deepen perspective in these matters.

*The Turks*

He first demonstrated that both Boué and Safařik had underestimated the strength and extent of the Turkish enclaves in the Balkans ; but on the other hand, his own reliance on Turkish figures led him to swing the scale too far in the opposite direction. His map, for example, marked a solid phalanx of Turks in north-eastern Bulgaria. He pointed out many interesting aspects of the Turkish settlement of the Balkans. He drew attention to the *Vardariotes*, pre-Osmanli Turks, who had become Christianized, had settled in the Vardar valley and then had been assimilated by the Slavs. (F. Meinhard distinguished a similar group in south-western Macedonia but referred to them as *Bardariotes*. See p. 127). The *Uzes* were another group of pre-Osmanli Turks who had settled near Lake Ohrid in the eleventh century and had also lost their identity to the Slavs. The *Konariote* Turks, called after the town of Konia in south-western Macedonia, had also been very early arrivals, probably associated with the nomadic herders who had composed the first wave of the Turkish onslaught. They had eventually settled down to become a peasant community and had managed to preserve their Turkish character. This group Lejean marked on his map. Also grouped with the Turks were the *Yüruks*, referred to in 1821 by O'Etzel (see p. 11). The name in Turkish signifies nomad. Lejean claimed that they were of pure Turkoman stock and that they dwelt in the plain of Sérrai with extensions as far as Dráma on the one side, and the mountains north-east of Salonika on the other. These also he marked on his map.

*The Greeks*

Lejean held that Greek territory in Thrace and on the Aegean coast was not nearly so extensive as was universally believed in his day. He reduced their distributions mainly to the advantage of the Turks and Bulgarians. The idea that the Greeks were preponderant on the mainland had been steadily losing ground between 1821 and 1861. The fact that Müller had preferred the term 'Pelasgian' in

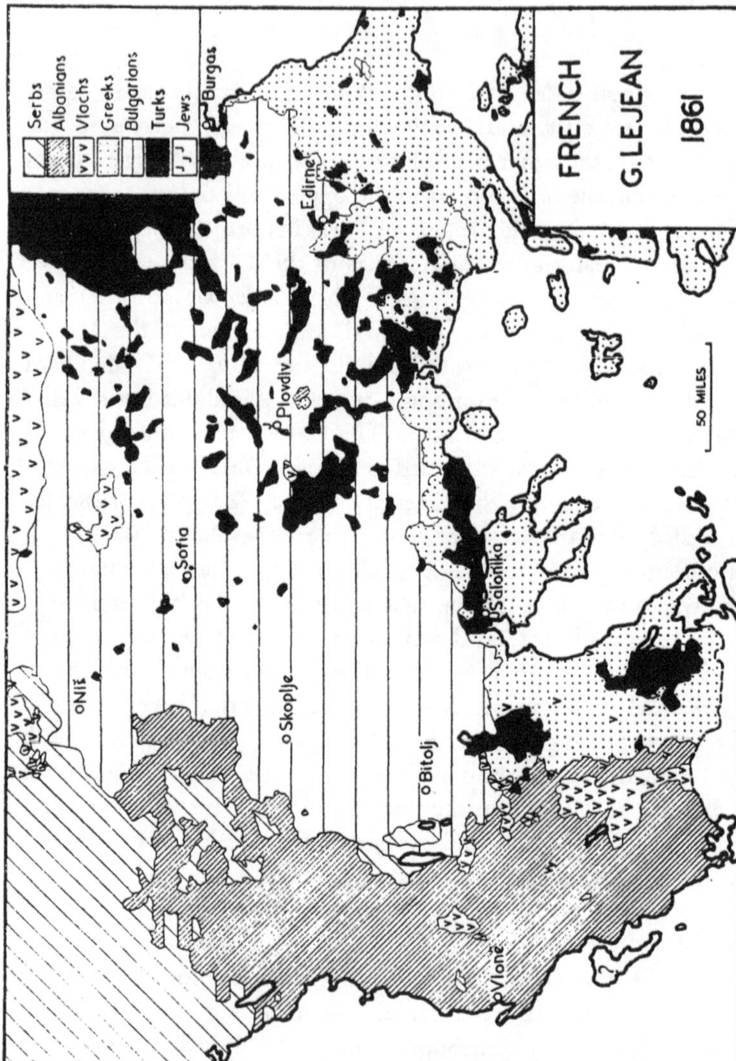


FIG. 11

Turkomans and Yüruks, separately distinguished in the original, have been included within the Turkish category

1842 had been a significant indication of the trend of opinion.[1] Fallmerayer's thesis, that nothing remained in this part of Europe of the original Hellenic type, had dealt a shattering blow at the idea of a Greek Balkans. Towards the middle of the century constant rumours of the venality of the Greek clergy had engendered a feeling of contempt on the part of the West for the Greek Church, a feeling which steadily grew until the term *Phanariote* connoting, amongst others, a high dignitary of the Greek Church, came to have a distinctly unsavoury nuance. Lejean depicted Thessalía and Khalkidhikí as the real strongholds of the Greeks outside the archipelago. He showed them also along the coast of the Aegean and along the coast of the Black sea from Salonika to Burgas, but he stressed that here they were, for the most part, mariners and fishermen with no vital interest in the interior. The Greeks were at this time (1861), still confined politically to their Archipelago, and they did not dispute Lejean's findings, but at a later date, when the full implication of his thesis was apparent, they subjected him to heavy criticism. In 1919, V. Colocotronis, for example, accused him of religious prejudice and even went so far as to maintain that Lejean carried an incidental commission from the Bulgarians interested in the formation of the Exarchate.[2]

*The Slavs*

Lejean believed with Safařik and Boué, that the Bulgarians prevailed over the greater part of the Balkans including four-fifths of Macedonia and that the geographical limits of the Bulgarians comprised the Danube as far west as the river Timok, Niš, Prizren, Ohrid, Kastoría, Náousa, Salonika, Edirne and Sizeboli, the Black sea coast as far north as Burgas, Sliven and Razgrad. Outside these limits there were also groups of Bulgarians to be found amongst the Albanians, amongst the Romanians and amongst the Greeks ; they were represented also in Bessarabia, in the Dobrudja and even in Asia. Lejean apparently knew more of the origins and history of the Bulgarians than did his

[1] 'Pelasgian' was the name used by nineteenth-century classical historians to refer to the inhabitants of the Greek mainland, in contradistinction to the term 'Hellenes' used of the population of the archipelago (see p. 20).

[2] V. Colocotronis, *La Macédoine et L'Hellénisme* (Paris, 1919).

predecessors in the ethnographic field. He pointed out that the Bulgars were of Ugrian origin, that they had been mentioned in Armenia 600 years before their arrival on the banks of the Danube in 484 A.D. For a time they had occupied Moldavia and Wallachia before crossing the Danube and raiding as far as Constantinople. They formed their trans-Danubian Empire in 679 and remained a power until suppressed by the Byzantines in 1019. Thereafter they united with the Vlachs to form another Empire in the twelfth century. Two hundred years later, they were finally overwhelmed by the Turks at Kosovo. (The Serb historians were later to claim Kosovo as the last great battle fought by the Serbs. Kosovo figures in many Serb traditional ballads. It is very probable that the army confronting the Turks at Kosovo included all types of Slavs, as well as many alien mercenaries).[1] In spite of their origins, remarked Lejean, the Bulgarians were essentially Slavonic from a cultural point of view. They had spoken Slav since the ninth century when they had been Christianized and their leaders had been made *boiards* (dukes).

Lejean differed little from Boué and Safařik on the all-important question of the existence of Serbs in Macedonia. He followed Boué in showing small Serb exclaves around Lake Ohrid and Lake Prespa but these were isolated by Bulgarian territory from the main mass of Serbs, north of the Sar mountains. Lejean envisaged the entry of the two great Slav branches, the Serbs and the Croats, into 'Illyria' about the seventh century ; probably the Serbs were derived from the *Soraben* (Sorbs) of Lausitz, whilst the Croats who came a little later were originally known as the *Horvat*. They were associated with other tribes such as the *Zachlumi* from Chelm (Za-chlum). The latter became known later as the Hercegovinians. Nearly all Slav tribes were initially named after their geographical locality, for example, the *Narentari* after the name of the river, the *Diocleaten* after Dioclea in Montenegro. The Serbs, according to Lejean, actually pushed as far south as Durrës before being driven back by the *Skipetars* (the Albanians) ; Lejean then advanced the interesting theory that the Serb exclaves around Lakes Ohrid and Prespa, and in northern Albania, dated back to the ninth century and that they marked the

[1] H. W. V. Temperley, *History of Serbia* (London, 1917).

southern limits of Serb expansion. Alternatively, he suggested, these isolated Serbs could be the remnants of the *Zupani* or colonies of Serb officials, established as outposts of the mediaeval Serb Empire. The fact that Lejean did recognize these Serb elements in the population of Macedonia is both interesting and important. When the Serbians began to claim a sphere of interest in Macedonia, they stressed the evidence of their historical colonization of much of Albania and Macedonia. Lejean, however, regarded the Serb remnants as purely of academic, and not of political interest.

Lejean believed in the close affinities of the Slavs of Turkey, north of the Sar mountains, and agreed that they formed a composite 'Serbo-Croat' group which included five divisions.

(1) The Serbs.

(2) The Bosnians of Turkish Croatia, who differed considerably from the Serbs on several counts. Whereas the Serbs constituted a free society, Bosnian society was essentially feudal. The Christian peasants laboured under an oppressive Moslem aristocracy which was intensely conservative. Travellers in Bosnia often mistook upper-class Bosnians for Turks.

(3) The Rascians of Novi Pazar.

(4) The Hercegovinians.

(5) The Montenegrins who had preserved their independence since 1504 under their bishop princes. Their small population contained numerous Serbs, who, having fled from the Turk, found refuge in this lofty mountain kingdom.

All five branches were closely related. Their differences of language were negligible and although drawn from varied stocks, such as the Avars and the Illyrians, the process of 'slavization' had been very effective. Even the *Morlachs* of the northern Adriatic coast had been slavized. They were the remnants of the pre-Slav inhabitants and Lejean thought they may have been latinized Illyrians.

### *The Vlachs*

Lejean referred to the Vlachs as *Zinzares* and speculated on their origins. They themselves, wrote Lejean, claimed to be the descendants of Roman soldiers, who had conquered Macedonia. He thought this was hardly likely as their language was not pure Latin but rather was derived from 'latinized Dacian' (i.e. Romanian). They probably represented descendants of Dacians transferred from Moesia. Their language was understood by the Romanians, from whose language however it differed. Of every eight Vlach words, only three were derived from Latin, two were imported Moslem words—Greek, Turk, etc.—and three belonged to an unknown basic stock similar to Albanian. Lejean increased the distribution of the Vlachs in south-western Macedonia largely at the expense of the Greeks.

*The Albanians*

In his distribution of Albanians Lejean made a substantial departure from Boué's map. He ascribed a great extent of territory to the Albanians in Old Serbia and in northern Macedonia. This departure was partly due to the influences of the researches of J. G. von Hahn, an Austrian railway engineer who maintained, as a result of personal surveys made in 1858, that a greater part of the population of the region between the Drin and the Vardar, hitherto regarded as Serb and Bulgarian, was in fact Albanian. Hahn's thesis was that a great wedge of Albanian territory stretched from the bend of the Drin valley to the valley of the Morava and more or less effectively separated the Serbs from the Bulgarians over a considerable zone. Lejean, whilst his map was influenced, as indeed were most subsequent maps prior to 1913, by von Hahn's discoveries, did not accept all von Hahn's claims for the Albanians. Large tracts of the Morava valley were, however, acknowledged by Lejean as Albanian, and he justified this innovation by reference to that large-scale replacement of Serbs by Albanians which had, in 1861, been in operation for over 200 years. This process had been encouraged by the Turks themselves for obvious reasons, since the Albanians co-operated with them, whilst the Serbians were a constant source of annoyance. The northward and eastward migration of Albanians had reached such proportions that Lejean was able to record outpost villages of Albanians at *Novo Crdo* near Vranje and at *Arnaut-Koi* near Tiranë (Arnaut was the Turkish name for Albanian).

Like Boué, Lejean emphasized the religious differences between the

Albanians but he did not, as Boué had done, show religious distributions on his map. He depicted Orthodox, Catholic and Moslem Albanians as one ethnic group, at the same time stressing that they were by no means united. They were organized on a tribal basis and, furthermore, were divided by the river Shkumbin into two major groups—the *Guegues* (Gegs) in the north and the *Toskes* (Tosks) in the south. They all called themselves *Skipetars* (a word which signifies mountaineer in Albanian). They were, Lejean believed, the remnants of the hellenized Illyrians, who, in the classical era, had occupied all the southern Balkans (*cf.* Weigand's view on their origins, Appendix A).

*Jews and Gypsies*

Lejean, in his notes on the ethnography of this region, made many references to minorities which, although not numerous enough to warrant depiction on the map, nevertheless deserved some mention. The *Mammins* or Moslemized Jews of Salonika constituted such a group.[1] Together with the Orthodox Jews, they formed the bulk of the population of the port. Most of the Jews of Salonika were the descendants of refugees from the Inquisition, expelled from Spain during the reign of Isabella, and later of Philip II. They still spoke Espaniol or *Ladino*. The Gypsies he also distinguished as a separate group ; they were known variously as Bohemians, *Gitanos, Zigeuner, Tsiganes, Tsigani* or *Tchinghench* ; Lejean described them as Hindus ejected from India by the Mongols in the fifteenth century, although the *Sigynnes* mentioned by the ancient Greek geographers, Strabo and Herodotus, appeared to be the same people. They still led a nomadic life, but numbers of them had settled as skilled iron workers. Lejean made a reference to gypsies in the High Maritsa. Müller had also made a special point of representing the *Zigeuner* in this area in 1842 (Fig. 6).

*Conclusion*

A. Petermann praised Lejean's map as 'epoch-making,' although in actual fact it did not differ in its broad classifications from the maps of Boué and Safařik. What Lejean did do was to cement together

[1] Sometimes referred to as *Domnes*.

FIG. 12
In the original, the Vlachs of Macedonia and Albania are not depicted but one may infer their distribution from the blank spaces left on the map.

the mosaic loosely assembled by his predecessors, and modify some distributions, especially those of the Turks and Albanians. He went into the question of origins and made fuller use of place-name evidence and historical references. It is difficult to find any contemporary criticisms of his work. One senses a feeling in the literature of the period that Lejean's study had, to all intents and purposes, settled the vexed question of ethnographic distributions in Turkey-in-Europe. Safařik had put forward the idea of the Balkans being essentially Slavonic, Boué had corroborated his thesis by personal observation, and then Lejean had substantiated the idea with a wealth of historical evidence.

## The Map of Miss G. M. Mackenzie and Miss A. P. Irby, 1867

Strangely enough, it was due to the efforts of two ladies, Miss G. M. Mackenzie and Miss A. P. Irby, who had travelled 'unescorted' through the Balkans, that the essentially Slavonic nature of the peninsula was made known to the British public. Their map—the first English ethnographic map of the Balkans—was published in 1867 (Fig. 12). It was based largely on Lejean's map but incorporated minor modifications arising partly out of their own personal observations and partly from the criticisms of an Austrian railway engineer (who may have been von Hahn). A reference to the rather delightful account of their journey leaves no doubt that these two ladies were extremely well informed of conditions in the Balkans. Their work impressed Gladstone who wrote:

> I do not mean to disparage the labours and services of others when I say that, in my opinion, no diplomatist, no consul, no traveller, among our countrymen, has made such a valuable contribution to our means of knowledge of this important matter [the question of the Ottoman Empire], as was made by Miss Mackenzie and Miss Irby, when they published in 1867, their travels in some of the Slavonian Provinces of European Turkey.[1]

[1] *op. cit.* (Preface to 2nd edition, 1877).

*The Turks and Greeks*

Miss Mackenzie and Miss Irby first emphasized that the Turks formed a very small minority in the Balkans, a necessary prelude to any discourse on Turkey-in-Europe in 1867 since A. K. Johnston's atlas, as late as the edition of 1856, was still utilizing G. Kombst's map showing the greater part of Thrace and southern Macedonia as Turkish territory (see p. 25). Secondly, they emphasized that throughout Turkey the name Greek was used to denote a Christian of the Eastern Church, and pointed out that the idea that the population of Turkey was Greek by ' race ' had lately been dissipated by persons careful to insist on the antipathy that really divided Slav from Hellene. According to their text :

> The line of the Roman " via Egnetia " was roughly the boundary between the Bulgarians and Greeks, in that part of it which runs between Salonica and Ochrida [. . . ] though it leaves some Bulgarian country to the South and some Greek settlements to the North. It would be more difficult to draw a S.E. boundary from the coast of the Black Sea to the Gulf of Salonica for the Slavs do not at any point touch the Bosphorous or the sea of Marmora. In Ţhrace, Adrianople may be taken as a boundary city for the Bulgarians [and Greeks].

*The Slavs, Vlachs and Albanians*

Miss Mackenzie and Miss Irby were convinced of the preponderance of the Slavs over all other groups, and of the fact that Macedonia was, in the main, Bulgarian territory (the Bulgarians they regarded as a branch of the ' Yugo-Slavi'). They did not attempt to show the distribution of Macedonian Vlachs on their map and they believed that they were Greek in sympathy. They wrote of the Greeks in Bitolj, for instance, as being represented by the crafty ' Tzintzars ' (Vlachs). Their distribution of Albanians was much the same as that of Lejean.

*Conclusion*

It might have been expected that the work of Miss Mackenzie and Miss Irby would directly influence public opinion in the British Isles.

But the British belief in the Hellenic character of the Balkans was not so easily shaken. Perhaps it was that ethnographic opinion tended to be very conservative. If so the British public was not the only offender in this respect. In 1855, the second edition of M. A. Denaix's atlas still carried an ethnographic map of the Balkans showing the Greeks as a majority, and in 1873 there appeared in Paris F. L. Passard's map which actually favoured the Greeks with an ethnographic frontier on the crest of the Balkans. As late as 1877, E. Stanford published a map in the British Isles likewise depicting an extensive distribution of Greeks (see p. 70).

## Two Slav Maps, 1867-8

### *M. F. Mirković's Map of 1867*

Russia's assiduous cultivation of the idea of a great Slav brotherhood began to bear fruit about the same time as the work of Miss Mackenzie and Miss Irby appeared. It took the form of a great Pan-slav Conference in Moscow in 1867. M. F. Mirković's map, showing the ethnographic distribution of Slavs in Europe, was one of the chief exhibits. The conference was intended to draw attention to the political significance of the recent researches which had so profoundly modified the ethnographic map of eastern Europe.

Mirković based his Balkan distributions mainly on those of G. Lejean. He differed from Lejean in two important respects. First of all, he broke up Lejean's solid Turkish groups in north-eastern Bulgaria into mixed Bulgaro-Turkish exclaves. Secondly, he extended Bulgarian territory in Thrace well to the south of Edirne, depicting many Bulgarians where Lejean had indicated only Greeks and Turks. Otherwise his map was very similar to that of Lejean, more especially in the western Balkans. The Serbo-Bulgarian frontier remained unchanged and the Albanian distributions in Old Serbia were not disputed. That these distributions should pass unchallenged in 1867, at a Slav conference, was a tribute to Lejean's work, and at the same time an indication of the fact that all the Slavs, including Serbs and Bulgarians alike, were in agreement with the conclusions embodied in Mirković's map. It is true that much later, in 1906, J. Cvijić under-

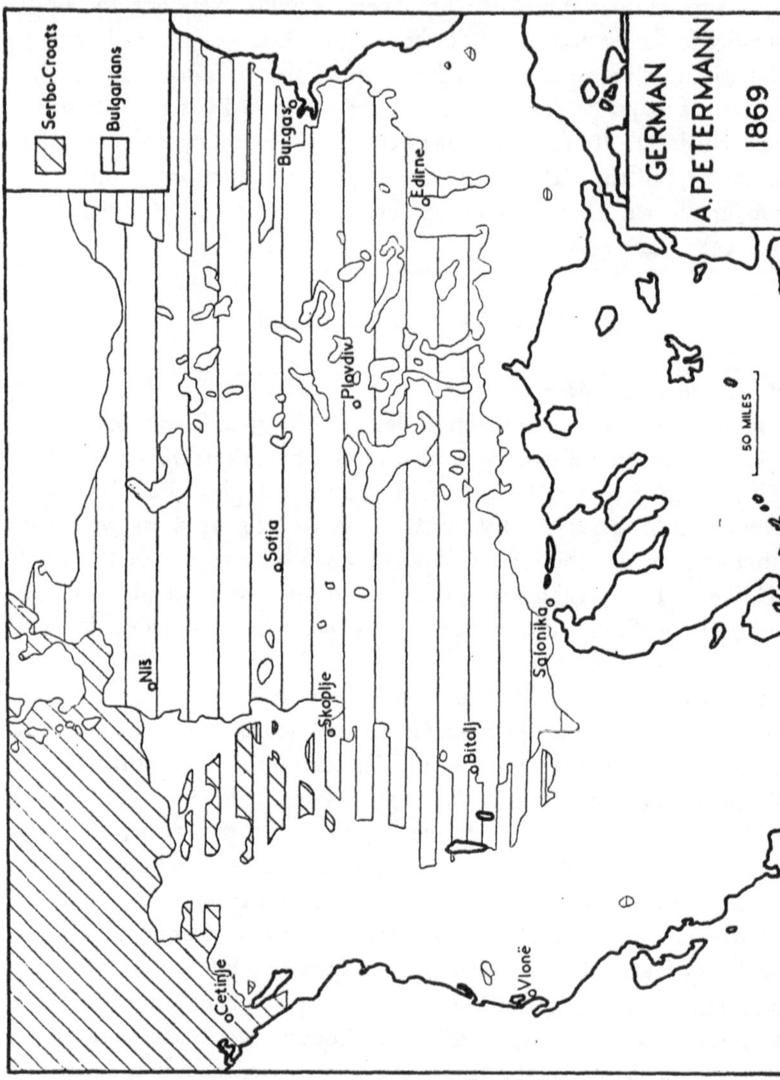
Serbo-Croats
Bulgarians
GERMAN
A. PETERMANN
1869
Burgas
Edirne
Plovdiv
Sofia
Niš
Skoplje
Salonika
Bitolj
Cetinje
Vlonë
50 MILES

FIG. 13

took to criticize Mirković's map, but he failed to draw attention to the fact that he was referring to the third edition (1877) of a map, which, when it was first published, contained a fair reproduction of the views held at the time by all Slavs, including the Serbs.[1]

### *J. Erben's Map of 1868*

J. Erben's ethnographic map of 1868 was on rather a small scale. It was important only in so much as it represents the Czech opinion of the period. Erben cited Safařik, Czeornig, Boekh, Lejean, Koeppen and Mirković as his sources. He followed Mirković's distributions in the Balkans. His only major modification was a reduction of the Albanian element in Old Serbia to the advantage of the Serbs.

## A. Petermann's Map of 1869

In 1869, A. Petermann compiled a composite map which was intended to be a summary of the latest researches into the ethnographic distribution of the Slavs in the Balkans (Fig. 13). A. Petermann was the editor of the German geographical periodical, *Petermann's Geographische Mittheilungen* which, during the period of his control, acquired a world-wide reputation. There was hardly a geographer of repute of any nationality whom Petermann did not know personally. Ethnographic maps interested him and in 1861 he had made Lejean's map the occasion of one of his famous supplements.

The theme of Petermann's map and of the article which accompanied it, was the importance of the 'new idea' of Balkan ethnography. F. Bradaška, a Croatian professor responsible for the article, pointed out that ideas on the nature of the population of Turkey-in-Europe had been revolutionized in the thirty years between 1840 and 1869. Initial over-estimates of Turks and Greeks had been due to lack of Slav studies and also to the fact that Moslem Slavs and Albanians had been wrongly declared Turks. He pointed out that the Slavs, practising agriculture, were passive and inarticulate whilst the Greeks, engaged in commerce with the outside world, overshadowed them, and that the maritime connections of the Greeks gave them control

[1] J. Cvijić, *The Ethnography of the Macedo-Slavs* (London, 1906).

of the coast and created the impression that their hold on the interior was equally secure ; the Greeks were also the heroes of neo-classical sentiment and above all they controlled the schools and the churches of the Slavs. Finally, he noted that it had been the deliberate policy of the Turks to make concessions to the Greeks and to the Albanians, who were given a free hand to suppress the Slav elements of the population. All these factors had been instrumental in blinding observers and scholars to the fact that the Slavs and not the Greeks, formed the majority of the inhabitants of Turkey-in-Europe.[1]

*The Slavs*

The change in ethnographic ideas in favour of the Slavs, which had been initiated by P. G. Safařik and had been confirmed by the views of A. Boué and Lejean, considerably impressed Petermann. He realized that the Pan-slav movement, which had been a great force in central Europe, had now taken on an ominous significance in the Balkans. What impressed him was, not the differences which might exist between the Slavs in this region, but the similarities—the fact that their alleged affinity might indicate a common political future. It is true, he distinguished Serb from Bulgarian on his map by a line from the Sar mountains to the Timok river, through Prokoplje, thus including Vidin, Niš and Leskovac within Bulgarian territory: he also indicated an isolated Serb exclave on the banks of Lake Prespa. But he marked all the Slavs, both Bulgarian and Serb, in the same colour, in order to present their hegemony in south-eastern Europe with greater force.

The preoccupation of Petermann with the political significance of the Slav groupings was symptomatic of a foreboding universally held at this time. This was the fear that all the Slavs would ultimately unite into one political unit—an idea based to a large extent on the postulate that only superficial cultural and linguistic differences existed between their various branches. In particular, there was a definite tendency to regard the relationships between 'Southern Slavs,' including Serbs and Bulgarians,[2] to be so close as to warrant a common

[1] F. Bradaška, "Die Slaven in der Turkei," *Petermann's Mittheilungen* (Gotha, 1869).

[2] 'Southern Slav' was used later to denote the Serbs, Croats and Slovenes, and excluded the Bulgarians.

political fate. Petermann had coloured them as one to illustrate this idea. Miss Mackenzie and Miss Irby had referred to Serbs and Bulgarians as the 'Yugo-Slavi.' Indeed, the supposition that a common political fate was in store for both Serbs and Bulgarians was one of the principal reasons why the Serbo-Bulgarian ethnographic boundary was not regarded as a crucial issue before 1870. Before that date the Serbo-Bulgarian frontier was purely an academic question. Until then Serbs and Bulgarians were members of the 'Southern Slav' family fighting for freedom and recognition. It mattered not whether the inhabitants of Macedonia were regarded as Serbs or Bulgarians. The important thing was that they were regarded as Slav.

*The Albanians*

Petermann was not primarily concerned with the non-Slav groups, but the distribution of the Albanians of western Macedonia may be inferred from his map. He was of the opinion that here they formed a considerable minority, west of a line from Skoplje to Kastoría, through Veles and Bitolj. H. Kiepert later incorporated this distribution of Albanians in his map of 1876 (see p. 66).

## Chapter IV

# BULGARIANS AND GREEKS, 1870-1878

### The Bulgarian Exarchate

In 1870 an event of far-reaching importance occurred—the Turks established the autonomy of the Bulgarian Church by appointing a Bulgarian Exarch. The Bulgarian national revival had necessarily been slower and later than that of the Serbs or Greeks. The fact that the Bulgarians were in closer physical proximity to Constantinople was one reason why they were not able to express themselves so freely. South of a line drawn from the Danube through Vidin, Niš and Prizren lay the real core of the Turkish Empire in Europe. In this area Turkish settlement was closer than it was farther north. Here the Turkish *begs* were established in their *cifliki* or large farms on which the Slav *rajas* laboured. In this region Turkish reaction to any separatist movement was strong and immediately effective. In Bulgaria and Macedonia, Turkish power was thus more firmly established than it was in the Serbian provinces farther north, or than it was in the Greek archipelago. For this reason the marginal zones of the Turkish Empire—Pelopónnisos, Serbia, Wallachia—witnessed scenes of national revival, whilst the Bulgarians still lay inert under the weight of Turkish administrative tyranny. But in spite of greater opposition the Bulgarian national movement had been slowly progressing. The first work printed in the vernacular had appeared in 1824 and the first Bulgarian school had been founded in Gabrovo in 1835. In the 1840's many schools were formed and Bulgarian literature blossomed forth, while the wealthier Bulgarians began to send their sons to be educated in Russia, Bohemia and France rather than in Constantinople as formerly. Bulgarians, who even until late in life had written the Greek language, persisted in learning to write their own. It was the obstacle presented to such efforts of natural self-expression by the hierarchy of the Greek Church that finally brought the Bulgarians to the resolution of freeing their Church from the control of Phanar (the district in Constantinople where the Greek bishops were quartered).

After 1840 the Bulgarian movement began to meet ever-stiffening opposition from the Greek Patriarch and the Greek prelates. Any national awakening on the part of the Bulgarians militated as much against the privileged spiritual position of the Greek Church in south-eastern Europe, as it did against the temporal power of the Turk. The Greek clergy not unnaturally used all their enormous influence to restrain Bulgarian nationalism, in order to retain the monopoly of ecclesiastical offices which had been theirs since the abolition of both the Old Serbian Patriarchate of Peć and the former Bulgarian Patriarchate of Ohrid in the eighteenth century. Ecclesiastical liberation was therefore the first step towards the development of Bulgarian nationalism and since 1854 the Bulgarians had been pressing for a restoration of ecclesiastical independence. In 1860 they had even negotiated with Rome for the establishment of a Bulgarian Uniate Church which while preserving Orthodox ritual would have been Roman in allegiance. Between 1860 and 1870 Bulgarian opposition to the Greek clergy had precipitated open insurrection on more than one occasion. In 1862, " A series of scandals took place throughout the provinces. Churches were closed in order that the Greek liturgy might not be read therein. When the Greek bishops returned from their revenue-gathering progresses they found their palaces locked and were conducted beyond the city walls. If they entered a church to officiate, no Bulgarian priest would take part in the service : when they departed the floor was ostentatiously swept, as if to remove traces of impurity. In Sofia, when a new bishop was expected, men, women and children filled the palace and blocked it up, till, unarmed as they were, they had to be expelled by Turkish soldiers."[1] The sympathetic attitude of the Russians towards the Bulgarians and their

[1] G. Muir Mackenzie and A. P. Irby, *op. cit.*

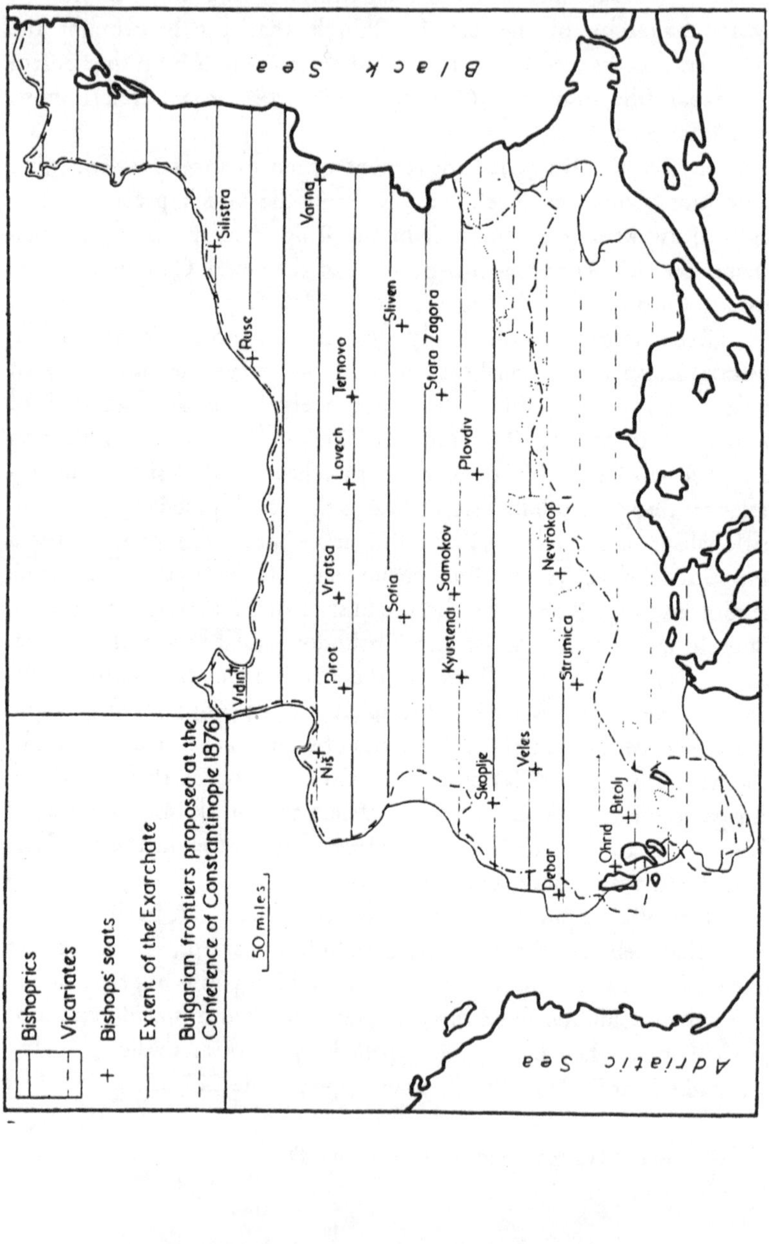


FIG. 14. THE BULGARIAN EXARCHATE, 1870–1912 (AFTER J. IVANOV).

grievances encouraged them to an open boycott of the Greek bishops. Russian influence in Constantinople was finally instrumental in obtaining a concession in favour of the Bulgarian point of view, and the establishment of the Bulgarian Exarchate between 1870 and 1872 guaranteed ecclesiastical independence and, at the same time, marked a new phase in the development of the Bulgarian national revival.

The creation of the Exarchate was an immensely important political concession. The bestowal of spiritual benefits upon the Bulgarians was one of its lesser functions. The Church played a part in the worldly affairs of eastern Europe which often transcended matters of mere spiritual significance. The newly-independent Church of Bulgaria, whilst it stood for religious freedom and maintained Bulgarian cultural traditions, was pre-eminently a political force. Its chief objectives were to combat the process of hellenization at work in Bulgarian territory and to prepare the populace for political independence. Its members were at once spiritual and political adherents of the Bulgarian cause. Its inauguration marked the end of that unholy alliance between the Sultan and the Greek bishops, which had jointly controlled for so long the destinies of the Slavs in the Balkans.

The most significant feature of the Exarchate was its extraordinary extent. Its limits were established as a result of a plebiscite organized by the Turks in 1872. According to the Bulgarian geographer, J. Ivanov (whose work appeared in 1917 and 1918), the extent of the Exarchate corresponded approximately to the distribution of Bulgarians on Lejean's map (Fig. 14). It included Vidin, Niš and Leskovac (which were ceded to Serbia in 1878), Macedonia and Thrace. The Exarch's influence was apparently not so strong in the south because he was there represented only by vicars, and not by bishops. The relative weakness of the Exarchate over much of southern Macedonia and Thrace would seem to suggest that the Slavs of these regions did not desire to be emancipated from Greek control at this time. The fact that large numbers of Slavs preferred the Greek to the new Bulgarian Church led eventually to the Greek hypothesis of the existence of *Slavophones* or *Bulgarophones* (i.e. Slav-speaking Greek nationals. See page 71). One further point needs stressing: according to the Serbians there were large numbers of Slavs in the

Skoplje, Niš and Vidin areas who embraced the Bulgarian Church, not because they were Bulgarians, but because the new Church offered a respite from Graeco-Turkish oppression.

The significance of the establishment of the Exarchate was not lost upon the Great Powers of Europe, nor upon the Balkan national states already in existence. The Great Powers, of course, believed that the concession had been wrung from the Porte by Russian pressure. The Greeks and Serbians excitedly remarked that a Bulgarian colossus had been born overnight and that Bulgaria would undoubtedly become the paramount power in the Balkans, if the domain of the Bulgarian Exarchate was any indication of the ultimate political extent of Bulgaria. It was from this date that the Serbians began to have certain misgivings about their Bulgarian brethren and the Greeks felt cheated of their rightful legacy, the mantle of the Byzantine Empire, so boldly lifted from their shoulders by the upstart Slav. The unanimity of opinion on the Serbo-Bulgarian ethnographic frontier, so remarkable a feature of all the maps so far considered, was to be shattered by this new approach, with its accent on the dissimilarities between the Slav groups, rather than on the qualities they had in common. The immediate effect of the creation of the Exarchate, however, was to emancipate the Bulgarians from Greek control.

## The Conference of Constantinople, 1876

In 1876 a Conference of the Great Powers took place at Constantinople with the objective of formulating a programme of reform in the Balkans and of forcing the Sultan to implement such reforms as he had already promised. The revolt of the Bosnians and Hercegovinians in 1875, and of the Bulgarians in 1876, had lifted the veil on the deplorable conditions still to be found in the Turkish Empire. The massacre of Bulgarians which followed the insurrection drew attention to their cause in the British Isles and invoked Gladstone's thunderous indictment. At the Conference, the universal dissatisfaction with Turkish maladministration was unquestioned. The problem was—who was to replace the Turk in Europe? The Conference revealed a clash of ideas regarding the nature and extent of the new political formations necessary to supplant the Turk in his European domain.

Ethnographic maps were about to play an important role in the re-assessment of the political situation. Major ethnic groupings were suggested at the Conference as a useful basis for the political reorganization of the Balkans. It would be wrong to imply, however, that any of the participants at the Conference were inspired by the principle of self-determination. Russia, at the time, was engaged in the systematic decimation of her nomadic minorities in the Ukraine; Great Britain was adamant in her refusal to recognize Irish nationalism; whilst Germany had recently added tens of thousands of Frenchmen to her population. The suggested use of the ethno-political boundary was largely a Russian idea which, had it been utilized, would have resulted in Slav hegemony in the Balkans. It was Ignatyev, the Russian Ambassador, who suggested the creation—"of an autonomous big Bulgaria stretching from Bourgas [ . . . ] to Dedeagatch [ . . . ] then to the Lake of Ochrida [ . . . ] Albania [ . . . ] Nish [ . . . ] Vidin and the Danube."[1] He produced a map showing such a state based on his own ethnographic data, and its boundaries coincided roughly with the Bulgarian ethnographic frontier as drawn on Lejean's map and also with the limits of the Bulgarian Exarchate (Fig. 90). *Petermann's Mittheilungen* reported that St. J. Verković's ethnographic data was also made use of at the Conference. He was a Serb but his conclusions on ethnographic distribution favoured the Bulgarians in Macedonia.[2] This was incidentally a further indication that the Serbo-Bulgarian ethnographic quarrel had not then developed.

Great Britain and Austria, however, were opposed to the idea of a great Slav state. The prospect of a political Pan-slavonia stretching from Prague to Moscow was still a dreadful possibility to the Great Powers of peninsular Europe. The fact that Russia had played the leading role in promoting Bulgarian national consciousness was taken to mean that the new state would merely represent an enlarged sphere

[1] M. D. Stojanović, *The Great Powers and the Balkans, 1875-1878* (Cambridge, 1939).

[2] "Statistisch-ethnographische Daten des Sandschaks Seres, gesammelt von St. J. Verković," mitgetheilt von Fr. Bradaška. *Petermann's Mittheilungen*, Bd. 24 (Gotha, 1878).

of Russian influence and thus the coincidence of Russian aims and Bulgarian national aspirations prevented either the Austrians or the British from taking a sympathetic view of Bulgarian plans. Ignatyev's proposals were therefore opposed by "[ . . . ] both Britain and Austria [ . . . ] in accordance with their traditional policies. The British proposed a division into two parts with the southern boundary moved much farther north to the Adrianople-Monastir line."[1]

The Conference achieved no concrete results. The Turks were allowed to remain in possession of Bulgaria and Macedonia, provided that they put into execution the much needed reforms. Their failure to do so led two years later, in 1878, to the Turko-Russian war. But if the Conference achieved nothing else, it did create an interest in ethnographic maps because it became clear that through the medium of ethnographic ideas the Bulgarians had gained a moral ascendency over all the other peoples of the Balkans. This may have been due in part to the widespread sympathy felt for the most oppressed of the Sultan's subjects, but even more so was it due to the simple fact that for thirty years the greater part of the territory between the Danube and the Aegean, between the Macedonian lakes and the Black sea, had been coloured as Bulgarian on scores of ethnographic maps. The terms Balkan and Bulgarian had become practically synonymous and this had been achieved solely by that subtle emotional appeal which emanated from ethnographic maps. The Greeks realized too late the magic of such maps. In the simple flat colours were to be seen the hopes and aspirations of a nation. The trend of events was already pointing to the maxim that if a people were to lay political claim to a territory, they must first establish their ethnic rights. The ambassadorial meeting at Constantinople brought home the principle of the ethnographic frontier to the peoples of Europe with all the force of a thunderbolt. It dramatized its significance and invested it with a propagandist value which the Balkan peoples were not slow to grasp.

The very idea of using ethnographic distributions to plot political boundaries was so revolutionary in its conception that the Greeks were horrified by the trend of events and made all haste to improve their

[1] Stojanović, *op. cit. Cf.* also *Turkey, Blue Book No.* 13 (1878), which includes a map of the proposed Bulgarian provinces.

moral position by a re-statement of their ethnographic claims. The British also, the French, the Austrians, even the Turks were startled by the grim possibilities of this primitive application of the principle of self-determination to a solution of the Balkan problem. All nations began to take an avid interest in ethnographic distributions and a spate of ethnographic maps appeared in the next year or so.

## H. Kiepert's Map of 1876

The most famous map of this period was compiled by the foremost geographer of his day—H. Kiepert, professor of geography in the University of Berlin (Fig. 15). Kiepert had always had a very strong regional interest in the Balkans, more especially in Greece. He was well acquainted with the history of the Balkan region and moreover was an able cartographer. His sources included such maps as had already appeared, in particular those of G. Lejean, M. F. Mirković and A. Petermann. His ethnographic map incorporated into a composite whole what he thought were the better points of each map. In addition, he introduced several innovations, based very largely on the writings of a Greek historian, P. Aravantinos, whose studies had been published in Athens in 1856-7. He also spent some time in Constantinople enquiring into official sources. Altogether he recognized six nationalities—Turks, Greeks, Serbs, Bulgarians, Romanians and Albanians.

### *The Turks*

As a result of his use of Turkish statistics, his map re-emphasized the Turks in Thrace at the expense of the Greeks. The tendency to reassess the numbers of Turks in the Balkans had begun with Lejean's map of 1861. A special map by Petermann of the Turks in Europe, produced early in 1876, had greatly underestimated their importance. Kiepert literally put them back on the map. In so doing he probably included a number of Moslem Bulgarians in his Turkish classification. The difficulty of distinguishing Moslem Slav from Turk had always been great, especially when, as was sometimes the case, the Slavs had lost their language.

GERMAN
H. KIEPERT
1876
Serbs
Albanians
Vlachs
Greeks
Bulgarians
Turks
Burgas
Edirne
Plovdiv
Sofia
Niš
Skopje
Bitolj
Salonika
Vlonë
50 MILES

FIG. 15

*The Greeks*

If he reduced Hellenic territory in Thrace, he increased it considerably on his map in south-western Macedonia and Ipiros. He believed that the Greeks stretched as far north as Edhessa, or Vodena as the Slavs called it. This was the town which Miss Mackenzie had referred to in 1867 as "the Bulgarian City of the Waters." His justification for classifying the inhabitants of most of southern Ipiros as Greek rather than as Albanian, was that the process of hellenization had proceeded to such an extent amongst the Tosks of southern Albania, that in many cases it was impossible to distinguish Greeks from Albanians, particularly from those Albanians belonging to the Greek Orthodox Church. It is of interest to note that he produced a map of Ipiros later, in 1878 (Fig. 19).

*The Slavs*

The distribution of Slavs on Kiepert's map had been taken from Petermann's map of 1869 and so the Bulgarians were shown as inhabiting most of Macedonia but not so widely distributed as they had been on the maps of Lejean and Boué. No Serbs were shown south of the Sar mountains. Kiepert removed the Serb minority round Lake Ohrid which had been a feature of some of the earlier maps.

*The Vlachs and the Albanians*

In Kiepert's view the Vlachs were so sparsely distributed throughout southern Macedonia as scarcely to justify recognition except in the Píndhos region, and he grouped the smaller enclaves with the Greeks. He believed that the Albanian element in Ipiros had been overestimated in the past, but he still showed the Albanians in strong force in both western Macedonia and in Old Serbia.

*Conclusion*

Kiepert's map had a great vogue. His work was known to the Tsar who was very impressed by his scholarship, and to Bismarck who expressed himself thus:

> La situation ethnographique de la Bulgarie, comme je le sais de source authentique et comme il résulte de la meilleure carte que nous

connaissons, celle de Kiepert, est celle-ci: les limites de la nationalité [Bulgarian] descendent à l'ouest à peu près sans mélange jusqu'au delà de Salonique et vont à l'est, avec un peu de mélange d'elements Turcs, jusqu'à la Mer Noire, tandis que la Conférence de Constantinople, comme on peut le voir par ses délibérations, s'est arrêtée dans la Bulgarie orientale, un peu au nord des limites de la nationalité, est en revanche, a peut-être à l'ouest, rattaché à la Bulgarie un peu plus qui le territoire habité par un population exclusivement bulgare.[1]

Kiepert's map was used by the delegates at the Congress of Berlin. Its popularity was largely due to the fact that Germany appeared at that time (1878) to be the only Great Power not directly interested in the issue of the Balkan dispute and Kiepert's map was regarded as part of Bismarck's 'honest brokerage.' Maps based on Kiepert appeared in *Markham's Geographical Magazine* for October, 1876, in E. G. Ravenstein's *Universal Geography* of 1877, in the second edition of the travels of Miss Mackenzie and Miss Irby (1877), and it was resurrected by Kiepert's son in 1898 and re-issued in 1910. Rizov incorporated it in his atlas of 1917 and J. D. Bourchier, the famous pro-Bulgarian correspondent of *The Times*, reintroduced H. Kiepert's map to the British public in 1921.

## English Maps of 1876 and 1877

### *Wyld's Map of 1876*

The year 1876 marked the climax of the pro-Bulgarian ethnographic fashion, and the supremacy of the Bulgarians throughout the whole of the southern Balkans, including Bulgaria, Macedonia and Thrace, was universally accepted. The fashion was challenged neither by the Greeks nor by the Serbians. England, which was the home of the Greek cause, welcomed the appearance at this time of three ethnographic maps, all of which adhered to the pro-Bulgarian fashion and which indicated that even in England the trend of public opinion had gone against the Greeks. The first of these maps—Wyld's map of 1876—was rather an interesting combination of the maps of Boué and

[1] A. d'Avril, *Les Négotiations relatives au traité de Berlin* (Paris, 1886).

Petermann. The distribution of the Bulgarians was based on Petermann; the division of the Albanians into their religious sects, and the location of the Vlachs, were derived from Boué. Wyld's map did make a few concessions to the Greeks, representing, for example, all Turkish exclaves of southern Macedonia and Thrace as Greek. The Byronic tradition died hard.

### *The Maps of Bacon and Miss Muir Mackenzie*

Bacon's map which appeared in 1877 was no more than Lejean's map simplified. However, it did not separately distinguish Serbs and Bulgarians but marked both as 'Southern Slav.' The third map was that of Miss Muir Mackenzie and Miss A. P. Irby, which was re-issued in 1877. The 1877 map differed from that of 1867 in so much as the western part of the map was modified in accordance with Kiepert's views. Thus it showed more Albanians and Greeks, but the Slavs, particularly the Bulgarians, were still given a most liberal distribution, and more Serbs were shown in the Peć district. The edition of 1877 contained a preface penned by none other than W. E. Gladstone, whose speeches, directed against the Ottoman Empire in Europe, had done so much to popularize the Slav cause.

## Pro-Greek Maps, 1877-8

The Greeks were well aware that the views popularized by P. G. Safařik and A. Boué, by G. Lejean and H. Kiepert, had become fashionable in all parts of Europe. The idea that the vast majority of the Sultan's subjects were Greek had been dispelled, and in its place the idea that they were Bulgarians had been established. Moreover the re-emergence of a Bulgarian national Church had deprived the Greeks of a valuable instrument of hellenization. The growth of Bulgarian nationalism had not only crippled the spiritual supremacy of the Greeks in the Balkans but it threatened also their commercial and administrative control and so reduced the possibility of a revival of the Byzantine Empire under the auspices of the modern Greeks. That the pro-Bulgarian ethnographic map was being widely published even in the British Isles was an indication of the bleakness of Greek prospects.

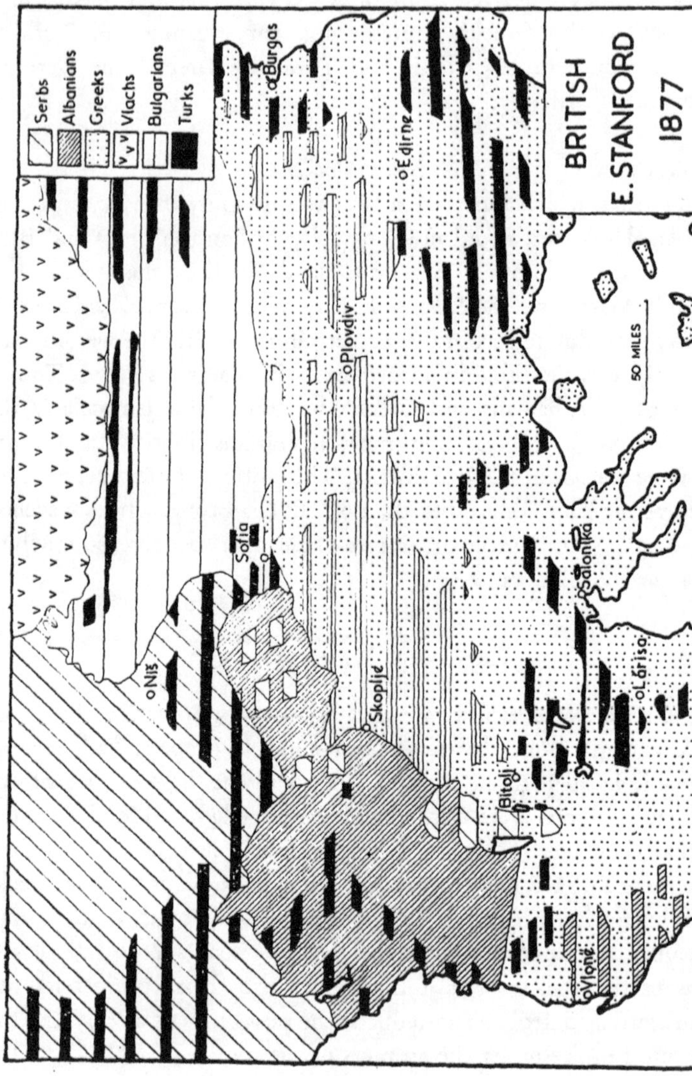
Serbs
Albanians
Greeks
Vlachs
Bulgarians
Turks
BRITISH
E. STANFORD
1877
Burgas
Edirne
Plovdiv
Sofia
Niš
Skoplje
Bitolj
Salonika
Lárisa
Vlonë
50 MILES

FIG. 16

The new ethnographic ideas had been instrumental in completely undermining their moral position.

### *E. Stanford's Map of 1877*

Under the circumstances the Greek reaction, so long delayed, was violent but unavailing. In 1877 three maps appeared in rapid succession disputing the validity of H. Kiepert's map and incorporating the Greek ideas of Balkan ethnography. Of these the best known was Stanford's map, derived from Greek sources and published both in Greek and in English (Fig. 16). On this map, the whole of the peninsula between the crest of the Balkans and the Aegean, including central and southern Albania, was coloured as Greek territory. The anonymous Greek author dismissed language as an unreliable criterion on which to base an ethnographic map. He maintained, however, that the Greek language was widely understood by all the peoples of the Balkans, that the Vlachs were Greek nationals, and that large numbers of the so-called Slavs were in fact *Bulgarophone* Greeks (i.e. Greek nationals who happened to speak Bulgarian). Above all, the Balkans, he argued, were culturally and historically an unalienable part of the Hellenic world. These views, elaborated and recast, formed the core later of the theses of C. Nicolaides in 1899, of N. Kasasis in 1903, of S. P. Phocas Cosmetatos and V. Colocotronis in 1919 and those of a host of other Greek scholars who continued to labour for the next fifty years in a desperate effort to revitalize the old idea, that the Balkans constituted an Hellenic province.[1]

### *F. Bianconi's Map of 1877*

The second of the pro-Greek series to appear in 1877 was F. Bianconi's map. Bianconi was a Frenchman who had been at one time the chief engineer and architect of the Ottoman Railways. His map was based on Turkish statistics. Such statistics were derived from rough surveys

[1] V. Colocotronis, *La Macédoine et L'Hellénisme*, *op. cit.*, contains the following : " Mais il existe aussi dans certains districts Macédoniens de telles populations Slavophones, qui, en outre ont conservé une conscience nationale purement grec." C. Nicolaides, *Makedonien* (Berlin, 1899). S. P. Phocas Cosmetatos, *La Macédoine : son passé et son présent* (Lausanne, 1919). N. Kasasis, *L'Héllenisme et la Macédoine* (Paris, 1903).

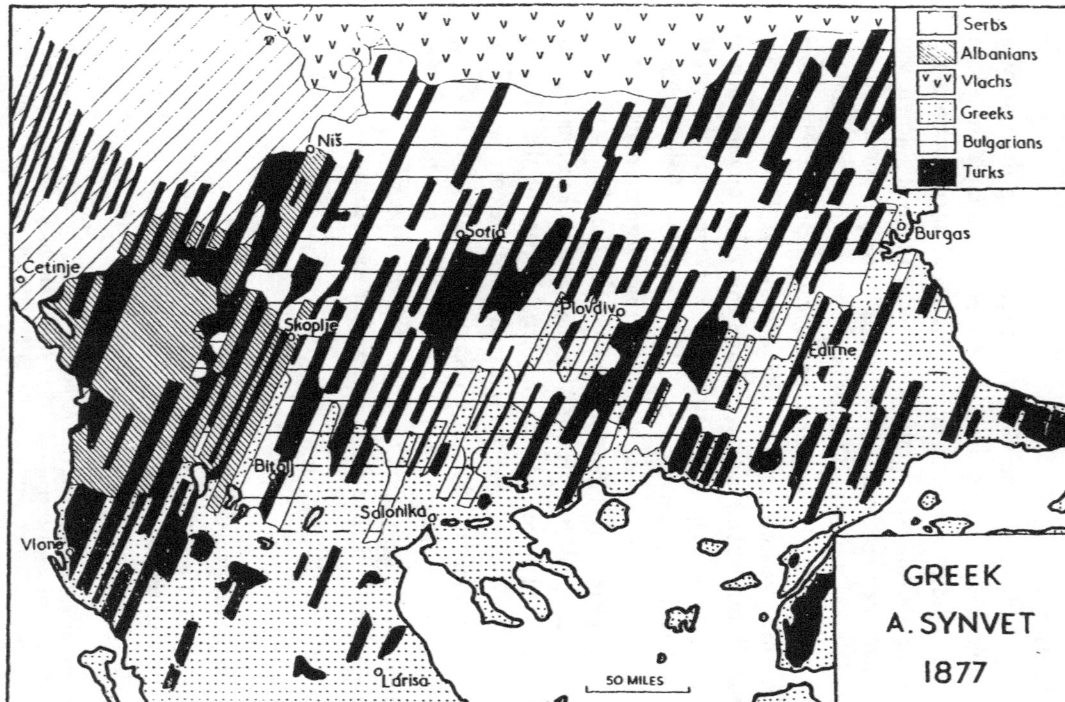


FIG. 17

In the original, the Turkish category is referred to as *Musulman*. The Graeco-Bulgarian category, which is depicted above by an overlap of Greek and Bulgarian shadings, is separately distinguished but broadly classified with the Greek in the key.

made for tax assessment purposes; they were not uniform from one place to another, nor were they complete. 'Heads of households' only were counted; women and children were disregarded. Only Turks and non-Moslems were separately distinguished and non-Moslems were often referred to as Greeks (see p. 42). In consequence Bianconi's map portrayed the whole of 'Roumelia,' Macedonia and southern Albania as Greek. It may be regarded, in company with Stanford's map, as part of the reaction to the idea of the Slav domination of the Balkans.

*A. Synvet's Map of 1877*

The Greek reaction was further reinforced by Synvet's map of 1877 (Fig. 17). Synvet was a Greek schoolmaster from Constantinople and he adopted a novel method of belittling the Slav claims. Only in Serbia proper, in Romania and in a small part of Bulgaria did Synvet distinguish any *compact* ethnic groups. Elsewhere, over the whole of the Balkans, his distributions were drawn to stress the heterogeneous character of the population.

*The Turks.* All the Moslems on his map were shown as Turks, an interpretation which incurred the wrath of later Slav scholars, but which in 1877 might have been justified because the Moslem religion did have political significance. A great many Bosnians, Serbs, Albanians and Bulgarians (Pomaks) had adopted the Moslem faith and were closely allied to the Turkish cause. These Moslems often had a great deal more in common with the Turks than they had with their compatriots. They enjoyed privileged positions which were regarded as vested interests and often they themselves formed the very bands of *bashi-bazouks* whose savage reprisals put an end to many a national demonstration against the Turk. Synvet's map was a vivid reminder that the Moslems were still a factor to be reckoned with in the Balkan political situation.

*The Greeks.* By recognizing the Bulgarophone Greeks in Macedonia and Roumelia and by classifying the Vlachs as Greeks, Synvet considerably extended the Hellenic ethnographic frontier but he did not make the extravagant claims for the Greeks, put forward in the maps of Stanford and Bianconi. His was rather a reaffirmation of the Graeco-Turkish viewpoint. It was welcomed by the Greeks in 1877.[1]

Outside Great Britain the three pro-Greek maps had a very poor reception[2]. The Slavonic character of the Balkan peninsula had been impressed upon the mind of Europe so thoroughly by the Lejean school, that any radical re-interpretation enjoyed no immediate vogue. Nevertheless, the Bulgarian ethnic hegemony had been challenged. The

FIG. 18. H. KIEPERT'S ETHNOCRATIC MAP OF 1878

British welcomed the representation of the Greek claims, which lent some support to British foreign policy in the eastern Mediterranean.

### *H. Kiepert's 'Ethnocratic' Map of 1878*

The Greek case was strengthened by the appearance of another of Kiepert's maps in 1878. This was not an ethnographic map, but it was based partly on ethnographic data (Fig. 18). The point that cultural

[1] This map, although welcomed by the Greeks in 1877, was rejected by them in 1919. Colocotronis (*op. cit.*) went so far as to accuse Synvet of prejudice. He alleged that Synvet used as his base map, "La Bulgarie d'après le Prince Tcherkasky," traced by a Bulgarian professor from diplomatic reports.

[2] Karl Sax for example described Stanford's map as unworthy of consideration. *Mittheilungen Geographischen Gesellschaft*, Bd. XXI (Wien, 1878).

groups did not necessarily have common political aspirations, and that ethnographic frontiers were not suitable political boundaries was evidently appreciated by Kiepert. He therefore produced this new map, at the request of the Greeks, to offset the political significance which events had given to his ethnographic map of 1876 (see p. 65). It was entitled *Tableau ethnocratique des Pays du Sud-Est de l'Europe*. In the explanation accompanying the map Kiepert outlined the difficulties inherent in the production of an ethnographic map and he maintained that the use of such maps for drawing up political boundaries was a malpractice which no geographer ought to countenance. He therefore attempted to produce a composite map, embodying all the factors which he believed should be taken into account in the framing of political boundaries. His criteria included race, language, religion and historical associations; and he added the proviso that such boundaries should be drawn in accordance with the realities of physical geography. As a result of all these considerations Kiepert's *ethnocratic* map was very different from his ethnographic map. He depicted on the former the Balkan crest-line as a suitable northern boundary for an enlarged Greek state, which was also to include southern Macedonia and southern Albania. Bulgaria, he confined to the lands between the Danube and the Balkan range but including northern Macedonia, the Dobrudja and the Niš district. Serbia he extended to include Bosnia, Hercegovina and Montenegro and he made allowance for an independent Albania which incorporated Old Serbia. Later, in 1919, V. Coloctronis claimed that Kiepert's later map invalidated his earlier one, but the two were meant of course to be complementary.[1] C. Nicolaides in 1899 went so far as to base his *ethnographic* map on Kiepert's *ethnocratic* map (see p. 122).

## K. Sax's Map of 1878

The Greeks and the British were not alone in the campaign, which aimed at destroying the idea of Slav supremacy in the Balkans. Austria

[1] V. Colocotronis, *op. cit.* (p. 484), states : " Or, ce même H. Kiepert lorsqu'il fut chargé d'Athènes de publier une nouvelle carte ethnographique de Balkans n'hésita pas un instant a désigner comme frontière de Hellénisme la crête même de l'Hemus. Décidément Kiepert ne voulait mécontenter personne."

had always had a vital interest in Balkan affairs. Since her defeat at the hands of Prussia, Austria's hope of achieving the leadership of a Germanic central Europe had vanished. Austria was left with her Slav possessions and the possibility of expansion in the Balkans. She had no desire to see Slav national aspirations successful anywhere in Europe. Serbia, the one Slav state of consequence in the Balkans, was a constant source of annoyance to the Habsburgs. The triumph of the idea of the *Serbo-Croat* group, which had become a feature of all ethnographic maps since the time of Šafařik in 1842, spelled danger for an Empire which relied on Slav loyalty to German rule. Moreover the existence of Serbia set up a pole of attraction for all the Slavs of the western Balkans, not only for those Slavs already within the Habsburg domain, but perhaps even more important, for those Slavs in the very regions where Austria wished to consolidate her political power, namely in Bosnia, Hercegovina, Old Serbia and Macedonia.

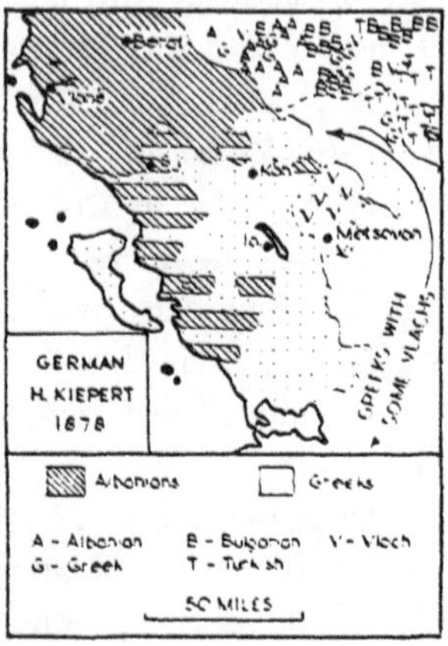


FIG. 19. H. KIEPERT'S MAP OF IPIROS, 1878

Abbreviations of towns are as follows: **G.** Gjinokastër; **Kón.** Kónitsa; **Io.** Ioannina. The distribution of Albanian, Bulgarian, Vlach, Greek and Turkish villages in south-western Macedonia is depicted by the letters given in the key above.

In the light of Austria's attitude, the map compiled by an Austrian official, Karl Sax, which appeared in 1877, is worth consideration (Fig. 20). For many years Sax had served the Austrian Empire as a consul in Adrianople. He was well acquainted with Balkan affairs and had evidently cultivated an intellectual interest in ethnography, both as a pastime and as part of his official business. His sources were reputable for he was acquainted with the maps of Lejean, Kiepert and Synvet and with the travel works of Blau, Kanitz, Hahn, Roskiewicz, Thömmel, Hochstetter and Boué. He was also able to make use of a Bulgarian estimate of the population of Thrace and of the official reports of

Austrian officials in Shkodër (Scutari), Prizren, Skoplje, Bitolj, Salonika and of other smaller towns. It must be remembered that the Austrians, by virtue of their geographical position and their immense experience in dealing with diverse nationalities, commanded sources of information not available to the French or British or even to the Russians.

Sax adopted a different set of criteria from that used hitherto for compiling ethnographic maps and as a result the picture he obtained differed considerably from that of Kiepert. He argued that since the time of Safařik and Lejean too much emphasis had been placed on linguistic, and not enough on religious data which in the Balkans were extremely important. Furthermore, everybody had ignored what Sax called 'group consciousness'; he did not mean mere folklore, or racial similarity or any of the more obvious and facile cultural traits; he meant a deep-seated feeling of community which often united peoples of diverse culture. He called it *das eigene nationale Bewusstsein*.[1] Sax went on to point out that reliance on linguistic criteria had given to the ethnographic map a simplicity which in real life did not exist. Many of the Slavs, for example the Bosnians and the Serbs, could be represented linguistically as one group, whereas in actual fact religious differences in their case were of a higher order than language. Thus the Bosnians, he declared, were invested with an entirely separate sense of nationality from that of the Serbians.[2]

*The Turks*

As a result of the application of new criteria, Sax produced a very complex ethnographic map of the Balkans. He was much more liberal in his Turkish distribution than Kiepert. He recognized a considerable minority of Turks in central Macedonia, particularly in

[1] In the original : "Ich spreche gar nicht von der Völkergeschichte, vom physischen Typus, von den Gebräuchen und derartigen selbstverständlichen, aber ferner liegenden Merkmalen."

[2] "Die bosnischen Mohammedaner, deren Muttersprache die serbische ist haben sich niemals Serben nennen lassen, sondern " Türken," sie sind doch wohl kein Türken, aber sie dürfen auch nicht einfach unter die Serben gerechnet werden, weil sie sich zu dieser letzteren Nation selbst nicht rechnen und sogar in den schnoffsten Gegensatz stellen."

the valleys of the Vardar and Struma, and in the regions intervening. The zone bounded by Skoplje, Veles, Dojran, Sérrai, Melnik and the Osogovska mountains, he depicted as half-Turkish.

*The Greeks*

He was kinder to the Greeks than Kiepert and he showed strong Greek minorities extending as far north as Bitolj, Devdelija on the Vardar, Melnik and the Rodopi mountains; in addition, he showed a strong Greek minority in the valley of the Maritsa as far north as Edirne. The Greeks he pictured in Thrace, as at least equal in importance to the Turks and Bulgarians.

*The Slavs*

The Slav groupings on Sax's map were neither so simple nor so widespread as they had been on previous maps. He distinguished not two, but nine branches of the Slavs in Turkey-in-Europe. They included the following 'national groups': Greek Orthodox Serbs, Catholic Serbo-Croats, Moslem Serbo-Croats, Serbo-Bulgarians (a mixed group), Exarchate Bulgarians, Hellenized Bulgarians of Greek Orthodox faith, Uniate Bulgarians (a very small group), Catholic Bulgarians (likewise a small group) and finally Pomaks or Moslem Bulgarians.[1] In this manner he endeavoured to annihilate the idea of both Serbo-Croat and of Bulgarian solidarity. How well he did so may be appreciated merely by a glance at his map. The Serbs proper, for example, he depicted as extending south to Novi Pazar and Pristina with very small minorities around Peć and Prizren. South of the Ibar and west of the Lin rivers, the Serbs gave way to Moslem Serbo-Croats. Thus the all-important Novi Pazar corridor between Montenegro and Albania on the south-west, and Serbia on the north-east, he

[1] (1) Serbische Stamme, griech-orthodoxer Religion (Serben, Cernagorcen, herzegovinische und bosnische Serben). (2) Serbokroaten, Katholischer Religion, oder bosnische und herzegovinische Lateiner. (3) Serbo-Kroaten mohammedanischer Religion oder bosnische Turken. (4) Serbo-Bulgaren oder mit Serben vermischte Bulgaren griechisch-orthodoxer Religion. (5) Bulgaren griechisch-orthodoxer Religion der schismatisch-bulgarischen Kirche. (6) Graeco-Bulgaren oder halb hellenisirte Bulgaren der griechisch-orthodoxer Kirche. (7) Griechisch-Katholische oder unirte Bulgaren. (8) Latein Katholische Bulgaren. (9) Pomaken.

represented to be populated by Slavs of a nationality alien to that of the Serbs. A significant innovation was the portrayal of the region immediately north of Niš as Serbo-Bulgarian.[1] He did not quarrel with the basic distribution of Bulgarians as represented on Kiepert's map. In fact, he showed them to be more numerous in Thrace than had hitherto been accepted, except by Mirković on his Russian map. Bulgarians, he marked as far west as Niš, Leskovac, Vranje, the Sar mountains and the Drin valley and as far south as Flórina, Konia, Véroia, Salonika, Sérrai and Dráma. But he robbed the Bulgarians of their former clear-cut majority in Macedonia by the recognition of numerous minorities, not only Turk, Albanian and Greek, but also Pomak, which Sax did not regard to be of Bulgarian nationality. The Pomaks he marked in the upper Struma valley, in the Edhessa district and in the whole of the Mesta valley eastwards as far as Komotiní.

*The Vlachs*

Sax took care to distinguish the Vlachs from the Greeks. He not only marked them in the Píndhos but also in eastern Thessalía, in the valleys of the Semen and Devoll of southern Albania, in south-western Macedonia around Kastoría and Flórina, and west of Bitolj.

*The Albanians*

Sax divided the Albanians into their religious groups. The Moslem Albanians, however, he indicated as the predominant group, extending from Lake Shkodrs (Skadar) to as far east as the Morava river in the vicinity of Prokuplje; he believed they formed the majority of the inhabitants of Old Serbia. He also indicated Moslem Albanians as a minority in western Macedonia, as far east as Skoplje, Prilep and Bitolj. South of the Shkumbin river, he indicated them as a minority in Northern Ipiros, particularly in the coastal region of the Kérkira Channel, and also in the Kónitsa and Kastoría areas. He also separately distinguished Greek Orthodox Albanians south of the Shkumbin river. They formed a compact group only in the Libohovë district; elsewhere, they were found in conjunction with the Moslem Albanians. His separate recognition of Greek Orthodox Albanians was a useful

[1] Shown as Bulgarian on most contemporary maps.

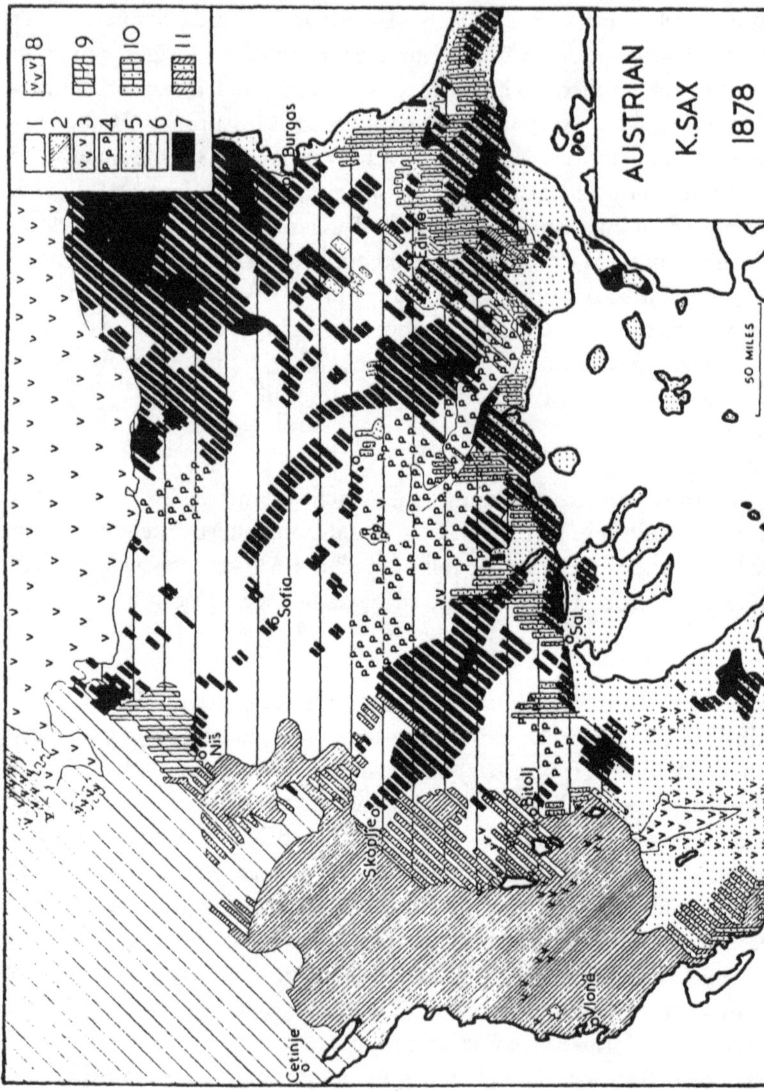
1
2
3
4
5
6
7
8
9
10
11
Burgas
Edirne
Sofia
Niš
Skopje
Sal
Bitolj
Vlonë
Cetinje
50 MILES
AUSTRIAN
K.SAX
1878

distinction as Greek claims in southern Albania were based largely on the assumption that these Greek Orthodox Albanians adhered to the Greek way of life, and were in process of becoming Greek nationals.

*Conclusion*

Summing up Sax's contribution, it might be said that he presented a good case for both Austrian and Greek intervention in Balkan affairs, in so much as he demonstrated that the idea of the existence of two major Slav groups, comprising Serbo-Croats and Bulgarians was a fallacy. He claimed that there were twenty-eight different nationalities in Turkey-in-Europe and so involved was their distribution that no possibility existed of granting political independence to each group. Macedonia in particular had a very heterogeneous population, and on this account there was no real reason why it should be linked with Bulgaria. The method of diagrammatic representation of minorities adopted by Sax accentuated the confusion of nationalities and emphasized the point he wished to make, that Lejean and Kiepert and their school of linguistic ethnographers had grievously erred in stressing the predominance and unity of the Slav groups in the Balkans (*cf.* Fig. 11). Karl Sax's map undoubtedly was related to Austrian policy in the Balkans in so much as it attempted to belittle the political significance of ethnic groupings, but a review in *Petermann's Mittheilungen* declared that Sax's map gave the truest picture until then of ethnographic conditions in the Balkans. W. Z. Ripley, who had a wide knowledge

---

FIG. 20

The references in the key are as follows : **1. Serbo-Croats,** including Greek Orthodox, Moslem and Roman Catholic (these sub-groups were separately distinguished in the original) ; **2. Albanians,** including Greek Orthodox, Moslem and Roman Catholic (also separately distinguished in the original) ; **3. Romanians**—Pindus Vlachs and Moldo-Vlachs ; **4. Bulgarian Pomaks** ; **5. Greeks,** including Orthodox and Moslem (separately distinguished in the original) ; **6. Exarch Bulgarians** ; **7. Circassians, Tatars, Turks and Turcomans** (each separately distinguished in the original) ; **8. Graeco-Vlachs** (separately distinguished in the original and classified as Romanians) ; **9. Serbo-Bulgarians** (separately distinguished in the original but classified as Bulgarians) ; **10. Graeco-Bulgarians** (separately distinguished in the original but classified as Bulgarians) ; **11. Graeco-Albanians** (separately distinguished in the original but classified as Greeks).

of European ethnography, incorporated Sax's map into his well-known work of reference, *The Races of Europe*, published in 1900.

## The Map of Nic. Densusianu and F. Dame, 1877

It is necessary to point out here that even in 1877 ethnic controversy was not limited to a discussion of the relative merits of Bulgarian and Greek claims in Macedonia. The existence of a Romanian speaking element in the population of various parts of Macedonia, Ipiros, southern Albania and Thessalía had, by that time, excited the close attention of the Romanians proper who had already achieved a measure of independence north of the Danube. However remote the relationship between the 'Macedo-Roumains' and themselves might be, they discerned it was real enough to be turned to advantage. Historical research even then had revealed the important role played by the Vlachs or *Blachi* in the political life of mediaeval Macedonia. By recalling the past and by drawing attention to the presence of the Vlachs in the south-western Balkans, interested Romanians sought to establish a case for political recognition of the Vlachs which might take the form of the revival of Great Wallachia or 'Megalo-Vlachie' as Densusianu called his proposed Vlach state. In the campaign for the emancipation of the Vlachs the education of public opinion concerning their history, culture and distribution was a primary necessity. The book on the 'Macedo-Roumains' written by N. Densusianu and F. Damé was dedicated to this end. The map it included was based on sources which gave emphasis to the distribution of the Vlachs; the works of F. C. L. H. Pouqueville, M. E. Picot, J. Thunmann and W. M. Leake were the most important of these.[1] The method used was to colour the map according, it would seem, to descriptions afforded by these authors. The results were interesting (Fig. 21).

### *The Turks*

Not much Turkish territory was shown on this map. Apart from some districts in the Rodopi mountains the Turks were confined to the

[1] The titles were respectively: *Voyage en Grèce* (Paris, 1826). *Les Roumains de Macédoine* (Paris, 1875). *Untersuchungen über die Geschichte der östlichen europäischen Völker* (Leipzig, 1774). *Travels in northern Greece* (London, 1835).

area immediately north of Lakes Korónia and Vólvi, stretching thence along the coast to Kávalla.

*The Greeks*

The Greeks were given a very meagre distribution in complete contrast to that favoured by Greek cartographers and much less than that suggested by G. Lejean and H. Kiepert. Only a small portion of

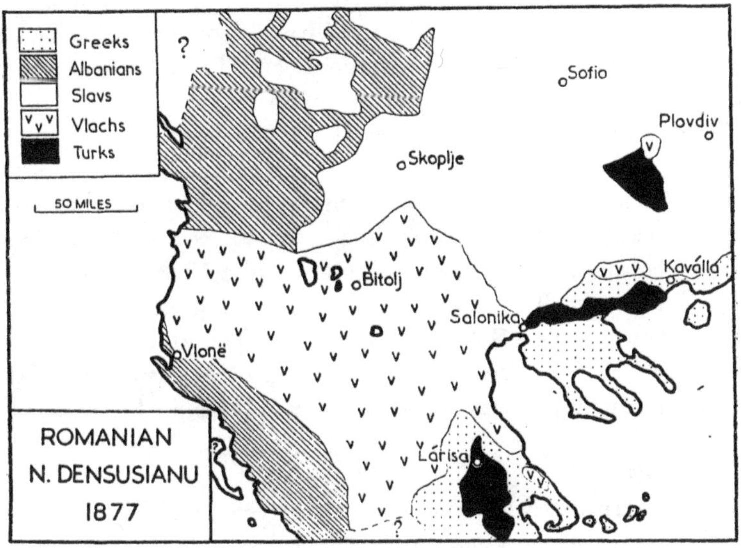


FIG. 21

Macedonia proper was shown as Greek and the Greeks were even excluded from the greater part of Thessalía.

*The Slavs*

The interesting classification adopted by N. Densusianu and F. Damé did not distinguish between Serbs and Bulgarians. All the Slavs were shown in one colour in complete contrast to Sax's map, an indication of the fact that at this time, at least in the Romanian view, the distinction between Bulgarians and Serbs was not important. Bulgaria, of course, had not yet been established as an independent state. The distribution of the Slavs consisted of enclaves in Old Serbia and a concentration in central Macedonia which, however, did not reach the Aegean.

*The Vlachs*

The Vlachs, or 'Macedo-Roumains' and 'Zinzares' as they were referred to in the text, were given a very wide and solid distribution (in contradistinction to the scattered distributions favoured by G. Lejean and H. Kiepert) which covered part of southern Albania, the greater part of south-western Macedonia and Thessalía. In addition three exclaves of Vlachs were shown outside this area.

There were four main branches of the Vlachs referred to in the text, the *Perhebiens,* the *Brouzi,* the *Massarets* or *Dassarets* and the *Boui* or *Boviens.* Their combined strength, according to the authors, totalled 1,200,000, of whom 450,000 were to be found in Macedonia, 200,000 in Thessalía, 350,000 in Ipiros and Albania, and 200,000 in Thrace. These estimates were something like four times as great as those given by G. Lejean and A. Boué. The Vlachs were, in the words of F. C. L. H. Pouqueville, reiterated by N. Densusianu, "the second most important nation after the 'Pelasgians' in Macedonia".

*The Albanians*

The Albanians were divided into two groups, one of which was found in southern Ipiros and the other, the larger of the two, to the north of the Shkumbin river extending into western Macedonia and Old Serbia as far, indeed, as the Morava river.

## THE TREATIES OF SAN STEFANO AND BERLIN, 1878

*The Provisions of San Stefano*

Between 1876 and 1878, the appearance of the pro-Greek maps, of the Austrian map, and of the Romanian map, all suggested that opinion on Macedonian ethnographic distributions was changing, and was being changed, largely at the expense of the Bulgarians. Two schools of thought were particularly in evidence during 1878 amongst the delegates at the important Peace Congresses held in that year. The one, the Lejean school, maintained that Macedonia was in the main Slav, by which was understood Bulgarian; the other emphasized the diversity of the Macedonian population. This conflict of opinion was to be given fuller expression in the events which followed the Russo-Turkish War.

After the breakdown of the Conference of Constantinople in 1876, war between Turkey and Russia could only have been prevented by close agreement between Russia and Great Britain, and when this was not forthcoming it needed only the excuse of a new Slav uprising to set the Russian armies on the move.[1] This came in June 1877, when Montenegro recommenced hostilities against the Turks and Russia declared war. Serbia and Romania joined the Tsar. The short campaign which followed enabled Russia to impose upon Turkey the armistice of San Stefano. The terms of the Treaty associated with the armistice, although they were not carried out, had enormous consequences for the political geography of the Balkans. The Russians had entertained grandiose schemes for the partition of the Balkans as far back as the reign of Catherine the Great (see p. 19). Then, in 1782, partition of the Balkans had been dependent only upon the co-operation of Austria. By 1878, however, any scheme for the partition of Turkey-in-Europe depended not only on Austrian but also on British co-operation. Since the time when Napoleon had threatened India by gaining a grip on Egypt, Britain had been interested in the development of events in the eastern Mediterranean. The purchase of shares in the Suez Canal in 1875 added to British commitments in this theatre. Any settlement in the Balkans therefore depended on granting satisfaction, both to Austria in the west and to Britain in the south. The Tsar obviously had such considerations in mind when he presented his peace terms to the Sultan. Those terms left Austria with a free hand in Bosnia and Hercegovina, in the Novi Pazar corridor and in Albania. Such a concession dealt a death blow to the old Pan-slav idea of a union of all the 'Southern Slavs.' Serbia was compensated by the gift of some 'Bulgarian' territory in the Niš and Leskovac area. The Tsar's decision broke the Serbo-Bulgarian accord and the result of his action

[1] W. N. Medlicott, *The Congress of Berlin and After* (London, 1938).

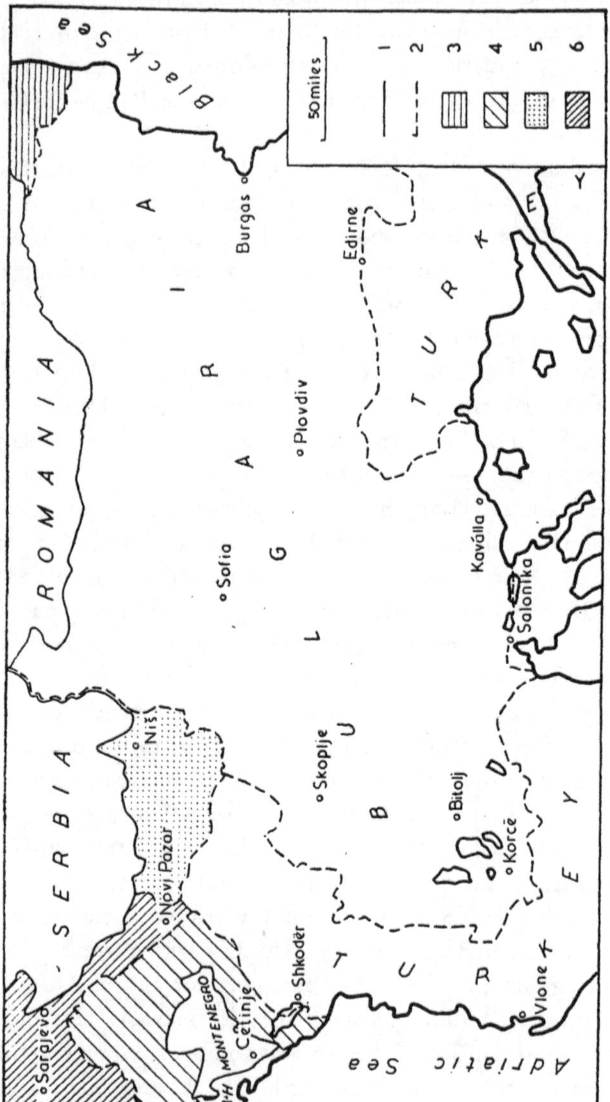


FIG. 22. THE TREATY OF SAN STEFANO (1878)

The references in the key are as follows: **1.** Boundaries of 1877; **2.** New boundaries envisaged under the terms of the Treaty; **3.** Territory conceded to Romania; **4.** Territory conceded to Montenegro; **5.** Territory conceded to Serbia; **6.** Territory acknowledged to be within the Austrian sphere of influence.

was to have far-reaching repercussions. In order to pacify Britain, he left Thessalía and Khalkidhikí in Turkish hands, a move which, in practice, left those territories open to Greek penetration. Thrace was also left to the Turks—a proof, which the Tsar hoped would be accepted, that the Russians had no designs on Constantinople.

Apart from these concessions, the main provision of the Treaty was the creation of an independent 'Greater Bulgaria' similar in extent to that suggested by Ignatyev in 1876 (Fig. 22). The creation of a Greater Bulgaria was, from the Russian point of view, a step in the right direction. Even if, as a temporary measure, many Serbo-Croats had to be thrown to the Austrian wolves, Russia had achieved her immediate objective—a Slav corridor into the heart of the Balkans. From another point of view, it invested Bulgaria with mythical frontiers to which the Bulgarians have continued to cling right down to the present day. The boundaries of the proposed Greater Bulgaria coincided closely with the ethnographic frontier laid down for the Bulgarians by H. Kiepert. They included some Albanians in the west and some Greeks in the south-west. On the other hand, many Bulgarians were left in Turkish Thrace and others were sacrificed to Serbia in the Niš salient.

*Austrian and British Proposals for Revision*

The Treaty of San Stefano was never enforced. Both Great Britain and Austria protested and threatened war unless new terms were reached which would be acceptable to all parties concerned. From an examination of Andrassy's proposals for a revision of the provisions of San Stefano, the real cause of friction between Austria and Russia may be deduced. His main proposals were: (1) that Austria should occupy Bosnia and Hercegovina as well as the territory between Serbia and Montenegro, after informing the Porte; (2) that the boundary of Montenegro should follow the line of the Konito, as far as the junction of the rivers Pipa and Tara, then to Belapolje (Bijelo Polje), crossing the river Lem (Lim) near Berani (Berane) and then to the Lake of Scutari. Austria was to annex the territory between the river Bojano, the lake and the sea. Montenegro would be guaranteed the liberty of navigation on both river and lake; (3) that Serbia should receive no aggrandizement in the west; but as compensation she would be given Vranje and Trn (Tren) in the east. That she should undertake to, or let Austria, construct a railway line through her territory and maintain the same tariffs as in Austria; (4) that Russia's right to Bessarabia should be recognized; (5) that the eastern frontier of Bulgaria between Kirk Kellessi (Kirklareli) and Constantinople should take the line proposed by the Conference of 1876, thus leaving Lule-Bourgas (Luleburgaz) and the Black Sea to Turkey; *her western frontier should be drawn from the Gulf of Orfano (Strimón) to Vranje; the district remaining beyond this limit should be given administrative autonomy independent of the Bulgarian Principality under the name of Macedonia, Salonika would be part of the new Province.*

Andrassy pointed out that his demands were intended to secure the railway line through Mitrovica to Salonika and free access to the Aegean sea for Austria.[1] It is very apparent from a reading of Andrassy's proposals that Macedonia was the real bone of contention between the two Emperors. It was here that Russian and Austrian interests came into conflict. Macedonia emerged as the key to the strategic control of south-eastern Europe.

British opposition to the Treaty of San Stefano was partly due to the fact that the Tsar had tried to settle the Eastern Question on his own. Lord Beaconsfield stated, "that the Treaty abolishes the dominion of the Ottoman Empire in Europe; it creates a large State which, under the name of Bulgaria, is inhabited by many races not Bulgarian[ . . . ] all the European dominions of the Ottoman Porte are [. . .] put under the administration of Russia".[2] The reference to the varied 'racial' composition of western Bulgaria was an indication of the influence of the ethnographic ideas incorporated in such maps as those of A. Synvet and K. Sax.

### *The Provisions of the Treaty of Berlin*

The Treaty of Berlin, concluded in the same year, finally settled the issue of the Russo-Turkish war (Fig. 23). Salisbury declared at this Conference that "Turkey should be freed from Russia's domination

[1] Stojanović, *op. cit.* (p. 239).

[2] J. A. R. Marriott, *The Eastern Question*, p. 339 (Oxford, 1917).

[. . .] This could be achieved by driving back the Slav State to the Balkans and substituting a Greek Province".[1] Wyld produced a second edition of his map in 1878 which showed the boundaries of the proposed 'New Helenic (sic) State.' It was to have included western Thrace (Dhitikí Thráki) and a considerable part of Macedonia and Ipiros. The final provisions of the Treaty of Berlin constituted a compromise between the Austrian, British and Russian viewpoints. The thorny problem of Macedonia was left unsolved as the region remained to Turkey. Serbia and Montenegro received some of the districts promised to them at San Stefano. Bulgaria was reduced, however, to the territory between the Danube and the Balkan range. Bulgarian territory to the south of the Balkan range was formed into the new province of Eastern Roumelia and handed back for administration to the Sultan. Thessalía, Thrace and Albania were also restored to the Turkish Empire. Bosnia and Hercegovina were handed over indefinitely to Austria, and Austrian troops were allowed to garrison Novi Pazar.[2]

*Consequences of the Treaty of Berlin*

The Treaty of Berlin denied Macedonia to the Bulgarians. Its other provisions concerning the Balkans lasted for hardly more than a few years. Thessalía joined the Greek kingdom in 1881 and Eastern Roumelia merged with Bulgaria in 1885. So the pattern which the Tsar had laid down at San Stefano re-emerged with the exception of the fateful Macedonia. That 'province' was destined to become a kind of no man's land in both a literal and a metaphorical sense. None of the Balkan peoples themselves was satisfied with the outcome of the negotiations of 1878. Serbia remained effectively divided from Montenegro and with her route to the Adriatic in the hands of Austria. The Serbo-Croats in Bosnia and Hercegovina had been saved from the Turks only to be given to the Austrians. The Serbians had been compelled to evacuate Prizren, which they had claimed as the ancient capital of Old Serbia. Blocked in the north and west, Serbia began to turn south to seek an outlet for her land-locked territory—an outlet which would

[1] Stojanović, *op. cit.* (p. 244).

[2] For a discussion of the problem of the Vranje-Pirot salient, see B. H. Sumner, *Russia and the Balkans*, 1870–80, p. 653 (Oxford, 1937).

FIG. 23. THE TREATY OF BERLIN (1878)

This map indicates, in addition to the changes effected under the terms of the Treaty of Berlin, some changes which took place subsequently to 1878. For the most part the latter changes were the result of negotiations between Turkey and the countries concerned.

only be gained at the expense of what was then believed to be Albanian and Bulgarian territory. Serbia's interest in the southern Balkans dated from this period. The Greeks, too, were disillusioned for in the end Beaconsfield had put his faith in Turkey rather than in Greece. Bulgaria gained independence only to have many Bulgarians left outside the boundaries of the new state. Nevertheless the constitution of an independent Bulgaria did mean that, with the exception of Albania, all the major ethnic groups of the Balkans had received some form of territorial recognition by 1878. From their several nuclei they now looked forward to expansion into those regions still remaining to the Turk. If they were prevented by the force or persuasion of the Great Powers, or by the vestiges of power still left to the Sultan himself, from extending their political boundaries, they found themselves free in the interim, at least to establish their moral claims to specific territories and to build up spheres of influence ; and of all the territories left to the Turk, none was desired so earnestly by so many powers, great and small, as Macedonia. And how might the smaller powers better establish their moral rights than through the medium of the ethnographic map ? The Treaty of Berlin had neatly delimited the areas where ethnographic speculation might continue and Sax had ably demonstrated that nowhere in these areas was the ethnographic issue so much in doubt as in Macedonia, and that it was possible to evolve new ethnographic mosaics by the adoption of fresh criteria.

## Chapter V

# THE SERBS, 1868–1891

### The Serbo-Bulgarian Schism

At a time when the great majority of scholars were indulging in the fanciful illusion that the Slavs were one great happy family, closely bound by ties of culture, language and history, C. Delamarre, a French professor, presented an ethnographic map to the Geographical Society of Paris in 1868. The map and its legend stressed the fact that Pan-slavism from a linguistic and literary point of view was a myth and, that there was no such thing as a single Slav language or a single Slav culture. Delamarre argued that there was a Polish, a Czech, a Serbian, a Bulgarian and a Russian language, but to regard each of these languages as so closely related as to be mutually intelligible was a fallacy which had unfortunately gained widespread acceptance. It had led, for example, to the establishment of the chair of Slavonic language at *Le Collège de France*. Delamarre pressed for its abolition since there was no one Slav language, but on the contrary, a plurality of Slav languages. He adopted the motto " Un Pluriel pour un Singulier et le Pan-Slavisme est détruit dans son principe." He inferred that the political fragmentation of ' Pan-slavonia ' was more likely on this account than any eventual unification. However, his point of view was not immediately appreciated. Delamarre was rather before the times in his ideas. But the differences existing between Slav and Slav were destined to become increasingly apparent in the next decade, and ethnographic ideas were to be considerably modified in the light of the recognition of those differences.

### *M. S. Milojević's Claims of 1873*

In 1873, five years after C. Delamarre had pointed out the crucial differences between the various Slav languages, and two years after the creation of the Exarchate, the Serbian Royal Academy was thrown into confusion by the contents of a paper read by M. S. Milojević. Its theme was that the Serb language and culture extended much further south than had hitherto generally been believed. Milojević argued in fact that Macedonia was Serb and not Bulgarian territory and he produced a map to support his contentions. It was a Serbian, Stoyan Novaković, a distinguished member of the Academy, who exposed the shallowness of his argument. In fact a great deal of the evidence submitted was proved to be forged. Milojević emerged as a cheap, mischievous chauvinist, ignominiously condemned by his fellow countrymen for having committed an unfriendly act against a good neighbour. But his advent was significant, if premature, for it meant that at least one Serbian had realized the political significance of the Bulgarian Exarchate. He was to be joined by many others as the mystic conception of Pan-slavism gave way to the intransigent nationalism of the component Slav groups. However, the clash between the Serbians and the Bulgarians in Macedonia was not to develop until Serbian expansionists, diverted by the march of events, turned their faces from the Adriatic towards the Aegean.

### *The Growth of Bulgaria*

The provisions of the Treaty of Berlin had severed the Bulgarians north of the Balkans from those to the south of them, but this rather artificial division did not last long (Fig. 23). Unionist movements sprang up immediately in both Bulgarian provinces. In 1885 the Bulgarians of Eastern Roumelia expelled their Turkish Governor-General and later in the same year effected a union with Bulgaria proper. The union of the two Bulgarias was ratified by the Turks in 1886. In the short time between 1878 and 1885 the Bulgarians had found themselves: they now controlled all the eastern Balkans between the Danube and the Rodopi mountains and, thanks to the active encouragement of Russian agents within its territory, the new state could already muster a formidable army from the ranks of the

Bulgarian nationalistic *sokols* (athletic clubs). Even the British at this time were inclined to favour a strong Bulgaria " which might screen the sick man (Turkey) from the fury of the northern blast." Everything in 1885 pointed to the fact that Bulgaria was destined to play a leading role in the future political geography of the Balkan peninsula.

*The Austro-Serbian Secret Treaty of 1881*

Meanwhile Serbia had watched the expansion of both Bulgaria and Greece, which had acquired Thessalía in 1881, with some alarm, for the Serbians even before 1885 had begun to covet the Macedonian territory of Turkey and it soon became evident that the interests not only of Austria, Greece and Bulgaria respectively, but of Serbia also, converged in Macedonia. The date when the Serbians first began to take an active, as apart from a friendly interest, in the Vardar valley may be fixed as the year 1881—the date of the secret treaty between Austria and Serbia, whereby Serbia renounced her claims in Bosnia, Hercegovina and Novi Pazar. Article II of that treaty read as follows :

> Serbia will not tolerate political, religious, or other intrigues, which, taking her territory as a point of departure, might be directed against the Austro-Hungarian Monarchy, including therein Bosnia, Hercegovina, and the Sanjak of Novi Pazar.
>
> Austria-Hungary assumes the same obligation with regard to Serbia and her dynasty, the maintenance and strengthening of which she will support with all her influence.[1]

Under the terms of the Treaty of Berlin (1878) Austria had taken over the administration of Bosnia and Hercegovina and had been allowed to garrison the Novi Pazar corridor. The strengthening of Austro-Hungarian influence in these areas put an end, for the time being, to any possibility of a union of the Serbo-Croats, and Serbian expansion was canalized from a westerly into a southerly direction. Article VII of the secret treaty made this clear :

> If, as a result of a combination of circumstances whose development is not to be foreseen at present, Serbia were in a position to make

[1] Dr. Alfred Francis Pribram, *The Secret Treaties of Austria-Hungary, 1879-1914*, Vol. I, Translation (Cambridge, 1920).

> territorial acquisitions in the direction of her southern frontiers (with the exception of the Sanjak of Novi-Pazar), Austria-Hungary will not oppose herself thereto, and will use her influence with the other powers for the purpose of winning them over to an attitude favourable to Serbia.[1]

When this Treaty of 1881 was prolonged in 1889, the following additional article was incorporated into the Prolongation :

> If the circumstances foreseen by Article VII of the Treaty of June 23, 1881, should chance to occur while this treaty remains in force and while Serbia has faithfully observed its stipulations, it is understood that Austria-Hungary will recognize, and support with other Powers, the recognition in favour of the Kingdom of Serbia of the territorial extension foreseen by Article VII above-mentioned, which extension may be carried out in the direction of the valley of the Vardar as far as circumstances will permit.[2]

The provisions of the secret treaty outlined above played an important part in encouraging Serbian aspirations in Macedonia and in promoting disagreement between Serbia and the newly-formed Bulgarian state.

### *The Serbo-Bulgarian War of 1885*

The freshly directed Serbian territorial interests were bound to bring Serbia into conflict with Bulgaria. But even before 1881 the relations between the two principal Slav groups of the Balkans had taken a turn for the worse. In spite of C. Delamarre's warning about the fallacy of Pan-slavism, the Serbs and the Bulgarians had remained bound closely together until 1878, in the belief that they were racially and culturally brothers and the fellow-feeling between them had been sustained by Russian propaganda. The formation of the Bulgarian Exarchate in 1870 had presaged the formation of a ' Greater Bulgaria ' but even then the Serbians had not been prepared to challenge that possibility. Nothing had illustrated the solidarity of the Serbo-Bulgarian accord better than the humiliation of M. S. Milojević in 1873. But in 1878 the Serbians had been alarmed by the Treaty of San Stefano, in spite of the fact that they had gained some territory. Until that time they had felt themselves to be the real nucleus of Slavdom in the

[1] *Ibid.* (p. 55). [2] *Ibid.* (p. 137).

Balkans. The Tsar's pre-occupation with the Bulgarians, and his willingness to sacrifice the Serbo-Croats of Bosnia and Hercegovina to Austria, had aroused their suspicions. Even when the Treaty of Berlin superseded that of San Stefano, Serbia's confidence in Pan-slavism was not restored, particularly as she had to evacuate Prizren, in order that Austria might retain a corridor to Macedonia. Friction which occurred over the boundaries between the newly-constituted Bulgaria and the enlarged Serbia of 1878, was the first indication of a growing hostility between the two Slav nations.[1] Therefore when the union of Bulgaria and Eastern Roumelia was achieved in 1885, the Serbians declared war against Bulgaria, in what would appear to be a premature attempt to establish themselves in Macedonia before the Bulgarians extended their influence there also. The cheers for Milan, "King of Serbia and Macedonia," with which the Serbians bade farewell to their monarch at the head of his armies in 1885, held a significant ring for the future.[2] The cry of the Serbian expansionists in 1885, sure of the support of Austria and behind Austria, of Germany, was 'to the Aegean,' whereas before that date it had been 'to the Adriatic.' Inevitably, the reorientation of Serbian expansionist policy completely broke the accord between the Slavs of the Balkans, and the Serbo-Bulgarian rupture was clearly reflected in the ethnographic maps of the period that followed.

## S. Gopcevic's Map of 1889

The ethnographic maps of two Serbians—J. Dragašević and M. Veselinović—were amongst the first of a series which the Serbians began to produce in and after 1885. They purported to embody some of the results of historical and philological researches in Old Serbia and Macedonia upon which the Serbians were then engaged. In these territories, claimed the two Serbian cartographers, the numbers of the Serbs in the population had previously been sadly underestimated and many Slavs hitherto regarded as Bulgarians were actually Serbs. They showed, on their maps, great extensions of Serb territory south

[1] G. P. Gooch, *History of Modern Europe, 1878-1919* (London, 1928).

[2] J. A. R. Marriott, *op. cit.*

of the Sar mountains, where earlier maps had indicated only Albanians, Bulgarians and occasional isolated groups of Vlachs.[1]

The first Serbian professor of geography in the University of Belgrade, Vladimer Karić, included an ethnographic map of Serbia and Macedonia in his text-book on Serbian geography which appeared in 1887. This also portrayed the greater part of Macedonia as Serb but showed Bulgarians in the south and south-west from Kastoría to Melnik. Karić's map had some influence later on one of his pupils, J. Cvijić, whose maps were to become well-known in western Europe.

It was S. Gopčević's map of 1889 (the date of the prolongation of the Austro-Serbian Treaty of 1881), however, that really brought the new Serbian ethnographic ideas to the notice of the rest of Europe (Fig. 24). S. Gopčević was a professional diplomat and a scholar of some repute. He had already produced, before 1889, several learned works on Balkan geography, although he had not previously concerned himself with Macedonia. His map was the first large scale ethnographic map of Old Serbia and Macedonia (1:300,000). It appeared in the original under the auspices of the *Militär.-Geographisches Institut k. und k. in Wien* (Vienna), an institution responsible for the detailed Austrian topographical survey of the Balkans. In the same year *Petermann's Mittheilungen* published the map on a reduced scale (1:750,000). Thus it received widespread publicity in two reputable journals. From a purely cartographical point of view Gopčević's map was far in advance of any publications of a similar nature. He had access to the latest Austrian surveys and for the first time an ethnographic map of this area was produced on a fairly reliable base-map. H. Kiepert's base-map, for example, had been so poor that members of the boundary commission of 1878, compelled to use it, had found it almost useless for the practical purpose of drawing boundaries. Gopčević himself corrected the Austrian maps on many details of a topographical nature. He made, in fact, no less than 2,000 corrections to the Austrian 1:300,000 series. Secondly, the scale of the map allowed great accuracy and clarity. Thirdly, Gopčević's methods of depicting ethnographic distributions were an improvement on methods

[1] J. Dragašević, *Carte ethnographique de la presqu'île des Balkans* (Belgrade, 1885).

used up to that time. As Sax did, he made an attempt to indicate religious differences, and he introduced a useful method of showing the ethnographic character of the larger towns (where the population was always mixed) by means of symbols. Altogether, he distinguished no less than twelve groups in Old Serbia and Macedonia—Christian Serbs, Moslem Serbs, Christian Bulgarians, Moslem Bulgarians, Albanians, Christian Albanians of Serb extraction, Moslem Albanians of Serb extraction, Turks (including Tatars and Circassians), Greeks, Vlachs, Gypsies and Jews.

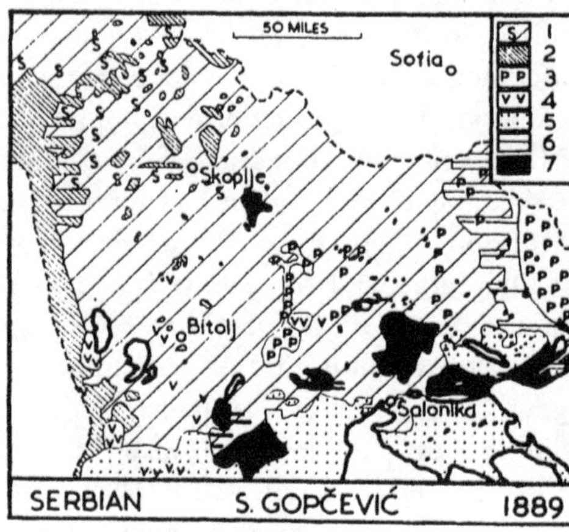


FIG. 24

The references in the key are as follows: **1. Christian** and **Moslem Serbs** (latter differentiated by an **S**) ; **2. Albanians** including Christians and Moslems of Serb descent ; **3. Moslem Bulgarians (Pomaks)** ; **4. Vlachs** ; **5. Greeks;** **6. Christian Bulgarians** ; **7. Turks** and **Tatars.**

In the original, ethnographic elements in the towns are indicated by symbols.

The territorial growth of Greece and Bulgaria between 1878 and 1886 had considerably limited Turkish territory in Europe. After 1886 ethnographers were no longer interested in producing maps of the whole of the Balkans but concentrated mainly on the part still remaining to the Turk ; it was into this territory that Austria, Serbia, Bulgaria and Greece hoped to expand. Turkish possessions in 1886 officially included Bosnia, Hercegovina and Novi Pazar, but as these regions were administered or garrisoned by Austria, and as Serbia had renounced her interests in them, they were usually precluded from the ethnographer's field of investigation until a later date. Research therefore was centred in the remaining Turkish territory—Albania, Old Serbia, Macedonia and Thrace. Gopčević limited his map exclusively to Old Serbia and Macedonia, and he made an attempt to define the boundary between the two. It may be seen from Figs. 1 and 2 that his definition of Macedonia was very limited. According to his interpretation, its northern frontier lay south of Bitolj and Strumica and the middle Vardar valley was part of Old Serbia. This was a radical departure from the more limited concept of Old Serbia favoured by Miss Mackenzie and Miss Irby in 1867, and even from the Serbian interpretation of 1878 (Fig. 89). In fact, before 1878 nearly all scholars had set the northern boundary of 'Turkish' Macedonia on the Sar mountains and the 'Kara Dag' (Crna hills).

*The Turks*

Gopčević depicted the distribution of the Turks in more detail than anyone had hitherto done. He showed them in all the main towns, in the Koniar district between Lake Vegorrítis (Ostrovon) and the Aliákmon river, in the Vardar valley, in the Lake Dojran depression and in the region between the lower valleys of the Mesta and Vardar rivers.

*The Greeks*

Gopčević did not agree that the Greeks were as widespread in southern Macedonia as Sax had represented them to be in 1877. He believed that even Kiepert had shown too many Greeks. He drew their linguistic frontier from the Píndhos mountains along a line north of Siátista to the upper Aliákmon river and thence to Véroia and the delta of the Vardar. Elsewhere in Macedonia he portrayed them as a rural population—in Khalkidhikí and around Lake Akhinoú. Sérrai, Kastoría and Edhessa were situated according to his map in Serb country.

*The Slavs*

Gopčević was responsible for giving academic support to the novel idea that the whole of Old Serbia and most of Macedonia were inhabited by Serbs, or to put it in another way, that all the Slavs in this territory, who had hitherto been regarded as Bulgarian, proved to be Serbian on closer inspection. He was accompanied by a Bulgarian servant on his Macedonian journey and he wrote that to their mutual surprise they had discovered that the majority of the Slavs of Macedonia belonged to the Serbian rather than to the Bulgarian stem.[1] But this was not all that Gopčević claimed on behalf of the Serbs. Hitherto nearly all maps had shown large numbers of Albanians in Old Serbia, and northern and western Macedonia, but Gopčević ignored older interpretations by marking these territories as Serb—either as populated by *Moslemized Serbs,* by *Christian Albanians of Serb extraction* or by *Moslem Albanians of Serb extraction.* The result of such a vigorous and revolutionary re-classification of old-established groupings resulted in an extension of the Serbian ethnographic frontier as far south as Nevrokop, Sérrai, Salonika, Véroia, the Grámmos Mountains and Kónitsa, and as far west as to include the whole of the Crni Drim and the western banks of the Drin-i-xy. Gopčević had proclaimed the Serbian cause in Old Serbia and Macedonia with a vengeance.

He substantiated these claims by reference to linguistic data, to the evidence of folklore and folk songs, and to historical evidence. The core of his argument was based on the linguistic factor because he declared that of all the travellers who had visited and described Macedonia hitherto, not one had been sufficiently qualified by his knowledge of Serbian and Bulgarian to pronounce on the nature of the language spoken by the inhabitants. Kiepert, Sax, Boué, Barth, Hahn, Kanitz, none of these scholars, argued Gopčević, had been acquainted closely enough with Slavonic languages to make a final decision on the Macedonian dialect. He then listed a number of similarities between the Serbian language and the Macedonian dialect. Because of their ignorance of the local languages, many ethnographers had jumped to the conclusion that because the Macedonian Slavs spoke *bugarske* and called themselves *Bugari* that they were Bulgarians. But the word *Bugari,* maintained Gopčević, merely meant peasant or *raja*, and had nothing to do with the word *Bolgari*

[1] "Die ethnographischen Verhältnisse Makedoniens und Altserbiens," *Petermann's Mittheilungen* (1889).

meaning Bulgarian. In this manner travellers had been misled and confused by the similarity between the two words.

According to Gopčević's representations, the Bulgarians occupied only a small part of Macedonia. He confined them in fact to the Mesta valley, the inhabitants of which he showed to be mostly Pomaks, but with some Christian Serbs and Bulgarians. Nor did he believe that the Pomaks were of Bulgarian nationality. He showed other groups of Pomaks in the Moglenítsas and Vardar valleys.

*The Vlachs*

Gopčević distinguished Vlachs in nearly all the towns of Macedonia. He also indicated them in the Píndhos and Grámmos mountains, in the Maloviŝte hills, on the south-western banks of Lake Ohrid and in the Kozuf hills.

*The Albanians*

Very few Albanians were, according to Gopčević, to be found either in Macedonia or in Old Serbia. They were found as a majority population only well to the west of the Drin valley. Ethnographers had erred in the past by depicting Old Serbia and part of western Macedonia as Albanian territory, he declared, because these areas were occupied not by Albanians but by *albanianized Serbs.*

*Conclusion*

Gopčević reckoned that the total population of Old Serbia and Macedonia was 2,849,050. Included in this total were the ethnic groups given in the table on the following page.

The national groups to his mind did not correspond to the ethnic groups given below. This was because 26,000 Albanians were adherents of the Greek Church and were Greek nationals. Another 10,000 Bulgarians had also become hellenized; 83,000 Vlachs could be regarded as Greek nationals, and the Pomaks were to all intents and purposes Turkish. Of the total Serb population, about 1,400,000 could be regarded as Serbs from a political point of view. The remainder had become Albanians, Turks or Bulgarians. This made up the non-Serb totals to 327,000 Greeks, 940,750 Turks and 82,000 Bulgarians.

Until the appearance of Gopčević's map, the controversy over the political affinity of the Christian population in Macedonia had involved only the Bulgarians, the Greeks and to a lesser extent the Romanians. Gopčević succeeded in introducing yet another element—the Serbs. Nobody before 1885, except for a few discredited Serbian chauvinists, such as M. S. Milojević, had believed that the Serbs were really an important minority south of the Sar mountains. Boué and Lejean had

*Population of Old Serbia and Macedonia after Gopčević*

| *Major Groups* | *Totals* | *Included Minorities* |
|---|---|---|
| Serbs .. .. .. | 1,830,100 | 418,500 Moslems |
| Turks .. .. .. | 269,000 | 6,200 Circassians<br>1,000 Tartars |
| Albanians .. .. | 189,250 | 138,150 Moslems |
| Bulgarians .. .. | 176,200 | 104,000 Pomaks |
| Greeks .. .. .. | 171,200 | 4,000 Moslems |
| Vlachs .. .. .. | 100,600 | 7,600 Moslems |
| Jews .. .. .. | 72,000 | 5,000 Moslems |
| Gypsies .. .. | 34,000 | — |
| Others .. .. .. | 6,000 | — |

shown a few small exclaves in a sea of Bulgarians. Kiepert had shown no Serbs in Macedonia at all. Gopčević admitted that he had himself been convinced, earlier in his career, that the Serbs could be guilty only of gross chauvinism in laying claim to Macedonia and he had even expressed such an opinion in writing before he embarked on his Macedonian travels.

His map immediately aroused a storm of criticism. A. Nehring in *Petermann's Mittheilungen* condemned the obvious political flavour of the map and the reviewer in *Deutsche Rundschau* was even more outspoken in his disapproval :

> This large and very clearly drawn map surprised us, not so much on account of its repudiation of all the well-known ethnographic maps of this region, as by its extremely detailed presentation of an uncommonly complicated ethnographic situation. When it is known by what devious ways Czoernig or Le Monier arrived at an understanding of the ethnographic map of Austria-Hungary, it might be concluded, that a short and hurried journey could not possibly provide the material for such a map.[1] Gopčević's methods are those of a dilettante rather than a professional ethnographer.[2]

These two shots, chosen at random from the barrage of criticism directed at Gopčević's work, indicated that the academic circles of Europe were not ready to accept the Serbian thesis at this stage. It is a firm axiom of the propagandist, however, that an initial failure may be turned into an ultimate success by the simple process of re-iteration. Gopčević at least provided the Serbs with their initial failure. Many Serbian ethnographers were to follow in his footsteps, although perhaps they displayed more caution in the presentation of Serbian claims. In time, their insistence on the presence of Serbs in Macedonia wore down the opposition and paved the way for Serbian political expansion southwards. To this extent Gopčević might well be called the father of Serbian political ethnography.

## N. S. Zaryanko's Map of 1890

### *The Niš-Leskovac Region*

The Serbians, although not initially successful in establishing their major claims, did exert some influence on European thought. An interesting example of their influence is to be found in the two editions of a Russian ethnographic map compiled by N. S. Zaryanko and published by V. V. Komarov in St. Petersburg in 1890. The first

[1] Czoernig, *op. cit.* F. Le Monier, *Sprachenkarte von österreichisch-ungarischen Monarchie* (Wien, 1888).

[2] Literatur-Berichte, Nr. 2464 in *Petermann's Mittheilungen. Deutsche Rundschau für Geographie und Statistik*, XII (1890). K. Oestreich also severely criticized Gopčević's distributions in " Die Bevölkerung von Makedonien," *Geographische Zeitschrift*, XI (Leipzig, 1905).

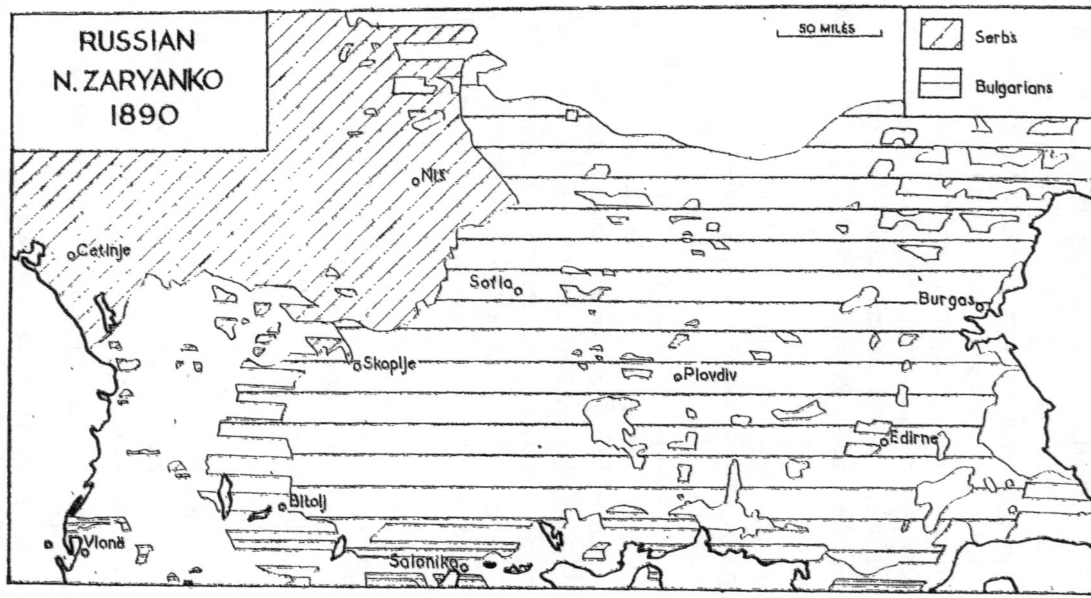


FIG. 25

Districts over 75 per cent Serb or Bulgarian are solidly shaded and districts over 50 per cent are shown diagrammatically by horizontal bars.

In the second edition of this map, published later in the same year, the Slavs of Macedonia are left undifferentiated.

edition purported to incorporate the latest researches on the distribution of the Slavs in Europe (Fig. 25). The region formed by the Niš and Leskovac districts was coloured as Serbian. Hitherto this region had been universally regarded as exclusively Bulgarian, and earlier Russian maps had stressed its Bulgarian character. The new interpretation which indicated the population to be wholly Serbian could mean either that the Slavs in the region had never been Bulgarians, or that the Bulgarians had lost their national traits within twelve years of incorporation into Serbia. Nevertheless, most ethnographic maps published after this date followed Gopčević and Zaryanko in recognizing Serbian territory in the Niš region. This concession in favour of the Serbs had important repercussions on the formation of the Serbian hypothesis of the amorphous character of the Macedonian Slavs (see p. 149).

*Macedonia*

In the first edition of the Russian map, Zaryanko also favoured the Serbs by reducing the Albanians in the area west of the Andrijevica-Prizren line. But he still countenanced some Bulgarians in Macedonia, going so far as to show strong Bulgarian minorities in the Vijosë valley of southern Albania and larger groups of Bulgarians in the Drin-i-xy valley and in the coastal districts of Kaválla and Dhitikí Thráki. The recognition by Zaryanko of the Bulgarian affinities of the Macedonian Slavs incurred the displeasure of the Serbians who protested to St. Petersburg that the Macedonian Slav population had been incorrectly classified as Bulgarian.[1] They had the satisfaction of seeing the Macedonian Slavs left uncoloured in the second edition of the map. If the Serbian ethnographers had not succeeded in establishing their claims in Macedonia, they had at least thrown doubt on the Bulgarian affinities of the Slav inhabitants of that territory.

But the Serbians were far from content with these limited moral successes. Western Europe was still sceptical. It was generally known that Russia had every reason for wishing to pacify Serbia. Now that the Slav family was growing up so quickly its members exhibited all the

[1] D. Rizoff, *Die Bulgaren in ihren historischen, ethnographischen und politischen Grenzen* (Berlin, 1917).

recalcitrance of adolescence. It was possible to read into the concession made in favour of the Serbs, nothing more than the Tsar's earnest desire to reconcile his allies in the western Balkans. So the Serbians pressed on with their research, eager to establish their case to the satisfaction of the rest of Europe.

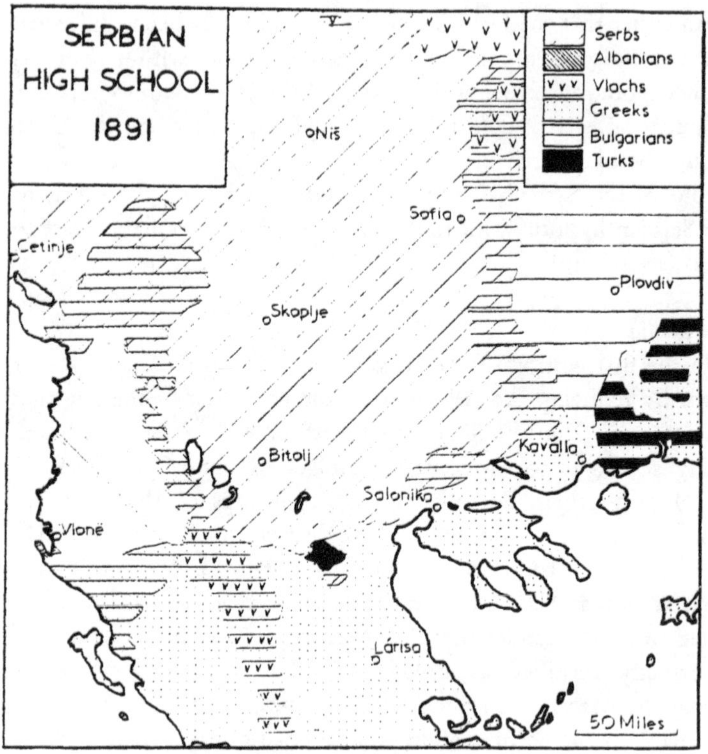


Fig. 26

THE SERBIAN HIGH SCHOOL MAP OF 1891

A Serbian ethnographic map was produced in 1891 by the scholars of the High School at Belgrade (Fig. 26).[1] This map was printed in both Serbian and French and was designed therefore for western

[1] It was reproduced also in 1903, edited by Prof. M. Andonović (Antonovitch).

European circulation. The anonymous authors of the map marked Old Serbia as extending from Novi Pazar to Prilep. Macedonia was given a narrow interpretation, being limited to the area south of Bitolj and Strumica.

*The Turks*

On this map, only the *Konariote* Turks were marked as a solid Turkish minority. Elsewhere, the Turks were depicted as a mixed population living principally in the hill country between the lower Struma and the lower Vardar valleys, in the lower Mesta valley, and in the region around the headwaters of the Arda river.

*The Greeks*

In eastern Macedonia the Greeks were given a more extensive distribution than on Gopčević's map, the Greek ethnographic frontier being placed on the peaks of the central Rodopi. Thus the Greeks gained at the expense of the Bulgarians. Where the Greeks had a common boundary with the Serbs, however, the ethnographic frontier was depicted as on S. Gopčević's map.

*The Slavs*

The Serbian authors of this map made even more striking claims for the Serbs than did Gopčević. Their map showed an enormous extent of Serbs over the whole of the western Balkans, from Shkodër down the Drin valley to Kónitsa in the west, to Kastoría, Véroia, Salonika and Dráma in the south, and in the east, as far as a line through Dráma, north to the Danube, including Sofia. There were no important minorities within their alleged Serbian territory, except for a few Albanians in the Drin valley and some Bulgarians in the valleys of the Mesta and Isker (Eskeje). Although it might be inferred from the legend that folklore and language had been taken into account in arriving at the limits of the Serbs, it is obvious that the distribution was based largely on historical criteria. In 1878, Spruner-Mencke's newly revised historical atlas[1] had incorporated a small map of the mediaeval Serbian empire of Dušan, an indication of the influence

[1] *Spruner-Mencke's Hand-Atlas für die Geschichte des Mittelalters und der neuen Zeit* (Gotha, 1871–8).

of the research of Serbian historians who had begun to take an interest in the origin and extent of pre-Turkish Serbian empires (Fig. 27). The Serbian High School map was little more than an attempt to invest Dušan's empire with an ethnographic significance which was extremely questionable. Thus the map was based on an inconsequential confusion of historical and ethnographic criteria. Its importance lies in the fact that it throws light on the psychology of its authors and is a measure of the desperation which prompted the Serbians in their attempts to

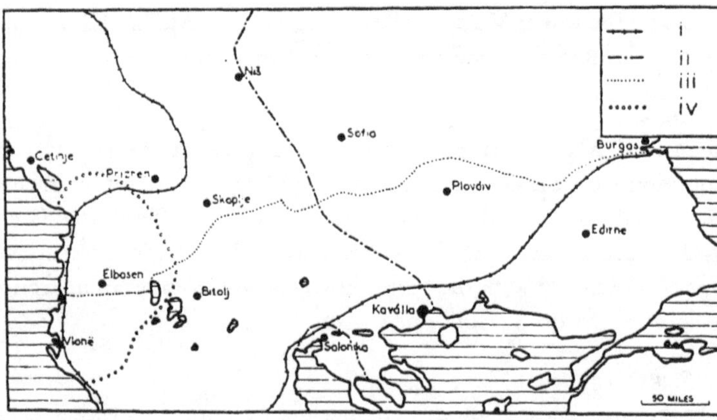


FIG. 27. MEDIAEVAL POLITICAL BOUNDARIES IN MACEDONIA
The references in the key are as follows : **i.** Extent of the first Bulgarian Empire, *c.* 900 A.D., based on S. Runciman, *A History of the First Bulgarian Empire* (1930) ; **ii.** Southern and eastern limits of the Serbian Empire in 1335—at the time of the death of Stefan Dušan, based on S. Stanojević, *Istoriski Atlas* (1934) ; **iii.** Northern limit of the Byzantine Empire in the 14th century, based on Spruner-Mencke, *Historischer Handatlas* ; **iv.** Limits of territory controlled by Skanderbeg, the Albanian, *c.* 1450, based on the *Enciclopedia Italiana* (1930).

secure a modification of the ideas on the ethnography of the western Balkans in their favour.

As for the Bulgarians, on this map they were practically excluded from Macedonia except for a minority indicated in the upper Mesta valley. Pomaks were not separately distinguished.

### *The Vlachs and Albanians*

The Píndhos mountains, the Grámmos and the districts intervening, were marked as Graeco-Vlach on this map. No other Vlach minorities were shown. The Drin valley and the plain of Kosovo were depicted as mixed Albano-Serb territory. Pure Albanian territory was limited to the region between the Mat valley in the north and a line between Korcë and the bay of Vlonë in the south. Northern Ipiros was shown as mixed Graeco-Albanian.

*Conclusion*

The impressions which these Serbian views had upon the British may best be appreciated by a reference to the work of Sir C. Eliot :[1]

> The history of the last fifty years in S.E. Europe is to a great extent the history of the disentanglement of the Slavonic Races from the Greeks and Turks, and to this is now succeeding the disentanglement of the Slavonic Races from one another. In the early part of this century all Christians in Ottoman Europe were called Greeks, and as late as 1878, the accurate and talented authoress of the " People of Turkey " no doubt faithfully reflecting local opinion, considers that the inhabitants of Veria, Doiran, Vodena and Strumnitsa are Greek rather than Slavonic. Ten years later, the progress made by Bulgarian schools and the Bulgarian Church persuaded most people who were interested in the question, without being prejudiced, that the whole of the Ottoman part of the Balkan Peninsula, west of Prishtina and of Ochrida, were for practical purposes Bulgarian. This view had hardly time to become commonplace before politicians put forward another idea, not without the support of scientific men, namely, that the districts in question are indeed Slav, but Servian—not Bulgarian. Those who knew Macedonia were astonished to hear that Servian consuls were appointed at Vodena and Seres to protect the interest of Servian Communities in those parts. Associations in Sofia and Belgrade did all that arguments and subscription lists could do to prove that the whole of Macedonia belonged to the nation which they represented, and the result of the Turko-Greek war of 1897 discredited still further the already weak Hellenic Cause and disposed those Slavs, who were officially called Greeks, because they belonged to the " Patriarchal Church of Constantinople," to call themselves Serbians.

[1] *Turkey in Europe* (London, 1900).

The above quotation illustrates the bewilderment which changes in ethnographic ideas were producing at this time, even on those members of the British public with a fair knowledge of the Balkans. There was certainly no suggestion that the Serbian view of Macedonian ethnography was accepted. There was in fact a tendency to ridicule their claims. The entry of the Serbians into the Macedonian arena resulted in a general intensification of Bulgarian and Greek propaganda in the Turkish provinces. The fate of Old Serbia and Macedonia had now become a crucial issue. The Bulgarians realized that they could not rest on their laurels, and that something more than G. Lejean's map was necessary to sustain their claims. The hostility of Serbia, and the possibility that the Serbians and Greeks might come to terms on spheres of influence in Macedonia, had placed Bulgaria's former secure position in Macedonia in jeopardy. The Greeks perceived that the Serbo-Bulgarian schism might be utilized to the advantage of the Hellenic cause, and they welcomed the opportunity of increasing the discomfiture of the Bulgarians in Macedonia by any means in their power, even by a rapprochement with Serbia. The flames of dissension were fanned by constant rumours of an impending Turkish collapse. Discontent with the Sultan's rule had become widespread even in Turkey proper. Turkish activities in Europe were restricted by guerilla operations to such an extent that the Turks had control only of main routes and big towns. In the capitals of Europe there were whispers of international action which would end once and for all Turkish rule in south-eastern Europe.

It was hardly surprising, therefore, that the interested parties redoubled their efforts to promote their respective national causes in the Macedonian theatre. The priest and the bishop, the school-teacher, the scholar and the brigand, all played their part in the struggle which now ensued. The battle was joined with all the fervour of a religious crusade and with all the ferocity of a blood feud. For the next thirty years in Macedonia the bandolier and the rifle were more often in a man's hand than the tools of his trade or the handles of his plough. Henceforth the ethnographic map assumed a much more complex character. Not only were ideas changing due to a re-examination of criteria but enforced proselytism, the persecution of minorities, abductions and educational coercion also began to have their effects in the modifying of ethnographic distributions. Greeks, Bulgarians, Romanians and Serbians did all in their power to strengthen their own respective elements and to weaken those of their rivals. The Great Powers, including Turkey itself, were not averse to exploiting the situation to their own respective advantages. Macedonia had become the problem area of Europe. The solution of the enigmatic character of the political affinity of its population constituted a challenge alike to the inventiveness of politicians and to the resources of scholarship. Interested and disinterested persons, scholars, diplomats and journalists, found increasing difficulty in resisting attempts at fresh interpretations of the ethnography of Macedonia.

## Chapter VI

# THE END OF THE CENTURY

### G. Weigand's Map of 1895

Gustaf Weigand's classical work, *Die Aromunen*, which clarified ideas on the ethnographic situation in south-western Macedonia in a remarkable manner, appeared in 1895. Since 1876, purely objective and dispassionate research into ethnographic distributions had been exceedingly rare, but Weigand as professor of Balkan languages at Leipzig enjoyed a reputation for impartiality and was well qualified to undertake ethnographic research, having had historical and philological training. He was a keen student of folklore and the author of Albanian and Bulgarian grammars and dictionaries.[1] His book included a map which was both attractive and useful, as its limited area and large scale (1:750,000) enabled distributions to be plotted with tolerable accuracy. Weigand attempted to re-define the ethnographic frontiers between Albanians, Greeks and Slavs, and at the same time to indicate those zones where the Vlachs or, as he preferred to call them, the *Aromunes*, predominated (Fig. 28).

*The Turks*

In the plains north of Salonika, Weigand indicated on his map large numbers of Turks, whose existence had hitherto been unacknowledged except on the maps of A. Synvet (1877) and K. Sax (1878). The particularly important routeway through the Lake Dojran locality was marked as Turkish. He also fixed the limits of the Konariote Turkish peasantry as Lake Vegorrítis in the north, and the Aliákmon river and the town of Kozáni in the south.

[1] See Appendix A.

*The Greeks*

Whilst he recognized Hellenic territory in Ipiros as far north as Delvinë and Kónitsa, thus accepting H. Kiepert's interpretation in favour of the Greeks, Weigand considerably reduced the Greek territory, shown on Kiepert's map in the Edhessa region. He depicted the Greek ethnic frontier here, as a line running south from the Grámmos mountains to the Valachades (Vlakh highlands), thence

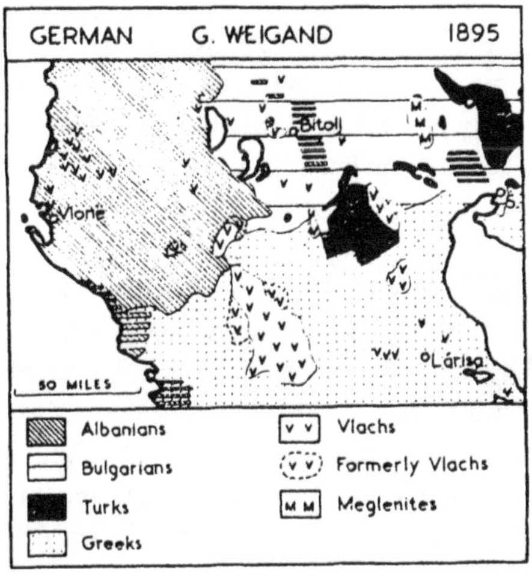


FIG. 28

In the original, towns with a Vlach element in their population are indicated.

to Kozáni and the Aliákmon river, with an extension north to include Náousa and Véroia. His Graeco-Slav frontier was almost the same as that of S. Gopčević's map with the notable difference that the Slavs were portrayed as Bulgarians and not as Serbs. He also agreed with Gopčević that the Albanian, Greek and Slav worlds met just west of Kastoría. Weigand's definition of the northern limits of the Greek ethnographic frontier in this region enjoyed widespread favour until the Graeco-Turkish population exchanges rendered it invalid ; it was, of course, rejected by the Greeks themselves.

*The Bulgarians*

The Bulgarians on Weigand's map were excluded from the Drin valley, except where that river enters Lake Ohrid. The northern banks of the Lake were indicated as Bulgarian. To the south the Bulgarian frontier was shown as joining the southern extremities of Lakes Ohrid, Prespa and Ventnok; thence it ran south to the Valachades, to include as Bulgarian the region around Lake Kastoría. From here it ran to Lake Vegorrítis and to the Gulf of Salonika, north of the mouth of the Aliákmon river. Salonika and the surrounding countryside were marked as Bulgarian. Weigand thus took an extraordinarily favourable view of the Bulgarian claims in Macedonia. He made no reference at all to the existence of any Serb exclaves around the Macedonian lakes. As his map did not include much territory north of Kruševo and Prilep there is no indication of where he set the northern limits of the Bulgarians.

*The Vlachs*

In his study of south-western Macedonia, Weigand had been concerned mainly with the delimitation of Vlach territory. Since the time when G. Lejean, and later K. Sax, had made bold attempts to map their distribution, the Vlachs had received scant attention from ethnographers, most of whom had decided that their distribution was too vague to warrant detailed treatment. The problem of mapping them had always presented numerous difficulties. Only a small percentage could be said to have a permanent domicile. The vast majority practised a nomadic mode of life, which led them from the lofty summer pastures of the Píndhos and the Rodopi mountains to the security of the Aegean and Iónian coastal plains in winter. Large numbers were itinerant traders and yet others constituted a merchant class, with representatives in all the main towns of the peninsula. Weigand was concerned with identifying those localities which had become the headquarters of the Vlachs—the regions where they foregathered in summer, where they accumulated their belongings, and where they had their more permanent habitations and regions, therefore, which had come to be regarded as exclusively Vlach territory, characterized by Vlach place-names. Apart from its economy, the most characteristic feature of the Vlach group was its language. It was upon this criterion that Weigand, as a philologist, built up his picture of the Vlachs as a distinctive group. Hitherto, although separately distinguished on many maps, the Vlachs had always been associated in an indeterminate fashion either with the Bulgarians or with the Greeks. Indeed many believed that their relation to the Greek people was analogous to that of the Albanian Tosks of Ipiros. The Vlachs, too, were for the most part Greek Orthodox in faith, although a small number had turned to the Moslem faith.

Weigand fixed about a dozen areas on his map where the Vlachs formed the bulk of the population. The more important of the Vlach elements are indicated below.

(1) The Vlachs of the Píndhos, who had been sufficiently important to figure on the maps of Boué, Lejean and Kiepert. Weigand stated that both historical evidence and the existence of place-names pointed to a more widespread distribution of the Vlachs in this region. Those on the flanks of the Píndhos were in process of being hellenized.

(2) The Vlachs of the Grámmos mountains. Here there was a tendency for the Vlachs to become albanianized.

(3) The Vlachs of the western slopes of the Olimbos mountains.

(4) The Vlachs of the mountains immediately west of Náousa and Véroia.

(5) The Vlachs of the Klisoúra region and the mountains immediately to the south.

(6) The Vlachs of the Flórina hills.

(7) The Vlachs of the hill country west of Bitolj.

In addition to these groups, he recognized many smaller exclaves throughout the Bitolj and Lake Ohrid region and also a strong minority of Vlachs in the plain of the lower Semen valley of Albania, between Berat and the coast. He also gave indications of areas once obviously Vlach (by virtue of place-name evidence) which, by 1895, had lost their Vlach character.

Weigand's work on the Vlachs promoted an interest in this extraordinary people who had so obviously at one time occupied a much greater area of the south-western Balkans. The Vlachs were apparently in process of being absorbed by the people amongst whom they lived; this was especially so where they were in close association with the Greeks. One of the consequences of Weigand's work was to invest the Vlach cause with fresh political significance. Not that Weigand himself had this aim in mind when he set about his work, but the advantages of a Vlach political movement in Macedonia had always been manifest to the Romanians, who seized upon Weigand's work to further their own scheme of promoting *Romania irredenta* in Macedonia (see p. 179 and Fig. 86).

*The Albanians*

Weigand limited the Albanians in the south, to a line from Sánta Quaranta on the coast of Ipiros, to Kónitsa, thence to the western fringes of the Grámmos mountains. Thus in Ipiros, he indicated the Albanians only as scattered minorities in the coastal districts, but in the east he showed them in the whole of the Devoll valley almost as far as Kastoría, in the greater part of the Drin valley, and in the Bitolj district where they formed scattered minorities.

*The Meglenites*

Weigand drew attention to a group of people hitherto unrecognized on any ethnographic map. He called them the *Meglenites* and they lived in the hill country west of the lower Vardar river and the plain of Mogléna (Karáj Ova). Apparently they had many peculiarities, which distinguished them from both the Greeks and the Bulgarians. They appeared to be the remnants of the Petchenegs (Patzinaks), who had swept into Macedonia from Asia long before the Osmanli Turks arrived. Apparently they spoke the Vlach language but by what mysterious process they had become Vlachs, Weigand was not able satisfactorily to explain.

*Conclusion*

Weigand's map was well received except by the Serbians and the Greeks. The great German geographer, A. Phillipson, welcomed his contribution and declared that it was all the more valuable as it was based on extensive personal experience of conditions in Macedonia, Albania and Thessalía. Reports in *Petermann's Mittheilungen* acclaimed his work as an example of painstaking research and as a real contribution to an understanding of the region and its nationalities.[1] Weigand's distributions were partially utilized later by J. Cvijić, by J. Ivanov and by many other authorities.

## The Maps of R. von Mach, 1899

The Greeks, however, were in no mood to subscribe to such sentiments about Weigand's map. The expansion of Bulgaria into Eastern Roumelia in 1885 had precipitated a domestic crisis in Greece and interest both in Crete and on the mainland had reached fever pitch. In 1894, a society popularly known as the *Ethnike Hetaireia* was formed with the object of furthering Greek claims in Macedonia, because with each move that the Slavs made on the mainland, the possibility of a restoration of the Empire of Constantinople was growing ever more remote. Gone for ever was the prospect of an Hellenic boundary on the Balkan peaks, but the possibility of accessions in Macedonia, Thrace and Asia Minor—scenes of ancient Greek glory—still remained. The possession of southern Macedonia was not only desirable in itself but it was the necessary prelude to any further Hellenic expansion on the mainland, as it was the link between Thessalía and Thrace. Nor without Macedonia was there any possibility of the Greeks aspiring to Constantinople.

The 'Thirty Days War' with Turkey in 1897 destroyed Greece's hope of attaining her objective by force and, somewhat chastened, the Greeks began to reflect on the attitude of the Great Powers whose policy of inaction during the brief conflict had substantially contributed to their failure. Greece could never hope to gain the sympathy either of Austria or of Russia; she appeared to have lost that also of the western Powers, who had failed at that time to support the Hellenic cause. In order to cultivate sympathy for their Macedonian aspirations, the Greeks turned towards the possibility of modifying opinion in their favour. They were fully aware of the need for a restatement of the Greek case. Weigand's map, since it imposed severe limits on Greek territory in Macedonia, was in the nature of a challenge which could not be ignored.

[1] *Hettner's Geogr. Zeitschrift* (1896). Literatur-Bericht Nr. 139, *Petermann's Mittheilungen*, XLV Band (1899).

*The Language Dilemma*

Since the production of E. Stanford's Greek map of 1877 (see p. 70), the Greeks had made no further attempts to modify prevailing ethnic ideas by the production of more pro-Greek maps. Their main difficulty arose out of the deference accorded to the criterion of language by the ethnographers of the day. Unfortunately for the Greeks, the idea had gained universal acceptance that Slav, Romanian and Turkish were together spoken more widely in Macedonia, than Greek. Since the linguistic factor still loomed large in any consideration of ethnographic distributions likely to impress western Europe, the Greeks had to consider carefully ways and means of re-emphasizing the importance of the Greek language in Macedonia.

G. Weigand had noted that before the advent of the Slavs in the sixth and seventh centuries A.D. the Greek linguistic frontier had corresponded roughly with the peaks of the Balkan mountains. The Slavs, after they had overrun most of the territory between the Danube and the Aegean, had been strongly influenced by Hellenic culture, but they had managed to retain their own language, until the Christianization of the Balkans gave to the Slav vernaculars an alphabet, and thereby the means of preserving and evolving their distinctive written languages. The Slav written languages were deeply rooted in the vernacular but the political importance of this healthy relationship between the spoken and the written word only became apparent towards the end of the nineteenth century. The Greek language, however, was never able to rid itself of its *damnosa hereditas*, the Attic tradition. The divergence between written Greek and the popular tongue (Romaic) has been a characteristic feature of the history of the language from ancient down to modern times.[1] It must explain in part the failure of the Greek language to maintain its hold over a much wider area of the peninsula than it did. The Greek poet, Solomós (1789-1856), was well aware of the danger of using for a national tongue "a language which nobody speaks, nor has spoken nor will ever speak," but in spite of his protests, traditions of a dead language rather than living Greek were preferred in 1832 as the basis of the Rook (official) language of Greece, and Koraïs' *Katharévouas*

[1] R. Byron, *The Byzantine Achievement* (London, 1929).

or expurgated Greek persisted until 1917 in spite of the opposition of many eminent men of letters such as Ionnis Psicharis (1854-1929).

The recognition of modern Greek on its own merits as the national language came too late to help the Greek cause in Macedonia. Even although the Greeks realized the political importance of language after 1876, their efforts to popularize Greek in Macedonia were not successful in spite of all the machinery of culture and education at their disposal. P. M. Chassiotis, in 1891, in his work *L'Instruction Publique chez les Grecs*, revealed in an interesting map that the expenditure on Greek public education in the so-called unredeemed territories was higher in proportion to the population than that within the Greek political boundaries. Greek schools manned by Greek teachers were to be found all over the southern Balkans. But their influence was negligible, as not only had the teachers to contend with Slav and Romanian rivals but all they had to offer in opposition to a living language was an abstract conception which nobody understood.

### *Educational Spheres of Influence*

The areas of Macedonia over which the rival educationalists had deployed their forces in 1899, was plotted by the German, von Mach, in a series of maps showing the extent of Greek, Bulgarian, Serbian and Romanian schools. The four maps gave useful indications of the respective spheres of influence of the contending parties. If the number of schools functioning in this region at the end of the century had been an indication of cultural progress, then surely Macedonia must have been a region of enlightenment and scholarship without parallel in eastern Europe.

In southern Macedonia, the Greek schools were extraordinarily strong and, compared with the numbers of the Greek speakers in these territories, the Greek concentration was higher than that of the Bulgarians. The Greeks had established numerous schools as far north as Bitolj, Ohrid, Strumica and Nevrokop (Fig. 88). In southern Albania, a network of Greek schools spread Hellenic culture as far as the Semen river. The Bulgarians had an average of over twenty schools per district (caza) over the whole of the traditional Bulgarian zone between the Sar mountains and the Kastoría-Salonika-Sérrai line

(Fig. 90). The Serbian schools were limited more particularly to the vicinity of Pristina and Prizren but there was an average of over five schools per district in the Tetovo, Skoplje, Kumanovo and Veles areas (Fig. 89). Finally there was a considerable number of Romanian schools in south-western Macedonia. Albanian schools were not separately distinguished, but von Mach's map showed many districts in the western Balkans where no Christian schools had been established, particularly in the highlands of eastern Albania.

It is apparent that throughout Macedonia, at this time, there was a considerable overlap of culture spheres of influence, so that children found themselves the objects of keen competition on the part of rival scholastic institutions. Schoolmasters were more often than not political agents, carefully trained for their task, and more than one I.M.R.O. leader subsequently hailed from their ranks (see p. 151). Competition indeed was so keen that a school would be maintained for only two or three scholars; conversely, village children would sometimes tramp twenty miles to attend a school of their national preference, whilst the village school remained empty. The art of enticing the young was assiduously practised, so that the peculiar situation would sometimes arise of 'Bulgarian' parents finding themselves in possession of 'Greek' children or vice versa.

## C. Nicolaides' Map of 1899

The aspirations of the Greeks in Macedonia were made clear by Nicolaides' book on Macedonia, the German translation of which was published in Berlin in 1899. It contained an ethnographic map, one of the first of its kind specifically limited to Macedonia (Fig. 29). Professor Nicolaides of the University of Athens was primarily an historian but he did not hesitate to supplement his views on the historical role of the Greeks in Macedonia with ethnographic evidence. His map provided just one more example of how ethnographic data might be manipulated in order to create an impression of conditions favourable to a particular thesis, and his criteria and methods are worth examination for this reason. It has been seen how the Greek case had been weakened by emphasis on linguistic criteria. Nevertheless

Nicolaides elected to base his map primarily on language but on 'commercial language' not 'mother tongue.' He maintained that, although Turkish might be a common medium over the Balkans generally, Greek, Slav and Albanian were the languages used in everyday relations in Macedonia. The Vlach language and the Turkish language were used only in small villages. Areas using the latter languages were therefore indicated on the map by means of symbols. His primary divisions, represented by flat colourings, were :

(1) territory where the Greek language was employed ;

(2) territory where the Slav languages were employed ;

(3) territory where both Slav and Albanian were employed.

Before examining the interesting distributions which Nicolaides obtained it might be as well to consider his definition of Macedonia. The definition of the frontiers between Albania, Old Serbia, Greece, and Thrace on the one hand and Macedonia on the other had by this time become an important issue. The political boundary of Serbia (1878), that of Bulgaria (1885), and the coastline in the south were the only definite limits to Macedonia. As the eastern limit Nicolaides adopted the administrative boundary between the *Vilayets* of Adrianople and Salonika, which utilized the lower Mesta river and then ran due north to incorporate most of the Rodopi mountain territory into Macedonia. In the north, he fixed the limits of Macedonia as the Sar mountains and the Crna hills. Thus he did not accept S. Gopčević's view, that Old Serbia extended south over the Sar mountains and, that the boundaries of Macedonia lay much farther south (Figs. 1 & 2). He adopted a very liberal view of Macedonia in the west, and included the great basin of Korcë and Lake Malik within the 'province.' He used the 1881 boundary of Greece as the southern limit of Macedonia. It is important to bear these limits in mind because both the Serbians and Bulgarians defined Macedonia rather differently and the Greeks themselves modified their conception of Macedonia at a later date (see p. 137).

*The Turks*

Nicolaides depicted the distribution of the Turks in such a way that he reduced their importance in south-western Macedonia and the Aegean coastal districts but increased it in the Mesta valley and the Rodopi mountains. He obviously regarded Pomaks as Turks, as a comparison with K. Sax's map will indicate, and he marked strong concentrations of Turks in the Struma valley north of Sérrai, in the Lake Dojran depression and around Stip, Kruševo and Bitolj. All these localities, he stated, were inhabited by a rural population composed of Moslem Turks. His evidence supported the idea that numerous

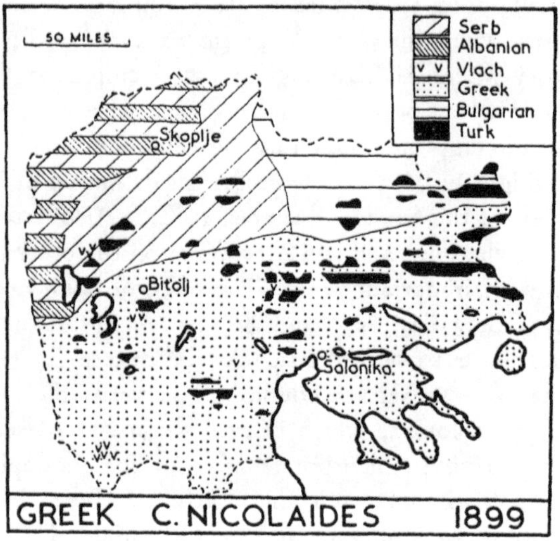


FIG. 29

In the original, the major linguistic zones of Macedonia are indicated as follows: 1. Albano-Slav; 2. Slav; 3. Greek. The Slav zone is sub-divided into Serb and Bulgarian (as above). Superimposed upon the main groupings are two other categories—'Turko-Moslem rural population' and 'Vlachs with a fixed domicile'—both of which are represented diagrammatically above.

Turkish settlements in Macedonia had been ignored by ethnographers of an earlier period too intent on rediscovering the Slavs.

*The Greeks*

Applying his criterion of Greek 'commercial language,' Nicolaides boldly portrayed the whole of southern Macedonia as Greek linguistic territory, pushing the frontier as far north as the Devoll river, the Galicia mountains between Lakes Ohrid and Prespa, and Strumica and Nevrokop. Within this frontier was included all that area which had been shown on von Mach's map to be covered by Greek schools. It bore no relationship at all to Weigand's Greek ethnographic frontier which had been much farther south. As Gopčević did in the case of the Serbs, Nicolaides staked all on a categorical affirmation of what the Greeks desired, rather than on what actually existed. His map defined the Greek sphere of cultural and commercial influence rather than Greek ethnic territory. Within his 'Greek territory' there were, he maintained, 594,500 Greeks, 324,000 Turks, 126,000 Slavs, 79,400 Jews, and 37,000 *Koutzo-Vlachs*. The non-Greeks in this zone, he declared, were minorities in a world where Greek language and culture prevailed.

### *The Slavs*

In northern Macedonia, Nicolaides made a distinction between (1) the Albano-Slav area, which included the Drin valley, the Tetovo and the Skoplje districts, and (2) the pure Slav area. In the pure Slav area the Greeks numbered only 57,000 compared with 268,200 Slavs and 182,000 Turks and Albanians. The frontier between Serbs and Bulgarians was drawn as a line from Strumica to Kriva Palanka. West of this line the Slavs were mostly Serbs and east of it mostly Bulgarians. This was by far the greatest concession to the Serbian thesis yet made by a non-Serbian. The Greeks had become aware that the Bulgaro-Serbian schism gave them a decided advantage in Macedonia and that the Bulgarian supremacy could be broken by insisting on the Serb character of its Slav population. Nicolaides, by splitting the Slavs into Serbs, Bulgarians and *Slavs-under-Greek-influence*, deprived any one of the Slav groups of a majority.

### *The Albanians*

The Albanians hardly figured at all on Nicolaides' map. In south-western Macedonia they were simply ignored, in spite of the fact that Nicolaides had pushed the limits of Macedonia well into territory which both the Turks and the western Europeans generally regarded as part of Albania in 1899. He did acknowledge the existence of 73,500 Albanians in the Albano-Slav zone but he insisted that in this case the Slavs were nearly as numerous as the Albanians. This idea, again, was typical of the views of Gopčević. Nicolaides' map thus initiated a period of close understanding between the Greeks and the Serbians as to their respective spheres of influence in Macedonia and his map may be regarded less as an ethnographic map than as a definition of the new territorial aims of the Greeks in Macedonia; these aims were much more restricted than had been the case in Stanford's day but more realistic insomuch as they took into account the growth of Serbian and Bulgarian nationalism. The areas claimed by Nicolaides as Greek territory in 1899 were precisely those areas occupied by the Greeks thirteen years later, during the period of the Balkan wars. It would seem from an analysis of this map, therefore, that its composition was prompted by three considerations: it was an attempt

(1) to gain moral support for Greek aims on the mainland by identifying Hellenic territory in those regions where expansion was desirable;

(2) to cultivate Serbian support for Greek claims by a modest recognition of Serbian claims in Old Serbia and northern Macedonia;

(3) to restrict Bulgarian expansion to north-eastern Macedonia.

However fantastic Nicolaides' ethnographic claims may have appeared to contemporary orthodox ethnographers, the fact remains that in many ways he anticipated ethnographic ideas commonly held twenty years later, and his ethnographic frontiers in Macedonia were almost identical with the later political boundaries which they ante-dated by over a dozen years.

*Conclusion*

After the appearance of Nicolaides' Greek map, it was obvious that the ethnographic situation was becoming more obscure with every new map that appeared. Weigand's map, it is true, had cast a shaft of light in one dark corner, but even reasoned analyses were soon obscured by the fog of propaganda produced by the rival schools. The claims of the Serbians in the north and the Greeks in the south had complicated the issue to such a degree that a proposal was put forward by O. Baldacci and K. Hassert, during the International Orientalist Congress in Rome in 1899, for an international enquiry into the ethnography of the Balkans, with the object of producing an ethnographic map on a scale of one in a million. Unfortunately the proposal came to nothing.

## F. MEINHARD'S MAP OF 1899

In the same year (1899), an effort was made by the Austrian, F. Meinhard, to reconcile the divergent points of view, in the form of a new ethnographic map of Macedonia (Fig. 30). Meinhard had spent a great deal of his life in Sofia as Director of the Bulgarian Railways and had therefore some knowledge of the Balkan peoples. Moreover, he was in close touch with both Bulgarian and Serbian sources of information. He was well acquainted with Weigand's work and he appears to have had a close knowledge of the older ethnographic maps of Turkey-in-Europe. His methods of representation were clearer than many of those adopted hitherto, because he did not use a relief map as a base but preferred to indicate the more important relief features by name—thus the colours on his map were not obscured by hachures. As Nicolaides had done, he confined his map to Macedonia and it is of interest to note the limits which he imposed upon the region. His definition was different again from that of Gopčević and differed slightly from that of Nicolaides. In the south-west, it excluded the Kónitsa and Korcë basins which Meinhard regarded as part of traditional Albania. In the south, it excluded the districts between the Kamvoúnia and Piéria mountains and the 1881 Greek boundary, which he understood to be part of Thessalía. In the east, it excluded the hill country on the far side of the Rodopi watershed and in the north adhered rigidly to the limits provided by the Sar mountains and Crna hills, thus excluding the Kačanik district. Meinhard's definition was more restricted than Nicolaides'. It was more favourable to the Bulgarian view insomuch as it excluded a Greek district in the south and a Serb district in the north (Fig. 1).

Meinhard believed that the religious subdivisions of linguistic groupings were significant, and thus his chief criteria included both language and religion, and he believed that history also should be taken into account. For example, he distinguished pure

Albanians from 'Albanians of Serb extraction.' Altogether he recognized seventeen different groups in Macedonia; the Austrian ethnographic picture was extraordinarily complex. Sax's map had shown similar complexity. There was no doubt that Austrian interpretations were closely related to political circumstances.

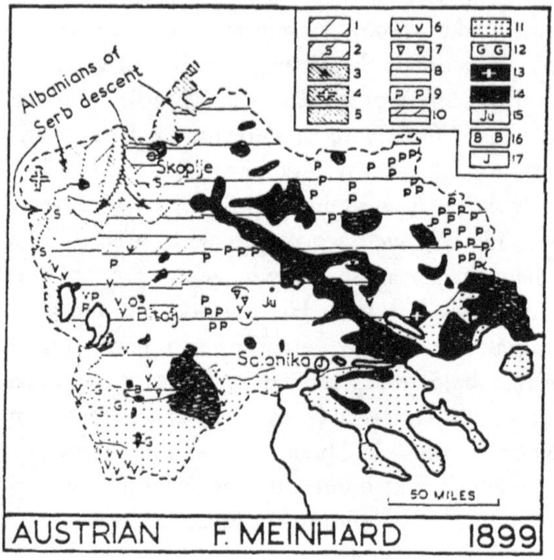


FIG. 30

The references in the key are as follows: 1. Christian Serbs; 2. Moslem Serbs; 3. Moslem Albanians of Serb descent; 4. Christian Albanians; 5. Moslem Albanians; 6. Christian Vlachs; 7. Moslem Vlachs; 8. Christian Bulgarians; 9. Moslem Bulgarians; 10. Bulgarians and Serbs, mixed; 11. Christian Greeks; 12. Moslem Greeks; 13. Christian Turks; 14. Moslem Turks; 15. Yürucks; 16. Bardariotes; 17. Spanish Jews.

*The Turks*

A feature of Meinhard's map was the re-emergence of the Turks around the Gulf of Kaválla and in the vicinity of the great transverse routes across Macedonia—the Dojran depression, Strumica and Stip. K. Sax and G. Weigand had already indicated the presence of Turks in these regions. Meinhard defined their distribution in detail and his views were accepted later by both V. Kānčev and J. Cvijić, the respective classical exponents of the Bulgarian and Serbian views on Macedonian ethnography. One interesting detail on the map was a reference to the Christian Turks on the borders on Lake Akhinoú—a description which in 1840 would have been regarded as a contradiction in terms. Meinhard believed that the *Yüruks* of the hill country around Páïkon still exhibited ethnic individuality. They had last figured on the maps of F. A. O'Etzel and M. A. Denaix (see p. 11), and G. Lejean had also mentioned them. The *Bardariotes* living in small groups south of Kastoría were also depicted on Meinhard's map. They had generally been classified as Turks. The reference to these small isolated remnants was a vivid reminder of the historical incapacity of Macedonia to absorb completely the many diverse groups which had converged upon this territory at different times.

### *The Greeks*

Meinhard's map marked a low ebb for the Greek frontier on the mainland because, although he retained Weigand's Graeco-Slav frontier, he limited the Greeks in the west to the line of the Píndhos mountains. The Greek frontier had receded slightly farther south, on each of the non-Greek maps published in the second half of the century, until the Greek population now appeared as small enclaves where before it had covered a broad band of coast. Little wonder that the Greeks were indulging in rather hectic attempts to strengthen their ethnic position. Meinhard's recognition of a group of Moslemized Greeks in the upper Aliákmon valley was an interesting detail.

### *The Slavs*

In dealing with the distribution and character of the Macedonian Slavs, Meinhard effected a compromise between Serbian and Bulgarian ideas. He recognized a strong Serb minority in north-western Macedonia which extended, mixed with Bulgarians and *Albanians of Serb extraction*, as far as the Vardar river in the east and the Lepenac river in the south. The Tetovo and Debar areas, he represented to be purely Serb. Thus it may be seen that the work of the Serbian ethnographers, combined with their educational crusade in north Macedonia, was beginning to influence opinion by 1899. Meinhard also considered that east of a line formed by the Pčinja, the Vardar, and the Crna rivers, mixed Serbo-Bulgarian territory gave place to purely Bulgarian territory. The Slavs around Lakes Ohrid, Prespa and Kastoría, he also depicted as Bulgarian. The Bulgarians, according to Meinhard's interpretation, still formed the largest single element in Macedonia, but they did not outnumber all the other ethnic groups put together. His recognition of so many Turks and Serbs had been at the expense of the Bulgarians. Although the Bulgarians were not yet depicted as a minority, the map indicated in a striking manner the decline of the idea that the Bulgarians formed the majority of the inhabitants of Macedonia.

*The Vlachs*

Meinhard adopted Weigand's distribution of Christian and Moslem Vlachs. He referred to them as *Kutzo-Vlachs*, a name not acceptable to the Vlachs themselves as *Kutzo* signified lame or halting and was said to have been derived from the halting mode of speech which the Vlachs adopted amongst strangers.

*The Albanians*

According to Meinhard the Albanians had been over-estimated in Macedonia. He showed fewer 'pure' Albanians in northern Macedonia because he elected to distinguish as separate groups the *Moslem Serbs* of the Drin valley and the *Moslem Albanians of Serb extraction*, who inhabited all the highlands of northern Macedonia. He also indicated a group of Christian Albanians in the Drin valley. He was more favourable to the Albanians in the south where he marked the Moslem Albanian frontier east of the Grámmos mountains.

*Conclusion*

Meinhard's map was generally regarded as providing a fair summary of the ethnographic situation in 1899. It was praised by Hassert as a useful contribution which gave shape to the ethnographic map of Macedonia.[1] The map bore many traces of the influence of S.

[1] Literatur-Bericht Nr. 396, *Petermann's Mittheilungen*, XLVI (Gotha, 1900).

Gopčević, K. Sax, and G. Weigand. Its complexity was an indication that ideas on the ethnography of Macedonia were beginning to swing away from pro-Bulgarian conceptions, to the view that the ethnography of Macedonia was altogether more complicated that G. Lejean or H. Kiepert had pictured, and that the strength of the Serbs had been underestimated in the past.

## V. Kancev's Map of 1900

Meanwhile the Bulgarians had not been unmindful of the changes in ethnographic ideas exhibited in recent maps of Macedonia. They were particularly concerned at the success attending Serbian propaganda, which had already resulted in the moral alienation of territory once regarded as Bulgarian. The Bulgarian position was all the more difficult since not only the influence of Serbia and Austria in the north, but also that of Greece in the south, had to be combated. So far the Bulgarians had not produced an ethnographic map of their own; they had relied on those of A. Boué, G. Lejean, and H. Kiepert, but consequent upon the appearance of the maps of S. Gopčević and C. Nicolaides, a statement from Bulgarian sources appeared imperative, all the more so since responsible quarters might interpret F. Meinhard's map as being the official Bulgarian view. Therefore, in 1900, in response to this need for a Bulgarian statement, appeared Kănčev's book on the ethnography of Macedonia. It contained an ethnographic map which for many years constituted the official Bulgarian interpretation (Fig. 31). Kănčev was a Bulgarian inspector of schools. His duties had interested him in the ethnographic situation in Macedonia where, as we have already seen, education and nationality were closely linked. He had access to Bulgarian statistics, which he was able to verify by personal observation. He was familiar too with most of the ethnographic maps produced in the years before 1900. The agents of the Bulgarian Exarchate were carrying out an ethnographic survey at this time and Kănčev was probably aware of their findings. Significantly, he ignored religious criteria and thereby was able to reduce his major ethnic groups to six. The simplification of groupings on the basis of language was bound to favour the Bulgarian element particularly when the Pomaks in Macedonia were included as Bulgarians.

His definition of Macedonia was the same as that of Meinhard, except in the west, where he excluded the Radomir district of the Drin-i-xy valley and the Bílishta district of the Upper Devoll valley. These areas he regarded as part of Albania.

*Turks, Greeks and Slavs*

Kănčev's distributions of Turks and Greeks were very similar to

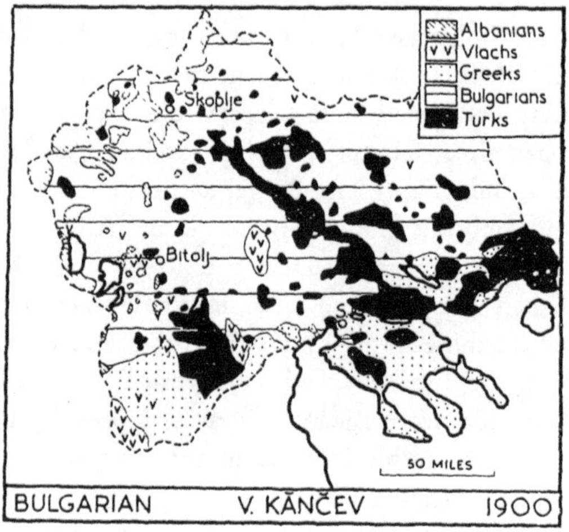


FIG. 31

The abbreviation **S** is for Salonika.

those on Meinhard's map. It may have been that both had access to the same Bulgarian source; they may have consulted each other's maps. His distribution of Slavs was also similar but they were all Bulgarians according to his representation, and he maintained that the regrettable tendency to acknowledge Serbs in northern Macedonia, either as Moslem Serbs, as Serbs mixed with Bulgarians, or as pure Serb minorities, dated back to the chauvinistic work of Gopčević in 1899. Such claims, declared Kănčev, were exaggerated or false and he insisted that the pre-1899 ethnographic maps bore ample testimony to the truth of his contention.

*Vlachs and Albanians*

Kănčev had nothing to add to Weigand's view of the Vlachs, the distribution of which he took from Weigand's map. He included both Christian and Moslem tribes in his Albanian group. He was not interested in Meinhard's 'Albanians of Serb extraction.' The Albanians according to Kănčev were an important minority and he depicted them in eastern and northern Macedonia in much more detail than hitherto—as in the occupation of all the higher land in north-western Macedonia, the Sar mountains, the Crna hills and the Golešnica mountains, the hilly country between Kruševo and Kastoría and the uplands between Gostivar and Kičevo.

*Conclusion*

Kănčev's map was reproduced in P. Miliukov's atlas in 1900 and later in the Carnegie Report, *Inquiry into the Causes of the Balkan Wars* (1914). In both cases it represented the official Bulgarian view of ethnic conditions in Macedonia. In the first place it re-emphasized Bulgaria's traditional rights in Macedonia, and secondly it gave weight to the idea that the Albanians were the most important minority in the north and west.

## The Bulgarian Exarchate Map of 1901

The agents of the Bulgarian Exarchate were also busy at this time vindicating the Bulgarian claims. They produced a map in 1901 on a large scale (1:200,000), covering the territory in dispute. Drawn up at the Bulgarian Institute of Cartography at Sofia, this map endeavoured to show the ethnic composition of the population by a new method, utilizing coloured symbols instead of flat colours and thus giving a better idea of the mixture of population which occurred in Macedonia. It exhibited a fair amount of agreement with Kănčev's map, as might be expected. The Bulgarians were still fairly confident about their superiority. In the words of the Bulgarian Exarch : "The Bulgarians need not be afraid of racial competition." According to his calculation there were over a million Bulgarians in the Roumelian provinces, of whom 800,000 were Exarchists and

300,000 Patriarchists. The Greeks numbered from 3-400,000 mostly in the southern Provinces, the Serbians, 200,000 mostly in Kosovo Vilayhet, the *Koutzo-Wallachs* (Vlachs), 60,000, and the Turks from 5-600,000.[1]

As the new century opened and events moved towards a climax the Bulgarians could congratulate themselves upon the success with which they had countered Serbian and Greek propaganda. Here and there some ground had been lost, but the greater part of Macedonia in the eyes of most of Europe still remained incontestibly Bulgarian. The standard atlases then in use in France, Germany and Great Britain, still subscribed to the view that Macedonia was Bulgarian territory. For example H. Berghaus's atlas (1892 edition), the *Times Atlas* (1900 edition), Niox's *Atlas de Géographie Géneral* (1899), the Vidal Lablache atlas, all contained pro-Bulgarian ethnographic maps of the Balkans.

[1] G. P. Gooch and Harold Temperley, *British Documents on the Origins of the War, 1898-1914*, vol. V, p. 103 (London, 1928).

## CHAPTER VII

# THE MACEDO-SLAVS, 1903-1910

## K. PEUCKER'S MAP OF 1903

THE events which took place in Macedonia in 1903 were of so violent a nature as to compel the attention of all Europe. Sporadic insurrections had been taking place ever since the Treaty of Berlin restored the region to Turkish maladministration and, after 1890, these disturbances, fomented by the intervention of Bulgarian, Greek and Serbian nationalists, grew in intensity. By 1903, the whole of Macedonia and most of Thrace were in a state of open rebellion which culminated in the Salonika outrages during which the Ottoman Bank was blown up by insurgents. The savage Turkish reprisals which followed these demonstrations aroused widespread sympathy for the rebels.

In the British Isles, a correspondence over Macedonia opened in *The Times* and it was never really closed until the intervention of the Great War in 1914. Sir Oliver Lodge made Macedonia the subject of an inaugural lecture at Birmingham. Bishops condemned the Turkish policy in Macedonia from their pulpits. The newspapers were full of the Turkish atrocities, perpetrated on the Christian subjects of the Porte. For the first time the British public became familiar with the name of Macedonia, as indeed did the general public of all Europe. Everywhere the question was heard—who are the Macedonians? Whilst interest in the welfare of the peoples of Macedonia was still white hot, numerous ethnographic maps made their appearance; all of them tried to answer that question. Serbian, Greek, Bulgarian, Italian and British views were amongst those expounded. The answer was not as simple as in 1870. The complete failure of the rival schools of thought, as represented by S. Gopčević, C. Nicolaides and V. Kǎnčev, to reach any agreement on the ethnographic distributions was taken to be an indication of a complexity which might never be resolved into order. The French, with a stroke of culinary genius, utilized the term 'macédoine' as the name for a dish in which various

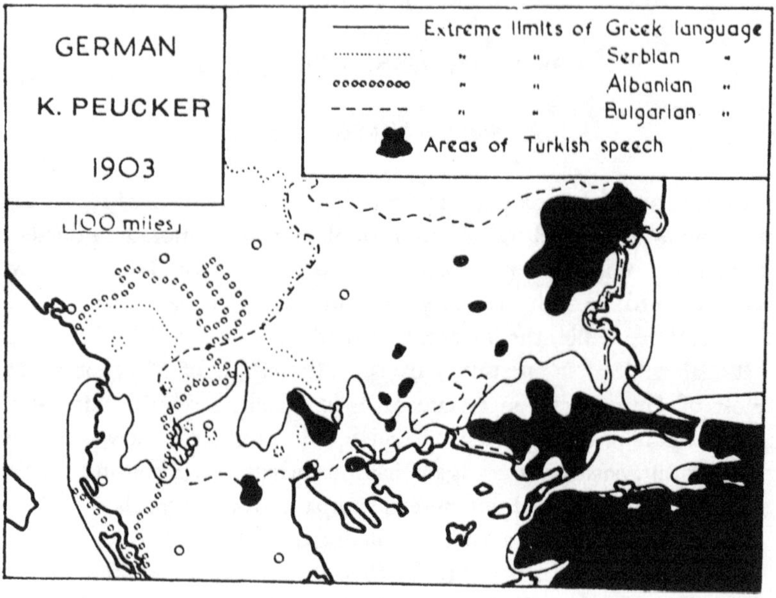


FIG. 32

ingredients, although mixed together, contrived to retain each its individual flavour.

*The Macedonians*

In 1903, Dr. K. Peucker of Vienna produced a map of Old Serbia, Albania and Macedonia, together with some relevant information of an historical, ethnographic and statistical nature bearing on the problem of Macedonia. He endeavoured to summarize the ethnographic situation by means of a small inset map showing zones of linguistic influence and zones of culture (Fig. 32). His information was derived from R. von Mach's map of 1899 and from the researches of J. Cvijić, a brilliant young Serbian geographer, who was ultimately to formulate revolutionary ideas about the ethnography of Macedonia. Briefly, Peucker's aim was to show that the admixture of Serbs, Albanians, Turks, Greeks and Bulgarians occurred in its most extreme form in Macedonia, that in fact Macedonia might almost be defined as an area of ethnic overlap. One might even infer from his map that Macedonia had an individuality of its own, which arose solely from the composite character of its population. Peucker also publicized Cvijić's view that the culture of Macedonia was Byzantine diluted by large, patriarchal communities of Serbian origin—a point of view which favoured both Greeks and Serbs at the expense of the Bulgarians. Peucker's map was important because it contained the germ of a new idea, later elaborated by J. Cvijić, that *Macedonians* might well be distinguished from Bulgarians, Serbs and Greeks alike. This idea was all the more remarkable because only ten years earlier Peucker had published a typically pro-Bulgarian ethnographic map. His changed viewpoint was an indication of the efficacy of Serbian and Greek propaganda in the interim.

## Greek Maps, 1903 and 1905

The Greeks continued to concentrate on the production of ecclesiastical and school maps which gave emphasis to the Hellenic spiritual and cultural role in the Balkans. M. Sullagos produced a map in 1903 which marked the political and ecclesiastical divisions of Turkey and indicated, by means of statistical tables, the number of adherents to the Greek Church in each division. In 1905, an anonymous map appeared in Paris which substantiated with a wealth of detail the preponderance of Greek schools in southern Macedonia. Lavishly produced, it claimed to show the distribution of Bulgarian, Serbian and Greek schools in Macedonia as well as the distribution of Greek Orthodox churches. Both maps indicated that Greek propagandists, after the disappointing reception of Nicolaides' map, were content to place their faith in the criteria which would reflect most credit on their case—on religion and education, rather than on language.

## D. M. Brancov's Maps of 1905

To give publicity to the Bulgarian viewpoint, the Exarchate map of 1901 was republished in Paris in 1905, and V. Kănčev's map reappeared in F. F. Voinov's work on Macedonia in 1905. The Bulgarians were still fairly confident of their ethnic superiority, the main challenge to which, they believed, came from the Greeks. The Serbian

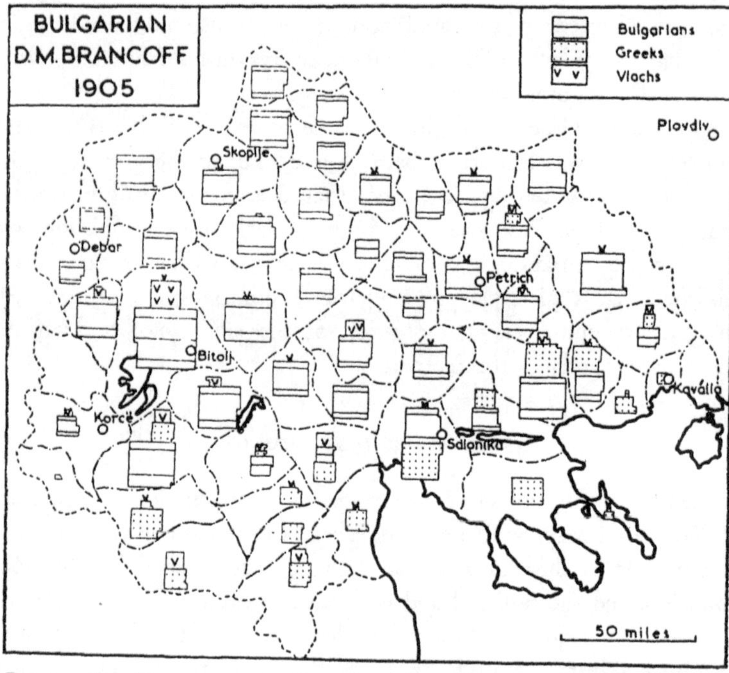


FIG. 33. THE DISTRIBUTION OF CHRISTIAN POPULATION IN MACEDONIA IN 1904
One millimetre squared represents 1500 persons.

claims they regarded as inconsequential, although they viewed the possibility of a Serbo-Greek understanding with some alarm. Insomuch as they represented the Bulgarian response to the Greek educational and religious propaganda, D. M. Brancov's maps are worthy of consideration. The name of Brancov (Brankoff) was a pseudonym adopted by the Bulgarian Exarchate's Secretary, D. Mischev. The maps were therefore semi-official in character (Figs. 33 & 34). The information they portrayed must have been derived from the Exarchate Survey of 1901. Brancov adopted a novel method of depicting ethnographic distributions. He used rectangular symbols to represent the proportional elements of Bulgarian, Greek and Vlach population in each of the districts of Macedonia, and in this manner indicated that only in the extreme south of Macedonia, in the districts of Grevená, Neápolis, Elassón, Kozáni, Sérvia, Kateríni, Véroia, Thessaloníki (a bare majority), Políyiros, Kaválla and Pangaíon (Pravitchta) did the Greeks outnumber the Bulgarians. Everywhere else the Bulgarians were in the majority. The Vlachs formed only a minority in each district ; since they were sometimes associated with the Greeks, it was Brancov's intention to draw attention to their relative unimportance.

His second map indicated, by the same method, the number of children attending Serbian, Bulgarian, Vlach and Greek schools in Macedonia. On comparison of the two maps some interesting points emerge. In the Tetovo district, for example, there were over 40,000 Bulgarians, but nearly all of them sent their children to Greek schools and also in Sérrai and in Dráma, the Bulgarians were in the majority, but most of their children went to Greek schools. Brancov thus emphasized the fact that the boasted superiority of Greek cultural establishments in Macedonia did not necessarily infer numerical superiority. A Bulgarian, although he might not belong to the Exarchate Church and although he might have been educated in a Greek school, was still a Bulgarian in his national sentiment.

### *The Limits of Macedonia*

Brancov's maps made clear the interesting division of opinion between Greeks and Bulgarians on the concept of Macedonia. It may be remembered that the Greeks excluded the Skoplje district but included the Korcë basin in their concept of Macedonia. The Bulgarians, stated Brancov, regarded Macedonia as a geographical unit and as such it should include Skoplje and exclude Korcë. This distinction between the Greek 'historical' Macedonia and the Bulgarian 'geographical' Macedonia was important for two reasons.

Firstly, the Bulgarians were interested in the acquisition of the whole and not part of Macedonia. For them, partition of Macedonia meant that the all important Salonika area might go to the Greeks or Skoplje to the Serbians. Secondly, the Bulgarians were anxious not to allow the growth of a Graeco-Serbian understanding which might even-

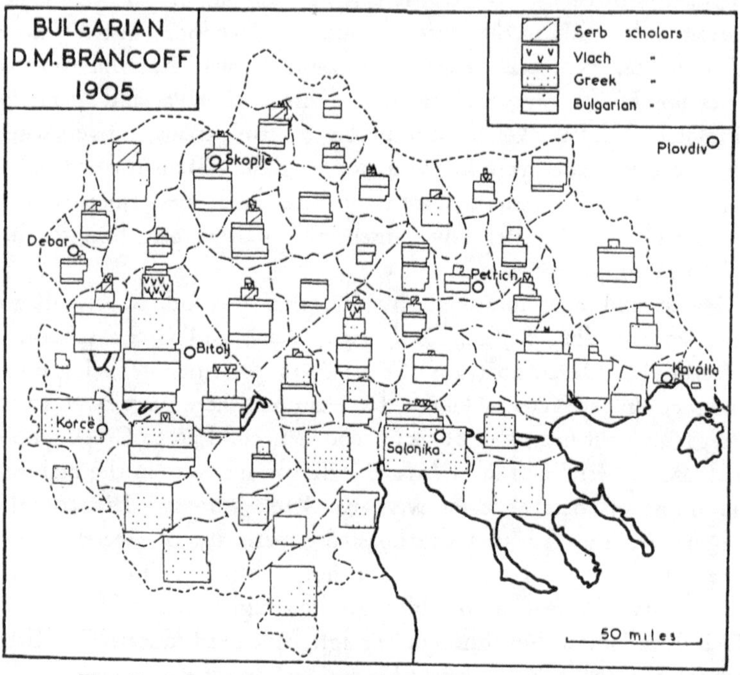


FIG. 34. THE DISTRIBUTION OF CHRISTIAN SCHOOL-CHILDREN IN MACEDONIA IN 1902
One millimetre squared represents 77 children.

tually eliminate Bulgarian interests. Already the Serbians had laid claim to northern Macedonia, which they had announced to be part of Old Serbia (see p. 100). Therefore the Greeks and Serbians had both acquired well defined spheres of influence which covered the whole of the region between the Sar mountains and the Aegean sea and bade fair to oust the Bulgarian interests. That is why Brancov was anxious to stress the geographical unity of Macedonia, a unity which, if respected politically, left the Bulgarians in a strong position with 51·4 per cent of the population Bulgarian, and left the Serbs and Greeks only as minorities. It is worthwhile noting that Brancov's ideas considerably influenced those of the American geographer, L. Dominian (see p. 208).

## H. N. Brailsford's Map of 1906

H. N. Brailsford's work on Macedonia published in 1906 made a valuable contribution to the understanding of Macedonian ethnography. It remains to this day one of the few books in English specifically devoted to the problem. Sometimes Brailsford is accused of pro-Bulgarian sympathy. In actual fact his views were widespread in the rest of Europe in 1906. He was, however, a firm believer in the Bulgarian cause in Macedonia and he maintained, then, that a Greater Bulgaria would exert a stabilizing effect on Balkan politics. He was one of the select band of students of Balkan affairs who, at the beginning of the century, grasped the significance of the future role of the Slav states in the Balkans. These men necessarily found themselves supporting either the Serbians or the Bulgarians and for a long time the Bulgarian cause was the more popular. In fact it was not until 1915 that hopes of a mutual understanding between Bulgaria and Great Britain were dashed to the ground by the Bulgarian entry into the War of 1914-18, on the side of the Central Powers.

*The Turks*

Brailsford did not attempt to delineate Macedonia on his map (Fig. 35) but one might infer from his text that he accepted the Bulgarian definition. He pointed out that the word Turk had no real ethnological significance and that there were very few genuine *Osmanlis* in Europe. Circassians, Levantines, Arabs and even Negroes passed under the name of Turk. He believed that the Moslem population for the most part consisted of renegade Slavs or *Muhadjirs* who had fled for some reason from the Slav states in the north. A great many Albanians, Gypsies and Jews were also Turkish in allegiance. Brailsford on his map depicted the *Kailar* (or Konariote) Turks and the Turkish enclaves between Sérrai and Dráma. He also indicated

Turks as a minority, usually composed of officials, in the towns of the Vardar valley. Nevertheless, he greatly reduced the widespread distribution of Turks shown on the maps of F. Meinhard and V. Kănčev.

*The Greeks*

He did not believe there was any evidence that the interior of Macedonia had ever been settled by a Greek rural population, and he

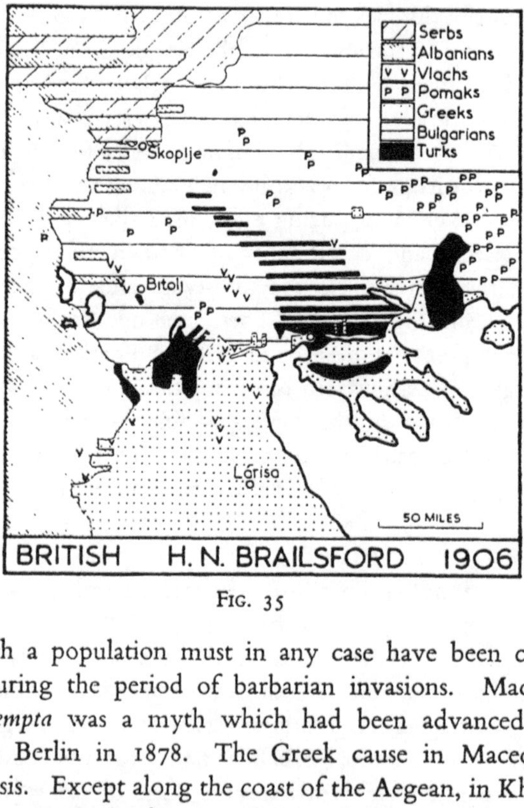


FIG. 35

thought such a population must in any case have been completely uprooted during the period of barbarian invasions. Macedonia as *Hellas Irredempta* was a myth which had been advanced by Lord Salisbury at Berlin in 1878. The Greek cause in Macedonia had no ethnic basis. Except along the coast of the Aegean, in Khalkidhikí, and in the extreme south-west of Macedonia near the Thessalían boundary, there were practically no villages in European Turkey where the mother tongue was Greek. He mentioned Kastoría as the frontier town between the Greeks and the Bulgarians in this area.

*The Slavs*

Brailsford did recognize some Serbs in Macedonia, but, only as a minority mixed with Albanians. He marked them north-west of the Vardar river between Skoplje and Kumanovo. The Slavs of Macedonia, he maintained, were Bulgarian and his map showed Macedonia as predominantly Bulgarian. He admitted, however, that the Slavs of Macedonia were of mixed blood. Bulgarian and Serbian conquests, Tartar (Tatar), Petcheneg (Patzinak) and Kuman (Cuman) raids had left their mark. He even allowed that these Slavs had no strong conception of nationality and that any strong Slavonic power might be able to impose itself upon them. But, he declared, it was not the Serbians, it was the Bulgarians who had managed to do this successfully. In Kosovo and north-west of Kosovo, the Slavs were definitely Serb ; in the region of Ohrid they were just as definitely Bulgarian. The nationality of the rest of the Slavs had been decided on political grounds and the victory had gone to the Bulgarians. "The Macedonians are Bulgarian today," wrote Brailsford, "because a free and progressive Bulgaria has known how to attract them. Serbia's cause is an artificial movement. As things are today the Serbian consuls are about as likely to win the Macedonians for Serbia as the American missionaries are to convert them to Protestantism." These words were written when the Bulgarian cause in Macedonia appeared irresistible—that is before the Balkan Wars. Only a few years later the Serbians were employing the logic of Brailsford's argument to prove that the Macedonians were Serbs ! The distribution of Bulgarians shown on Brailsford's map accorded roughly with that of V. Kănčev, except that Brailsford depicted much of eastern and southern Macedonia as Pomak, where Kănčev, himself a Bulgarian, had recognized only Turks.

There always had been a great deal of controversy about the Pomaks—those Bulgarian Moslems, often more fanatical than the genuine Osmanli Turks. It would be a most difficult task to attempt to define the national allegiance of the Pomaks at various stages throughout the development of Bulgarian nationalism. There were Pomaks who spoke both Turkish and Bulgarian, others who spoke only Bulgarian and others who were mainly Turkish-speaking.

but who retained a Bulgarian mode of life. Thus there were Pomaks to all intents and purposes Turkish and others just as definitely Bulgarian. Moreover, Pomak allegiance itself was very elastic. It vacillated between the Bulgarians and Turks but tended towards the Bulgarians, particularly after Bulgaria had acquired Eastern Roumelia in 1885.[1] Thus Brailsford, to a certain extent, was justified in his representation. He showed the Pomaks also in small groups in central Macedonia.

*The Vlachs*

Brailsford had little to say about the Vlachs. He estimated that they did not number many more than 200,000 souls, mostly living in the Píndhos region.

*The Albanians*

He regarded the Albanians as the most important minority in Macedonia, after the Bulgarians. He agreed with Barbarich that the whole of the territory between the Gulf of the Arta and Lake Shkodrs (Skadar) was the home of this 'unconquered race' (see p. 144). According to his representations, the population of the greater part of Ipiros, including the towns of Ioánnina, Kónitsa and Elbasen, was Albanian. From here their territory stretched as far east as Lake Ohrid and the hilly region on whose fringes lie the towns of Debar, Prizren, Peć and Shkodër. With the exception of small communities of Serbs in the towns, a few Jews and Greeks in Ioánnina, and some scattered Vlach settlements, the population within these limits was, in his opinion, the most homogeneous that could be found anywhere in the Balkans : it was pure Albanian. Nor were the Albanians confined to this region ; under the favourable circumstances afforded by the Turkish regime, they had expanded over the Píndhos into Thessalía, over the Drin valley into Macedonia, and through the hills and passes of the north-east into Old Serbia, which was, according to Brailsford in 1906, two-thirds Albanian, whilst whole districts around

[1] Interesting details of how, during their occupation of parts of Thrace in 1913, the Bulgarians carried out forcible conversion of Pomaks, are given in *Enquête dans les Balkans* (p. 139). Evidently some Pomaks did not regard themselves as Bulgarians at that time.

Bitolj and Skoplje were dotted with Albanian settlements (Brailsford's idea of Old Serbia was the same as that defined in the Introduction).

*Conclusion*

Brailsford's book appeared at a time when the name Macedonia was on everybody's lips and when sympathy for the Bulgarian cause was widespread and sincere. It suffered from the defects of its period insomuch as it over-estimated the influence of the Bulgarians in the southern Balkans. Its value is that it reflected fairly faithfully contemporary opinion. The British, for so long pro-Greek and pro-Turk, for a variety of reasons, had swung towards a sympathetic toleration of the Bulgarian cause.

## E. Barbarich's Map of 1905

Of all the major ethnic groups of the Balkans at the beginning of the century, only the Albanians had no part of their territory politically independent. They had at one time achieved a certain measure of unity under the renowned *Skanderbeg* (Fig. 27) during the fifteenth century but had made no further move towards political emancipation since that time. The reasons for this were not far to seek. They occupied some of the most mountainous and infertile territory in Europe. Knowing only tribal organization, they had never experienced the cultural renascence which had been the initial feature of Slav and Greek national movements. They had no schools of their own, no national Church, no storehouse of traditional literature, and so lacked the means of defending their cultural identity. Nor were they provoked into any desperate political activity by Turkish tyranny, for the Turks, having little to gain by a conquest of the Albanian mountains, had at a very early date entered into a free and easy relationship with the Albanians. Few Turkish (Osmanli) officials were to be found in Albania ; not only were the Albanians left to their own devices in the mountains but also their way of life was encouraged in the plains. The Albanians, in fact, had little to gain by political independence. Boundaries would have cut them off from the means of gaining a livelihood outside their own mean habitat.

*The Distribution of Albanians*

Since the Albanians had not exhibited any desire to become a free and independent nation, they had not produced any ethnographic maps of their own. But although they were politically inarticulate they did exhibit pride of ancestry and local patriotism and certain tribes were attached to certain districts. Their interests had hitherto been cared for by the Bulgarians, who looked to the Albanians for support against the Graeco-Serbian threat.[1] In 1905, however, the Albanian cause was championed from another quarter. An Italian, E. Barbarich, published a book about the Albanians which contained an interesting ethnographic map (Fig. 36). It was, of course, a map of Albania rather than of Macedonia, but Barbarich interpreted Albania in a very liberal manner to include the whole of Ipiros, the whole of the Drin valley and part of Old Serbia. He followed Marenin very closely in plotting the distribution of the Albanians, whom he represented as forming the majority of the inhabitants of Ipiros, western Macedonia and Old Serbia, and he took care to distinguish their various religious groups, because he believed that they had a particular significance. There had been no hesitation on the part of G. Lejean to include the Greek Orthodox Albanians with the Moslems in one

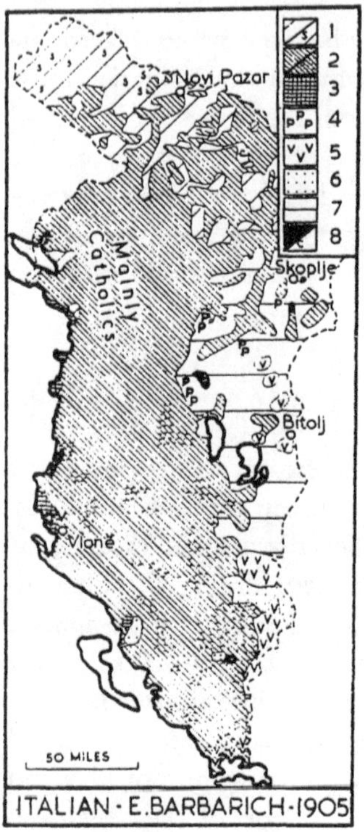


FIG. 36

The references in the key are as follows: **1. Serbs** and **Bosnians** (the latter differentiated by an **S**); **2. Roman Catholic Albanians** and **Moslem Albanians/ Greek Orthodox Albanians** (stippled); **3. Italians**; **4. Pomaks**; **5. Vlachs**; **6. Greeks**; **7. Bulgarians**; **8. Turks/Circassians.**

[1] An ethnographic map showing the distribution of the Albanians was produced by the Bulgarian, Dr. Marenin, in 1902.

ethnic group. But on later maps Albanian territory had been shrinking rapidly in the south because of the tendency to regard Greek Orthodox Albanians, particularly the Tosks, as Greek nationals, because they were highly susceptible to Greek influence. Both H. Kiepert and G. Weigand had agreed on the Hellenic character of a great many inhabitants of Ipiros formerly regarded as Albanian. One might even say that although the Albanians were admittedly an integral part of the Greek nation throughout the whole of the Greek archipelago, their inclusion within Greek territory in the Ipiros region had been a relatively recent phenomenon dating back only to the appearance of H. Kiepert's map in 1876.

A glance at a map of the Adriatic will immediately reveal the reasons for Italian interest in this part of Albania. The Ipiros region commands the Straits of Otranto, where the Italian and Balkan peninsulas approach to within fifty miles of each other. Both Greece and Italy were expanding too rapidly in the eastern Mediterranean to avoid conflict. Barbarich's work was a manifestation of Italian interest in the opposite shores of the Adriatic which was to involve Italy in an invasion of the Balkans on more than one subsequent occasion.

In the north, Barbarich pressed the Albanian claims in Old Serbia. Here the Albanians were represented principally by the Moslem Gegs. These Albanians had been particularly favoured under the Turkish regime, when they had been able to migrate from the highlands of Albania into the plains and basins of Old Serbia and Macedonia. As the Turkish power declined, however, the Albanians, having little sense of nationality, began to meet increasing opposition on the part of the Serbs. Their Moslem religion which might have given them a semblance of national feeling, meant very little because a great many Serbs were also Moslems. Moreover, S. Gopčević and his disciples had succeeded to a certain extent in modifying the idea that most of the inhabitants of Old Serbia were Albanians, by insisting on the Serb origins of the Old Serbian population. Barbarich repudiated this idea and re-emphasized the existence of at least a two-third majority of Albanians in Old Serbia.

Due to its limited circulation, Barbarich's map did not exert much influence on European ethnographic thought, but his distributions were used later by A. Dardano in preparing ethnographic maps for the Agostini Geographical Institute. The Italian and Bulgarian interest in Albania was not without its effects on the development of Albanian nationalism. Only a few years later, in 1908, a linguistic congress held in Bitolj evolved a suitable Roman alphabet for the Albanian language. Hitherto Italian, Cyrillic, and even Arabic alphabets had been in use. In this way one of the oldest of the Indo-European languages was safeguarded from further assimilation and dilution. It was not, however, until 1924 that Albanian or 'Shqip' began to be taught in schools. Whilst the Italians cultivated Albanian aspirations, there were exceptions: G. Amadori-Virgilj, for example, in 1908, ignored the existence of Albanians altogether on his map (see p. 159).

## J. Cvijic's Maps, 1906-7

Jovan Cvijić has already been referred to as the Serbian geographer whose views had so much impressed Peuker in 1903. Before he began to edit the series of monographs on the ethnography of the Serb lands—Насља Српских Земаља (Naselja Srpskh Zemalje)—the first volume of which appeared in 1902, Cvijić had already acquired a reputation amongst European scholars as a geologist, physical geographer and cartographer. He was to become even more renowned as a human geographer. As head of the Department of Geography in the University of Belgrade, he was well placed to carry out intensive research in the Balkans. Almost inevitably his work was affected by his country's political situation. Serbia was in a difficult position both from a strategic and from an economic point of view. Austria's hold on this small Slav inland kingdom had been tightening between 1903 and 1906. Serbia had suffered greatly from the 'Pig-war'—an economic trial of strength with Austria which had demonstrated only too well, that expansion to the sea would alone permit the existence of an independent Serbia.

### *His First Ethnographic Map, 1906*

Cvijić had made his first public statement on Macedonia in 1903, in the Viennese *Zeit*, at the time when the 'Mürzsteg Programme' had been undertaken by the Great Powers.[1] At that time he had been content to make a few remarks about the existence of Serb cultural traits amongst the population of Macedonia. In his *Outline of the Geography and Geology of Macedonia,* in Serbian, published three years later, Cvijić made further reference to the ethnography of Macedonia and Old Serbia.

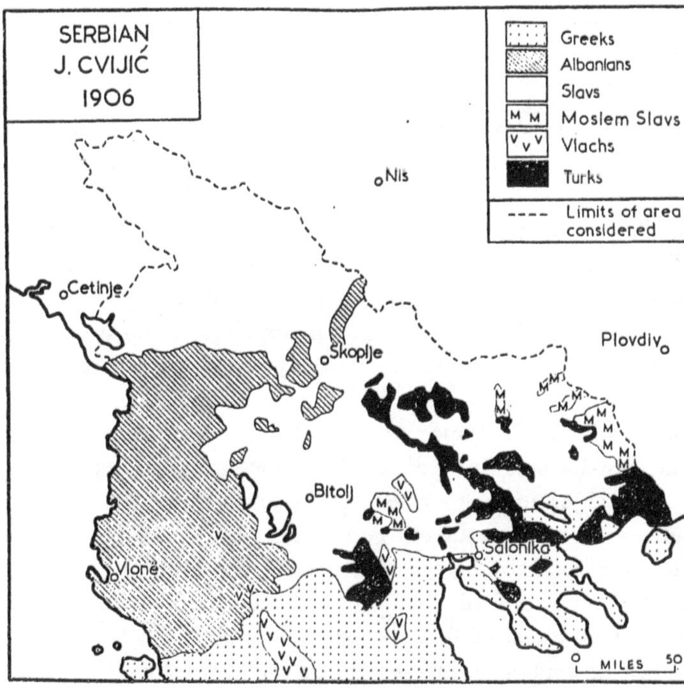


FIG. 37

In this work his first ethnographic map appeared, on a scale of 1:4,000,000 (Fig. 37). Compared with the detail which Cvijić produced in his later maps, the distributions he fixed on this map were only tentative (*cf.* Figs. 40 and 42). These distributions are examined more fully later. For the moment it is sufficient to draw attention to his

[1] In which Austria and Russia undertook responsibility for the policing of Macedonia, and the Sultan was forced to promise reforms, which were to be implemented with the aid of agents appointed by the Powers.

portrayal of extensive Albanian territory in northern Macedonia, and to the fact that he did not attempt to fix an ethnographic frontier between the Serbs and Bulgarians. He classified the Slavs of Old Serbia and Macedonia together in one group—*Serbs and Bulgarians.* The Moslem Slavs, who had generally been classed as Pomaks, he referred to as *Moslem Serbs and Bulgarians,* thus inferring the presence of Serbs in central Macedonia.

### *The Macedo-Slavs*

Cvijić was more explicit about the character of the Slavs of Macedonia in a pamphlet summarizing his views, which appeared in 1906 and 1907. Significantly, it was translated and published in Great Britain, France and Russia. It was obvious from the widespread publicity given to these views that the Serbians considered them to have some political importance. Indeed Cvijić had set himself the task of publicizing the Serb cause and making known the Serbian point of view where it was most likely to be appreciated. Russia could always be called upon for some support. Great Britain and France might be relied upon to support Serbia's aspirations also, since Serbia was the best fitted of the Balkan countries, by virtue of her geographical position, to offer resistance to Austrian and German expansion. Great Britain and France had been watching with apprehension the activities of the Germans in the Balkans. Wilhelm II made no secret of his designs on the Near East, the direct road to which traversed Serbia. 1906, therefore, was a propitious time for Cvijić to repudiate, in the eyes of the western world, Bulgarian claims to Macedonia, and to prepare the way with western aid for Serbian expansion southwards. In his pamphlet, *Remarks on the Ethnography of the Macedo-Slavs* (1906), he was introduced to his English readers in the preface as, " one of those individuals too rare, especially in the Balkans, who are able to subordinate their patriotism to the cause of scientific exactitude." Of his thesis it was remarked, " His treatise on the Macedonian Slavs is unique of its kind ; it gives a clear and unbiased estimate of ethnographic conditions in Macedonia." The rather strange belief that Cvijić was above patriotic fervour still prevails even to-day in the British Isles. It was a tribute to Cvijić's ingenuity that, in presenting the Serbian thesis, he avoided with consummate skill any hint of propaganda. His sources were impeccable and his conclusions always appeared to be modest. He pointed out in 1906 that Serbian research had been able to fix the limits of Old Serbia with accuracy. The view that the Sar Mountains and the Crna hills formed the southern limits of Serb territory was false. Skoplje and Tetovo had always been an unalienable part of 'Old Serbia.' He fixed the boundary between Macedonia and Old Serbia on the Baba mountains, the Babuna mountains and the Plačkovica mountains (Fig. 105). North of this line the Slavs were Serbs, but south of this line the Slav population had no sense of national feeling either Serb or Bulgarian. They exhibited cultural traits common to both Serb and Bulgarian. The fact that they called themselves Bulgarians had no political significance because the word *Bugari* was synonymous with *raja* and meant nothing more than a peasant. He declared that all previous ethnographic maps had been influenced by this fundamental misconception, and that all statistics produced to show the composition of the population of Macedonia were false because no census had ever been taken. An examination of the figures of V. Kănčev, C. Nicolaides, K. Oestreich[1] and S. Gopčević, revealed such startling discrepancies as to render all their figures invalid. Cvijić came to the conclusion that there was an element of truth, however, in all their claims because the Slavs of Macedonia were so mixed that, until their national feeling had been fixed by political adherence to one or another of the Balkan nations, they would remain in an amorphous state. If, declared Cvijić, the Macedonian Slavs were to be incorporated into the Serbian state, they would very soon become Serbian in nationality—a prophecy which was never fulfilled. Cvijić no doubt was thinking of the Slavs of the Niš and Leskovac areas who, before 1878, had been regarded as Bulgarians but who, after those districts became part of Serbia, had become good Serbians (see pp. 103—5).

The peculiar significance of Cvijić's conception of the *Macedo-Slavs* was that it neatly robbed the Bulgarians of their strongest claim to Macedonia—the claim that its inhabitants were mostly Bulgarians. Cvijić envisaged the *Macedo-Slavs* as politically neutral ; they might

[1] "Die Bevölkerung von Makedonien," *Geog. Zeitschrift*, XI (1905).

*become* either Serbs or Bulgarians. He did not agree with H. N. Brailsford that a quarter of a century of Bulgarian nationalistic propaganda in Macedonia had already had the effect of inducing Bulgarian national feeling amongst the Macedonian Slavs. He was anxious to promote the view that the *Macedo-Slavs* were potential Serbs even if they were at the same time potential Bulgarians. If this could be established then Serbia's claim to Macedonia would be on a par with that of Bulgaria. The beauty of Cvijić's conception was that on the surface it appeared to be a compromise between extreme Serbian and extreme Bulgarian ideas. The very fact that he criticized Gopčević's claims as extravagant lent substance to such a conclusion. Thus by its very moderation it commended itself to public opinion as impartial. It fitted in with the picture of overlapping ethnic groups in Macedonia as portrayed by Peuker in 1903.

### *Origin of the Idea of the Macedo-Slavs*

The idea that Macedonia might have a future independent of the Bulgarians had already arisen as far back as 1876, when schemes for a 'Macedonian Province' had been promoted at the Conference of Constantinople, but the idea had not been associated then with any differences between Macedonian Slavs and Bulgarians (see pp. 62-4). At that time the Bulgarian character of the Slavs of Macedonia went unchallenged. It was not until after 1885, when strained relations between Bulgaria and Serbia developed, that the Serbian claims in Macedonia gave rise to the speculation that the Slavs of Macedonia might be different from those of Bulgaria. In 1887, P. D. Draganov tentatively put forward this idea in *Les Nouvelles Slaves* (1887-8).

N. S. Zaryanko, the Russian cartographer, by omitting to colour the Slavs of Macedonia in the second edition of his ethnographic map of the Balkans (1890), likewise implied that those Slavs might be either Serbs or Bulgarians (see p. 105).[1] A powerful factor in promoting separatist tendencies had been the very interest which the Balkan nations and the Great Powers had themselves evinced in Macedonia.

[1]Zaryanko's revised ideas are reflected in a map compiled by T. Fischer for his *Länderkunde der drei Südeuropäischen Halbinseln* (Wien, 1893), entitled 'Skizze der ethnographischen Verhältnisse der Südosthalbinsel' (*n.s.*, 100 × 100 mm.) in which the Slavs of Macedonia are referred to as 'Serbs and Macedonians.'

Numerous secret societies began to flourish after 1890. Some were Serbian inspired, others were inspired by Greeks and Bulgarians, but at least one—the *Internal Macedonian Revolutionary Organization* (I.M.R.O.), established in 1893, with its headquarters in Salonika, was indigenous to Macedonia.[1] It aimed solely at achieving the political independence of Macedonia.[2] Although the chief enemy was regarded to be the Turk, I.M.R.O. resented interference in Macedonian affairs on the part of the Greeks, Serbians and even of the Bulgarians. It came into conflict on this account with the purely Bulgarian *Macedo-Adrianople Organization.* In the end I.M.R.O. did become more closely associated with the Bulgarians, but this only happened when it became increasingly clear that the danger to Macedonian independence came not from Turkey but from Serbia. The Bulgarians themselves were partly responsible for supporting I.M.R.O. and the idea of an independent Macedonia, because they believed that such a province would ultimately become part of Bulgaria, precisely in the same manner as Eastern Roumelia had become part of Bulgaria. The *Macedo-Adrianople Committee,* in 1895, addressed a memorial to the Powers in favour of an " autonomous Macedonia, with its capital at Salonika, to be placed under a Governor-General of the predominant nationality." The Graeco-Turkish war of 1897 fomented anti-Turkish strife in Macedonia and subsequently I.M.R.O. staged an abortive revolution at Salonika in 1903. In the same year Missinko, who described himself as a Macedonian, postulated, in an article published in Sofia, the existence of a Macedonian nation.[3] Possibly, Bulgaria was behind the Macedonian independence movement, which could be used to counter Greek and Serbian pretensions. Unfortunately for the Bulgarians, Cvijić seized upon the idea of the ' Macedonians ' and gave to what had originally been a political tag, an ethnic significance. A Macedonia which was ethnically as well as politically divorced from Bulgaria, and which was a field for Serbian expansion,

[1] S. Khristov, *Heroes and Assassins* (Gollanz, 1935).

[2] A history of the movement has recently been published in America—Ivan Mihailoff (one-time leader of I.M.R.O.), *Macedonia: A Switzerland of the Balkans,* translated from the Bulgarian (St. Louis, 1950). See also K. Anastasov, *The Tragic Peninsula. A history of the Macedonian movement since* 1878 (St. Louis, 1938).

[3] *Des Affaires Macédoniennes* (Sofia, 1903).

was not in keeping with the Bulgarian idea of a Macedonia in the shape of a 'Bulgarian Province.'

*The Reception of the Idea of the Macedo-Slavs*

In 1908, A. Išrkov, a Bulgarian geographer, demolished many of Cvijić's arguments in a pamphlet, published in France as a reply to Cvijić's statement.[1] He pointed out that the Macedonian Slavs had on numerous occasions associated themselves with the Bulgarians in nationalist uprisings against the Turks and that before 1878 the most fiercely nationalist of the Bulgarians had been those of Macedonia.[2] He also drew attention to the fact that Cvijić had entirely neglected the evidence of older ethnographic maps.

These criticisms were justified. There were many other weaknesses in Cvijić's argument. For example he compared the figures of Kănčev, Nicolaides and Gopčević but, since each had a different definition of Macedonia, no valid comparison was possible. Nor did he allow that Kănčev's work was widely accepted, whereas that of Gopčević and Nicolaides had been widely rejected. He was not justified in excluding the Slavs of Tetovo and Skoplje from his *Macedo-Slav* group because their national feeling was as much in dispute as, for example, that of the Slavs of the Lake Ohrid region, and he classed the Slavs of Skoplje as Serbs largely on historical grounds. Finally his claim that the *Macedo-Slavs* were amorphous from a national point of view entirely neglected to take into account the success which had attended Bulgarian propaganda since 1876.

The time was not yet ripe for the widespread acceptance of the idea of the *Macedo-Slavs*. Cvijić had not actually depicted them on a map. His suggested *Macedo-Slav* classification was tentative and it had yet to be taken seriously. He had addressed his views primarily to Great Britain, France and Russia in the hope that they would espouse the Serbian cause. His ideas no doubt impressed the Great Powers but the claims of the Bulgarians and the Greeks could not lightly be

[1] *Les Slaves de Macédoine. Résponse à M. Cvijić* (Paris, 1908).

[2] A Bulgarian, Matei Gheroff, included an interesting map in an article on Macedonia in the French *Revue Hebdomadaire* (Paris, 1915); it showed the distribution of Bulgarian uprisings against the Turks, and most of these insurrections had occurred in Macedonia.

dismissed at this time. The domestic affairs of Austria-Hungary also played their part in prejudicing Cvijić's thesis. The 'Trialists' were still powerful in 1906. Their policy was to bring the Slavs into the Austro-Hungarian Federation on equal terms with the Germans and Magyars, and if their policy were to prevail in Vienna then Serbia might range herself alongside Austria-Hungary. Whilst that possibility remained, the Great Powers did not wish to commit themselves too deeply in Serbia.

So Cvijić's initial venture into the realm of ethnographic polemics did not immediately revolutionize ethnographic ideas. He succeeded in impressing upon western Europe and Russia that the Serbs had a case in Macedonia and that the Slavs of the Skoplje and Tetovo regions were Serbs, but his idea of the *Macedo-Slavs* was not accepted. These conclusions may be justified by a reference to contemporary Slav maps.

## Russian and Czech Maps, 1907-10

### *T. D. Florinski's Maps, 1907*

These maps were the work of an eminent Slav scholar. T. D. Florinski was an authority on both Serb and Bulgarian history. It was mainly due to his efforts that manuscript evidence about the great Serb Empire of Dušan had been unearthed, and published at Kiev in 1888. He had also published details in 1899 of the Empire of the Bulgarian, Samuel (Fig. 27). These works dealt with the mediaeval Slav states of the Balkans and they had done much to stimulate both Serbian and Bulgarian nationalism. His greatest work, published in Kiev in 1907, was a study of the whole Slav family of nations. It was accompanied by two maps; one marked the extent of the Slavs in the ninth century and the other showed the modern distribution of the Slavs over Eurasia. On the latter map Macedonia was clearly shown as Bulgarian, with a Serb element in the Tetovo and Skoplje areas.

### *L. Niederle's Maps, 1910-11*

L. Niederle's enquiries into the origins of the Slavs were scarcely less consequential than those of Florinski. Niederle's famous work on the Slavs was translated into French in 1911, and his discoveries were also recorded in the Smithsonian Report of 1910. The map of

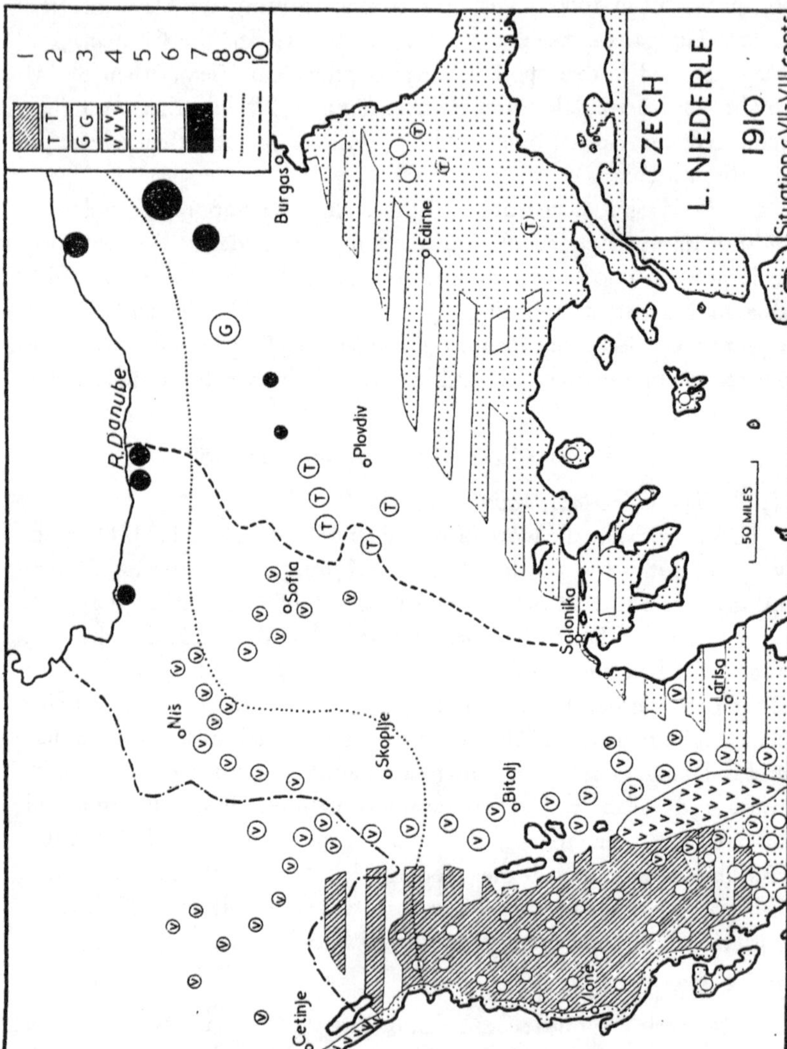
1
2
3
4
5
6
7
8
9
10
CZECH
L. NIEDERLE
1910
Situation c.VII-VIII cents
R. Danube
Burgas
Edirne
Plovdiv
Sofia
Niš
Skoplje
Bitolj
Salonika
Larisa
Vlone
Cetinje
50 MILES

the modern distribution of the Slavs which he produced was on too small a scale to be of practical use, but he followed Florinski in acknowledging the Macedonian Slavs as Bulgarians, except for a small Serb minority in the north.

But in addition to his map showing contemporary distributions Niederle produced a larger map in 1910, in which he endeavoured to reconstruct the ethnography of the Balkans in the seventh and eighth centuries, that is, before the Turkish invasions but after the arrival of the Slavs (Fig. 38). He believed that there were then eight main groups in the Balkans—*Slovanié, Romani, Rekové, Turkotataři, Sarmaté, Germani, Illyrové* or *Albanci*, and *Thrakové*. (Slavs, Romanians, Greeks, Turko-tatars, Sarmatians, Germans, Illyrians or Albanians, and Thracians). A note on their distribution helps to give perspective to modern ethnographic studies.

*The Turko-tatars.* They were marked on Niederle's map in north-eastern Bulgaria and in the Dobrudja. Obviously they derived from the northern steppe and had entered the Balkans from the north. Some had penetrated along the Danube but they nowhere formed a compact mass.

*The Greeks.* The Greeks were shown along the coasts of Albania, in Ipiros and southern Thessalía, in Khalkidhikí and south of a line, roughly from *Solun* (Salonika) to *Sozopolis* (Sozopol) and finally in Asia Minor.

*The Slavs.* The Slavs were all shown in one colour. They occupied, according to Niederle, the whole of the Balkans including the greater part of Macedonia and much of the Greek archipelago ; they formed a very important minority indeed in the Pelopónnisos, and were to be found also scattered throughout Albania. In Macedonia, Niederle distinguished the Slav tribes of the *Brsjaci, Strumenci* and *Smoléné*, the *Bulharové* (Bulgars proper) he confined to north-eastern Bulgaria and

---

FIG. 38. THE BALKAN PENINSULA IN THE SEVENTH AND EIGHTH CENTURIES A.D.

The references in the key are as follows : **1. Illyrians** (Albanians) ; **2. Thracians; 3. Germans; 4. Romanians** (Vlachs) ; **5. Greeks; 6. Slavs; 7. Turko-Tatars ;** **8.** Southern limit of the *Sto* dialect ; **9.** Limit of Greek and Latin provinces ; **10.** Western limit of the East-Bulgarian dialect.

the *Srbové* (Serbs) to the basin of the Drina around Raš. He also marked on his map two very significant boundaries : (1) The limit of the *Sto* dialect which later became the basis of the written Serbian language. It ran roughly from Antivari to Prizren and Pristina and ended on the Timok river ; (2) The limit of 'Old-Bulgarian' (*Východniho*) which ran, roughly, from Salonika north-eastwards to the Danube at *Hunnove*.

*The Vlachs.* Niederle distinguished the northern Romanians, *Romani*, from those south of the Danube, whom he referred to as the *Vlasi* (Vlachs). He gave the *Vlasi* a more widespread distribution than they had on modern maps. The Píndhos was the nucleus of their settlement but they were also found at this period in Ipiros, Thessalía, western Macedonia, Old Serbia, and in the Niš-Leskovac region.

*The Albanians.* Niederle believed the Albanians to be the direct descendants of the Illyrians and that in the seventh and eighth centuries they occupied the Albanian mountains between Peć in the north and Ioánnina in the south. There were few in Old Serbia or in western Macedonia.

*Sarmatians, Germans and Thracians.* The Sarmatians he marked north of the Danube ; the Germans also, but small German groups were apparently to be found in the eastern Balkans. However, they constituted only a minor part of the population. More numerous and of greater interest were the Thracians, remnants of whom were still to be found at this time, according to Niederle, in the mountains of the Balkan range and the Rodopi.

*Conclusion.* Niederle's reconstruction was based on an examination of manuscript evidence and on available Serbian, Greek, Bulgarian and Romanian sources. He made an attempt to sift the works of all the great Slav philologists and historians of the nineteenth century—Safařik, Belić, Vondrák, Novaković, Jireček, Jagić, Zupanić, Kănčev, Ivanov, Drinov, Zlatarski and many others figured in his bibliography. His synthesis was a most valuable contribution since he was neither Serbian nor Bulgarian but Czech. His ideas gave rise to the interesting speculation that a wedge of Romanian territory had once intervened between the Serb and Bulgarian lands. Neither of Niederle's maps gave much support to Serbian claims in Macedonia.

## G. Amadori-Virgilj's Map, 1908

Nor did Italian geographers at this time accept Cvijić's views. G. Amadori-Virgilj, for example, made little reference to the Serbian point of view in his work on the ethnography of the 'Rumelian' provinces. Amadori-Virgilj's work was unusal insomuch as he was pro-Greek although an Italian. A diplomat by profession, he had spent a considerable time in Macedonia. His book included an ethnographic map of the *Vilayets* of Jannina, Monastir (Bitolj), Salonika, Adrianople and parts of the *Vilayet* of Constantinople. He based his map largely on religious criteria and he insisted on the importance of religion as an index of nationality. The resultant distributions were very different from those produced, either by the Bulgarians on the basis of language, or by the Serbians on the basis of culture. He recognized the following elements in the population of Turkey-in-Europe: (1) Orthodox Greeks, (2) Exarchist Bulgarians, (3) *romanized* Vlachs, (4) Slavs under the influence of Serbian propaganda, (5) Musulmans, (6) Gypsies, (7) Armenians, (8) Jews. Of these the Vlachs, Gypsies, Armenians and Jews did not occupy any considerable territory (Fig. 39).

### *The Musulmans (Moslems)*

He believed that all Moslems formed one national group. Thus his Moslem group included not only genuine Osmanli Turks but some 350,000 Albanians of the *Vilayets* of Jannina and Monastir and some 185,000 Pomaks and *Mohadjers*. The latter were, as H. N. Brailsford had pointed out (p. 139), Moslem Slavs from the north who had migrated into the Turkish provinces. The distribution of the Moslems was very extensive. Particularly striking was the portrayal of the population of a large part of Dhitikí Thráki and eastern Macedonia as Moslem; in these areas they were mostly Pomaks. In thus excluding them from the Bulgarian national group Amadori-Virgilj favoured the Greeks because the main threat to the execution of the *Grand Idea* came, not from the Turks, but from the Bulgarians. If the Pomaks of these areas were classed as Bulgarians, a wedge of Bulgarian territory interposed between the Greeks of Macedonia and

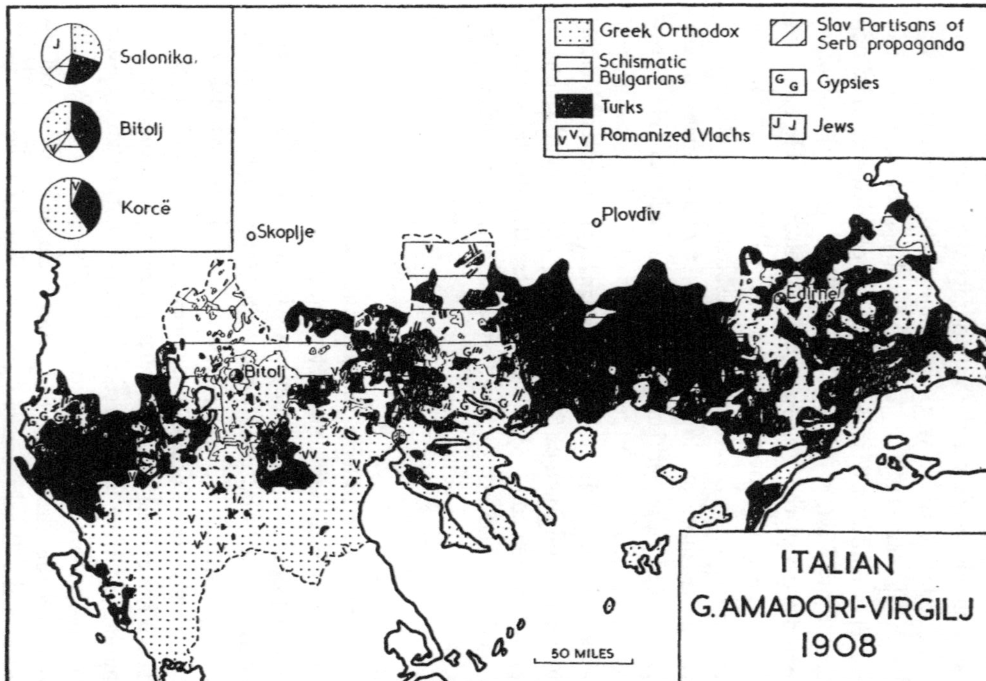


FIG. 39

In the original, the Turkish category is referred to as *Musulman*, Armenians are separately distinguished, and the individual ethnographic structure of each town is shown by symbols as exemplified in the inset above.

those of Thrace. Also striking was his portrayal of the territory between Berat, Vlonë, Korcë and Gjinokastër as Moslem or Turkish, particularly as Barbarich, also an Italian, had been eager to demonstrate its Albanian character.

*The Greeks*

On this map, Amadori-Virgilj portrayed as Greeks all members of the Greek Orthodox Church who were still faithful to the Patriarch. This meant the inclusion of a large element of Greek Orthodox Albanians, Vlachs and Slavs in his Greek national group. Hence the justification (1) for depicting the lower valleys of the Vijosë and Semen, in Albania, as Greek, (2) for extending the Greek ethnographic frontier into the Slav territory north of Bitolj, and (3) for ignoring practically all Vlach enclaves.

*The Slavs*

Nearly all the Slavs that Amadori-Virgilj marked on his map were classed as Exarchist Bulgarians, but he did recognize partisans of the Serbian party around Kičevo. He did not incorporate the districts of Debar, Prizren or Skoplje in his map, but his recognition of Serbs as far south as Kičevo seemed to indicate an agreement with Cvijić's claims made for the Serbs in 1907, even although Amadori-Virgilj did not recognize any *Macedo-Slavs.*

*Albanians and Vlachs*

Amadori-Virgilj marked no territory on his map as exclusively Vlach, but he did indicate what he called *romanized* Vlachs in the larger towns, the population of which he portrayed by means of symbols. He did not believe the Albanians to constitute a nationality. They simply did not figure on his map at all. The population of parts of Ipiros, often referred to as Albanian on other maps, he classed as either Turkish (Moslem) or Greek.

*Conclusion*

Amado ri-Virgilj's map was accompanied by a number of ecclesiastical maps, in which he attempted to show the distribution of Greek

Orthodox, Exarchate, Serbian national and Romanian Churches. These maps bore a striking resemblance to the pro-Greek ecclesiastical map that had appeared in Paris in 1905 (see p. 135) and must have emanated from the same source.

His statistics were evidently Turkish inspired and appeared to have been based on Turkish estimates of the numbers of peoples in the respective *millets*. These, at this time, comprised the *Rumi* or Greeks, the *Bulgari* or Exarchists, and the *Islami* (Moslems). Hence his figures for the area covered by his map were : Turks (Musulmans), 1,823,500 ; Greeks, 1,613,000 ; Bulgarians, 455,000 ; *romanized* Vlachs, 13,750 and Serb partisans, 16,550. It is interesting to note that Amadori-Virgilj's map was not altogether without influence on Cvijić's map of 1909, in which Cvijić was careful to insist on the influence of Hellenism in Ipiros and Macedonia (Fig. 40).

## Chapter VIII

# PRO-SERB MAPS AND THE BALKAN WARS 1909–1913

### J. Cvijic's Map of 1909

In 1908 Austria dramatically annexed Bosnia and Hercegovina. This move abruptly terminated any possibility of the entry of the Serbo-Croats into the Austro-Hungarian Empire on a federal basis. It meant that the Austrians had deliberately eschewed Slav co-operation and had set out to isolate Serbia. The Serbians until then had hoped that they might be granted an outlet to the Adriatic through those provinces, but they now found themselves compelled to look elsewhere for their 'open door.' There still existed three possible routes to the sea, but unfortunately all ran through territory not generally acknowledged to be Serb. The routes were:

(1) via the Drin valley to Shëngjin (San Giovanni di Medua)—this was the route of a projected *Adriatic-Danube* railway which was to have passed through Niš and Prizren;

(2) via northern Macedonia and central Albania to Durrës (Durazzo)—the latter port had greater possibilities than Shëngjin since it was possible to reach Durrës either via the Drin valley from Prizren, through Debar and Elbasen, or via Veles and Elbasen;

(3) via the Vardar valley to Salonika—Salonika was a better port than any to be found on the Adriatic coast; the route was much longer but easier.

From henceforth the whole of Serbian foreign policy was directed towards gaining control of one or more of these routes. In order to carry out their programme the Serbians looked for support to Great Britain and France, and to Russia, although in the latter case, the Tsar's refusal to accept the Austrian challenge on the issue of Bosnia and Hercegovina incurred Serbian mistrust and increased Serbian dependence upon the Western Powers.

For their part Britain and France were not unwilling to help Serbia in her quest for an 'open door.' The German Emperor had made no secret of his support for the Austrian drive to Salonika and only Serbia stood in the way of the Germanic Powers. Not only did the Serbians command the vital strategic routes to the Aegean but they were the nucleus of a potential Serbo-Croat state, the formation of which might have a disintegrating effect upon the Austro-Hungarian Empire.[1]

The mutual interests of Serbia and Great Britain in 1909 were reflected in the publication of J. Cvijić's statement on *Bosnia, Herzegovina and the Serb problem*, which contained another of Cvijić's maps, much more detailed than the one which had appeared in 1906. In this map Cvijić not only recognized Turks, Greeks, Vlachs and Albanians but he divided the Slavs into three groups, one of which had hitherto not appeared on any ethnographic map—the *Macedo-Slavs* (Fig. 40).

### *The Turks and Greeks*

On Cvijić's map, the distribution of the Turks was similar to that already given by F. Meinhard and V. Kănčev and need not be further discussed. The distribution of the Greeks in Macedonia was based on Meinhard's map and, in the case of Ipiros, on Weigand's map. That is to say, the Greek ethnographic frontier ran west-east from Sánta Quaranta to Kónitsa, Kozáni and Véroia and the Greeks on his map predominated in Thessalía, in Khalkidhikí and in the plain of Lake Akhinoú, with a small group on the coast, south of Kaválla. This distribution was considerably less extensive than the Greeks claimed but Cvijić overcame the difficulty of acknowledging the Greek claims by marking some Albanian and *Macedo-Slav* territory as *under Greek influence*. So territory *under Greek influence* stretched from Vlonë, in Albania, to Sérrai and Dráma in southern Macedonia, including Bitolj and the Lake Dojran area; it corresponded almost

[1] See R. G. D. Laffan, *The Guardians of the Gate* (Oxford, 1918).

exactly with the claims put forward by C. Nicolaides in 1899. Thus the map reduced both Albanian claims in Ipiros and Bulgarian claims in Macedonia.

*The Slavs*

The map exhibited revolutionary ideas on the distribution of the

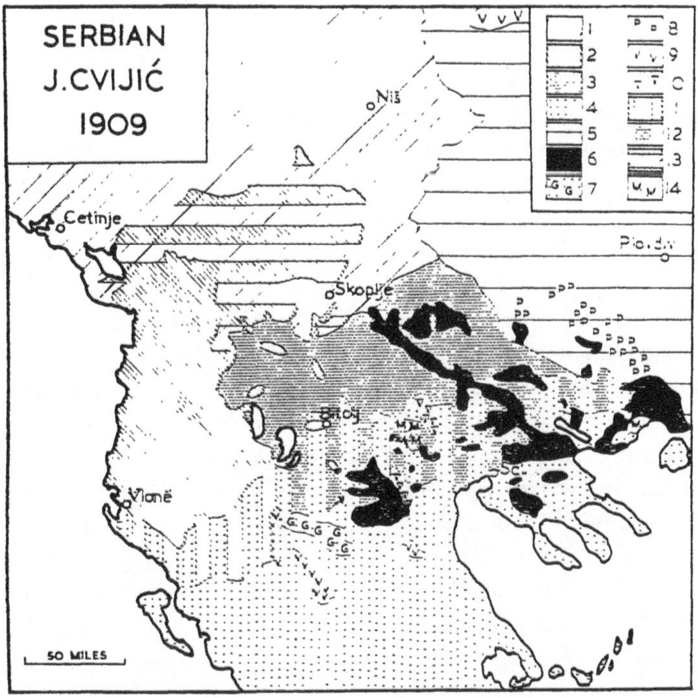


FIG. 40

The references in the key are as follows: **1. Serbo-Croats ; 2. Albanians ; 3. Macedo-Slavs ; 4. Greeks ; 5. Bulgarians ; 6. Turks ; 7. Moslem Greeks ; 8. Moslem Bulgarians ; 9. Vlachs** and **Romanians ; 10. Vlachs under Greek influence ; 11. Albanians under Greek influence ; 12. Macedo-Slavs under Greek influence ; 13. Albanian Serbs ; 14. Moslem Macedo-Slavs.**

Slavs in the western Balkans. The Serbs for example were depicted as far south as Peć, Mitrovica and Pirot, with a further extension south from Leskovac, to Vranje, Skoplje and Brod. The area north of and including Debar, Gostivar, Tetovo and the Crna hills, to the borders of Montenegro, to Peć and to Mitrovica, was represented as Albano-Serb territory. This mixed territory extended almost as far west as the Drin valley. Thus the Serbs were depicted in the majority north of a line through the Golešnica and Osogovska mountains. Apart from a few Serbians, only C. Nicolaides had marked such generous distributions of Serbs before this map appeared. F. Meinhard, although he had conceded a great deal of territory to the Serbs, had not shown them at all in the Kriva Palanka area, which he had judged to be Bulgarian.

The Bulgarians on this map were limited to Bulgaria proper, to Eastern Roumelia, and to a small extent of Macedonian territory in the upper Struma valley and the Nevrokop district. Only a limited number of Moslem Slavs were acknowledged as Bulgarian—those in the Mesta valley and the Rodopi mountains and a small group in the Struma valley. Thus both Moslem and Christian Bulgarians in Macedonia were shown as having a very limited distribution. In fact, the Bulgarians were all but excluded from Macedonia.

This novel and strikingly significant interpretation of Macedonian ethnography was accomplished by the simple expedient of distinguishing a third Slav grouping—that of the *Macedo-Slavs*. This was the first time that an attempt had been made to map their distribution, although Cvijić had postulated their existence in 1906. All territory between the Greek, Bulgarian and Serb ethnographic frontiers already mentioned, including the Crni Drim valley, was marked as *Macedo-Slav*. Thus classified as *Macedo-Slav*, were such strongly pro-Bulgarian Slavs as those of the Strumitsa valley and those around Ohrid.

The political significance of the idea of the *Macedo-Slavs* became manifest once their distribution had been plotted on a map. According to Cvijić the area between the Sar mountains and Salonika fell into two parts—the northern part was an extension of ' Old Serbia,' inhabited by Serbs, and the southern was largely inhabited by *Macedo-Slavs*. The *Macedo-Slavs* themselves were not homogeneous. Half of them in the south were *under Greek influence* and the remainder, from a casual glance at the map, appeared to have closer relations with the Serbian version of ' Old Serbia ' dominated by Skoplje than they did with Bulgaria on the far side of the Rodopi ranges. Anyone not well acquainted with previous ethnographic maps might well suppose after examining the Serbian map that, ethnically, the Serbs were in a dominant position in Old Serbia, and further, that those *Macedo-Slavs* who were not hellenized were a potential branch of the Serbo-Croats. Others more perfectly acquainted with the trend of ethnographic thought might conclude :

(1) that the large number of Serbs in northern Albania and Macedonia south of the Sar mountains was purely a Serbian interpretation based largely on the historical concept of the extended limits of 'Old Serbia,' and that it was more than a coincidence that Serb territory had been extended to cover the route to Shëngjin ;

(2) that the acknowledgement of the extent of Hellenic influence in southern Macedonia was an indication of Serbian willingness to reach agreement with the Greeks in the south.

It is important to note the limits of the Serbs, Bulgarians and *Macedo-Slavs* given on this map because in his later maps Cvijić varied the distribution of these Slav groups to suit expanding Serbian territorial ambition in Macedonia.

*The Vlachs and Albanians*

The Vlachs of the Píndhos, Grámmos and Klisoúra districts were marked on Cvijić's map. The distribution of the Albanians showed interesting modifications. Their ethnographic frontier in the south was drawn to include Kónitsa but all the Albanians in the Korcë, Kónitsa, Gjinokastër and Vlonë areas were marked as *under Greek influence.* In the north, purely Albanian territory was limited practically to the region south of the Drin. The whole of the area from the Drin valley to the Serbian political boundary of 1878 was referred to as a mixed zone containing three elements — Serb, Albanian and *Albanian Serb.* No Albanians were shown north of Peć and Mitrovica, so that the whole of the vital Novi Pazar corridor between Montenegro and Serbia was depicted purely as Serb. The distribution of Albanians in the north was thus a reiteration, in a modified form, of the ideas first promulgated by S. Gopčević. The Serb ethnic group was strengthened by the inclusion of *Albanian Serbs* who were, as Gopčević had maintained, Albanians of Serb extraction. Needless to say this interpretation was of doubtful validity. By the use of this device, the Albanian majority of two-thirds in the mixed areas of Old Serbia, and in the zone between the Montenegrin boundary of 1878 and the Drin valley, was turned into a minority, although Albanian speakers formed by far the most numerous element in the population. The mixed area included Shkodër and the mouth of the Drin, hitherto almost exclusively regarded as Albanian territory. This was the very area which had become so vital to Serbia as a possible outlet, since the Austrians had closed Bosnia and Hercegovina to Serbian penetration.

Cvijić's map of 1909 was reproduced in an unacknowledged form in a book compiled by A. Stead (one-time editor of *Review of Reviews*) and called *Serbia by the Serbians.* The appearance of the book was a reflection of the new interest being taken in Serbia by the British at this time and conversely of Serbian anxiety to propagate their claims amongst potential allies. There is very little evidence to suggest, however, that Cvijić's map wrought any appreciable change on English opinion at this juncture.

## R. W. Seton-Watson's Map of 1911

R. W. Seton-Watson may be regarded in many ways as one of the architects of the Jugoslav State. His opinions on the ethnography of Macedonia are therefore of considerable interest, more especially as they varied considerably between 1911 and 1917. His book on the Southern Slavs appeared in 1911 and it contained a map which showed the distribution of the Serbo-Croats, drawn up with the help of Dr. J. Smodlaka, who represented Dalmatia in the Austrian Parliament (Fig. 81). According to the map the southern frontier of the Serbs ran from the Bulgarian boundary near Kyustendil, through the Golešnica mountains to Kičevo, then due north along the Bistra mountains, thus excluding the whole of the Drin-i-xy valley, then it curved westwards to include Prizren and Dakovica. The Albanians in Old Serbia were marked as a minority. R. W. Seton-Watson's summary of the ethnographic situation in Old Serbia and Macedonia is worth quoting in full. It may be noted that he was, at this time, in agreement with H. N. Brailsford concerning the success of Bulgarian propaganda in Macedonia.

> In dealing with the Serb population, we are confronted by the complete absence of reliable statistics. The Kamidian régime did not trouble about censuses and such estimates as exist are almost avowedly based upon the wishes of their compilers, Greek, Serb or Bulgar as the case might be, rather than on the actual facts of the case. The Macedonian practice of forcible conversion of villages by the rival bands has still further complicated the problem, until it is by no means easy to form any definite judgement, even upon seemingly first-hand evidence as to the true nationality of many districts. Roughly speaking, the territory inhabited by Serbs comprises the whole Sanjak of Novi Pazar (which separates Servia from Montenegro and was from 1878 to 1908 garrisoned by Austrian troops), the district of Ipek, Jakova and Prisrend, from the Sanjak as far south as the river Drin : and the plain of Kossovo, from Mitrovica on the north extending through Pristina and Uskub to Istib on the south. South and East of this point there may be isolated Serb colonies, but if so they are doomed to rapid absorption by the Bulgar element. Even in the neighbourhood of Prisrend the Serbs are steadily losing ground at the expense of the Albanians. Since the accession of King Peter, Serbia had made more desperate efforts than ever to arrest the fatal process in Macedonia, which is destined some day to decide the struggle of races between Bulgar and Albanian and against Serb and Greek. But the efforts of the Serbian bands have not as a rule been successful. Whilst the Serbs talk and sentimentalize, the Bulgars act and shoot. . . . If the total population of Macedonia be reckoned at 2,500,000, the most liberal allowance cannot assign more than 400,000 of these (including 100,000 Moslems) to the Serb element.

R. W. Seton-Watson did not make any reference to the *Macedo-Slavs* and this seemed to indicate that at this time he did not take seriously the concept of the *Macedo-Slavs* recently put forward by J. Cvijić.

## The Liberation of Macedonia

### *The Formation of the Balkan League, 1912*

Meanwhile events in the Balkans had been moving towards a climax. Only the clash of interests in Macedonia had so far prevented the Serbians, Bulgarians and Greeks from rising together against their common enemy, the Turk. The Turk himself, however, with a singular lack of foresight, eventually forced the Balkan nations to act in concert. The annexation of Bosnia and Hercegovina in 1908 had precipitated a crisis in the domestic affairs of Turkey. The younger and more 'progressive' elements amongst the Turks had become extremely dissatisfied with the methods of administration and with the weak and vacillating foreign policy of the Porte. The revolution which followed significantly originated in Macedonia, and resulted eventually in the 'Young Turks' coming into power in Constantinople. The policy which the Turks had for so long successfully practised—that of sowing seeds of discord amongst the Balkan nationalities in order to maintain a delicate balance of power in their own hands—was cast aside. In its place 'Turkification' was inaugurated, particularly in Macedonia. Forced Moslemization and butchery of Christian subjects were carried out as part of a programme designed to restore the 'prestige' of the Turk in his European possessions and to create a Turkish national province in Macedonia.

This display of Turkish nationalism came as a shock to the non-Christian population, which had expected the new Turkish regime to inaugurate much needed reforms. The Young Turks even aimed at the economic isolation of Macedonia and went so far as to prevent the Greeks from engaging in commerce along the Aegean littoral; they refused, too, permission for the construction of a Greek railway from Thessalía into Macedonia. This policy eventually drove Serbia, Greece and Bulgaria to take common action, and in 1912 a number of Treaties between the various Balkan States was concluded. In March, 1912, Serbia and Bulgaria sank their differences sufficiently to sign a pact aimed against Turkey. In April, Bulgaria concluded a treaty with Montenegro and in May an alliance was drawn up between Bulgaria and Greece.

### *Territorial Agreements Affecting Macedonia*

These treaties and the negotiations which led up to them provide interesting evidence on the concessions which the rival parties in Macedonia were prepared to make to each other in 1912. No reference was made in the Graeco-Bulgarian Treaty itself, to the respective spheres of influence of Greece and Bulgaria in Macedonia, but according to D. M. Brancov, the Greeks at this time were willing to concede to Bulgaria the whole of Dhitikí Thráki and also the Turkish *cazas* of eastern Macedonia, provided that Greek claims in Crete were supported by the Bulgarians.[1] The Greeks themselves referred to the treaty as an 'impossible alliance' which could only have been brought about by unusual circumstances (Fig. 90).

The Serbo-Bulgarian Treaty differed from the Graeco-Bulgarian, because the Serbians and Bulgarians made a definite attempt to reconcile their differences in Macedonia by adopting a compromise solution in the form of a partition. The question of the partition of Macedonia had been broached by the Serbian Prime Minister during preliminary meetings with the Bulgarian Prime Minister before the Serbo-Bulgarian pact was concluded. At first, Skoplje and Kumanovo had been mentioned as the Serbian share. Public opinion in Bulgaria, however, was so strongly against any scheme of partition that the matter was dropped on that occasion. The Bulgarians preferred an autonomous Macedonia rather than partition. Finally the problem was half solved by the incorporation into the Serbo-Bulgarian Treaty of a Secret Annex defining the possible partition of Macedonia in the event of its successful conquest from the Turks. As neither party could agree on a line, the maximum claims of each party were marked on a map ; the overlapping area between these claims—the contested zone—was to be referred to the Tsar for final arbitration (Fig. 44). The Tsar apparently decided in favour of the Serbian line, after he had had a personal interview with J. Cvijić.

The Serbian line of partition was based upon J. Cvijić's latest ethnographic researches. By 1912 he had expanded Serbian claims in Macedonia far beyond the limits he had suggested in 1906-7 and beyond the limits laid down by his map of 1909. In his own words :

[1] D. M. Brancoff, *op. cit.*

> The frontier of Old Servia or the boundary which divides the Servian and Bulgarian spheres of influence starts from the Bulgarian frontier at Kustendil, with the dividing line between Petchine and Krilje, so that Kriva Palanka and Kratovo remain in the Bulgarian sphere, Uskub (Skoplje) and Kumanovo in the Servian. The southern frontier lies through Ovce Polje with the dividing line between Breganitsa and Ptchinje and it crosses the Vardar river north of Velles. From here it follows the offshoots of the mountains of Yakubitsa and by a further dividing line on the mountain of Baba, to the Lake of Ochrida, so that Prilep, Krushevo and Ochrida are in the Bulgarian sphere and Struga, Debar and Tchova in the Servian. A narrow strip of Old Servia opens on to the Adriatic sea near Scutari and Alessio. Thus we see that a territorial and ethnographical understanding has been arrived at between Serbs and Bulgars.[1]

This Serbo-Bulgarian divide as given by Cvijić, was identical with the limits of Serbian claims laid down in the Secret Annex. The terms of the Treaty had not then been published and Cvijić's statement was the first indication of the fact that the Serbians and Bulgarians had effected a compromise. Cvijić made no mention of his *Macedo-Slav* group at this time. The new limit he set for the Serbs in the Drin valley was an advance upon that suggested by him in his maps of 1909 and 1911. The inclusion of Debar and Struga in Serb territory prepared the way for a Serbian outflanking of the whole of northern and central Albania. The difficulties of railway construction in the mountains of northern Albania and Montenegro had become apparent in 1912 and the Serbians were eager to find new routes via central Albania. Serbia's need for Durrës as well as the port of Shëngjin had also become obvious to her economists. Hence the desire to gain control of the whole of the Drin valley—the gateway to Elbasen and Durrës. Cvijić himself made the Serbian aims on the Albanian coast quite clear in an article written for *Petermann's Mittheilungen* in 1912, in which he surveyed the possible railway routes from Serbia to the Adriatic.[2] The routes which he favoured in 1912

[1] *Review of Reviews,* vol. 46 (London, 1912).

[2] "Der Zugang Serbiens zur Adria," *Petermann's Mittheilungen* (Gotha, 1912).

were farther south than those he had suggested in 1909 and this fact apparently influenced his distributions (Fig. 41).[1]

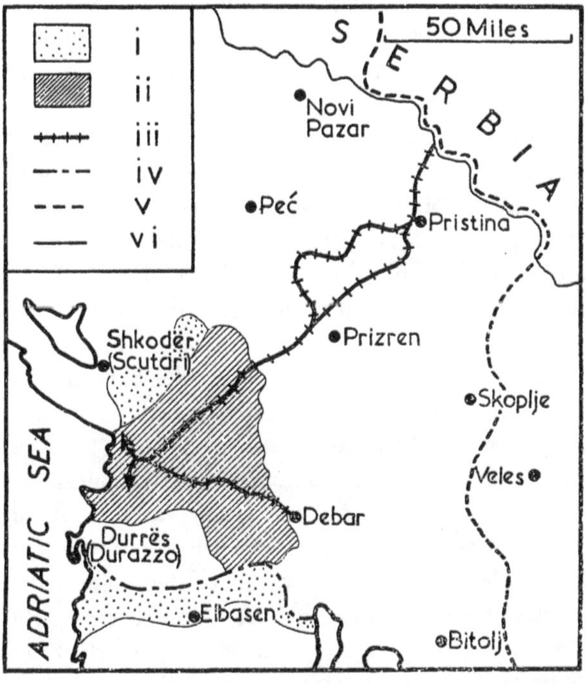


FIG. 41. SERBIAN ASPIRATIONS ON THE ADRIATIC COAST IN 1912 (AFTER J. CVIJIC)

The references in the key are as follows : **i.** Territory occupied but not required by Serbia in 1912 ; **ii.** Territory occupied and permanently required by Serbia for the purpose of railway construction ; **iii.** Projected rail-routes to the Adriatic ports ; **iv.** Southern limit of 'Old Serbia'; **v.** Eastern limit of Adriatic trade influences ; **vi.** Boundary of Serbia in 1912 (before the Balkan campaign).

### *The Balkan Wars, 1912-13*

The rash policy inaugurated by the Young Turks provoked even the Albanians into revolt. They achieved a great deal of success in northern Macedonia and their insurrection may be regarded as opening the Balkan Wars. The Porte was still occupied in 1912 with a war

[1] Cvijić first traced possible routes to the Adriatic in 1909 as follows : (1) through Hercegovina via Višegrad, Foča Nevesinje, Plana, Bileća and Trebinje to Dubrovnik ; (2) through Albania via Shkodër to Bar.

against Italy, and it appeared at one time that the Albanians, operating with the aid of renegade Turks, might obtain control of northern and western Macedonia. They even demanded the secession of Bitolj and Skoplje from the Turkish Empire as part of an independent Albania. Their success was an indication of Albanian numerical strength in these areas.

The Albanian movement was soon overshadowed by a declaration of war against Turkey by Montenegro. Soon after, Greece joined Montenegro and Serbia and Bulgaria presented the Turks with an ultimatum : by October 18th, 1912, the whole of the Balkan League was engaged in successfully driving the Turks from the Balkans.

The dispositions of the forces of the Balkan League consequent upon the defeat of the Turks were of some importance. The Bulgarians had been compelled to engage the bulk of the Turkish forces in Thrace, so that Greece and Serbia between them occupied the greater portion of Macedonia and stood fast. The Greeks even managed to occupy Salonika a few hours before the Bulgarians arrived. They also occupied southern Macedonia and southern Albania. The Serbians occupied Skoplje, Bitolj, the greater part of northern Albania, Old Serbia and Novi Pazar. The whole of the western Balkans was thus dominated by the armies of Serbia and Greece which were reluctant to evacuate any territory for the sake of their old mutual enemy, Bulgaria (Figs. 88, 89, 90).

## J. Cvijic's Map of 1913

Whilst the Serbians were in possession of most of Macedonia and before the Treaty of Bucarest had finally settled the new boundaries, J. Cvijić's third complete ethnographic map of the Balkans appeared in 1913 (Fig. 42). It was published simultaneously in the Proceedings of the Royal Serbian Geographical Society and in *Petermann's Mittheilungen* where it was sure of a wide circulation. This map reappeared with modifications in 1918. Subsequently, it formed the basis of practically all post-war ethnographic maps of the Balkans. Its influence on ideas of Balkan ethnography has been probably greater than that of any other single map.

The map was not confined to Macedonia ; it dealt with the whole of the Balkans. Cvijić was perfectly well acquainted with all previous ethnographic maps of south-eastern Europe. The method he adopted was to extract data from earlier maps and thus to build up a composite picture of Balkan ethnography with the additional aid of recent Serbian research. In particular the maps of Meinhard, Weigand and Gopcević influenced his Macedonian distributions. He recognized seven major groups—Turks, Greeks, Serbo-Croats, Bulgarians, *Macedo-Slavs*, Vlachs and Albanians. Each group, however, he divided into a number of sub-groups, generally on the basis of religion.

*The Turks*

Grouped with the Turks were Tatars and Gagauzi (Christian Turks). Their distribution in Macedonia was strikingly similar to that on the maps of Meinhard and Kănčev. The Greek Orthodox Turks shown on Meinhard's map around Lake Akhinoú also appeared on Cvijić's map. Cvijić's distribution of Turks was, however, more broken than that of Meinhard.

*The Greeks*

Two sub-groups of the Greeks were recognized by Cvijić—Greeks proper and Moslem Greeks. The latter had appeared for the first time on Meinhard's map but Cvijić showed a more extensive distribution of this group in the upper Aliákmon valley. It is interesting to note that these Moslem Greeks were later classed as Turks under the Graeco-Turkish population-exchange scheme of 1923. Cvijić marked his distributions of Greeks in south-western Macedonia, after Weigand, and in southern Macedonia, after Meinhard. He departed from his map of 1909 in disregarding elements *under Greek influence*, such as the Bulgarophone Greeks. Cvijić in 1913, therefore, supported the view that the Greeks formed only a minor element in the population of Macedonia. The only large groups of Greeks on his map were to be found in the Aliákmon valley, in Khalkidhikí and in the Struma-Angítis valley. The total territory controlled by the Greeks according to this interpretation amounted to a small fraction of Macedonia—less than 15 per cent of the total area.

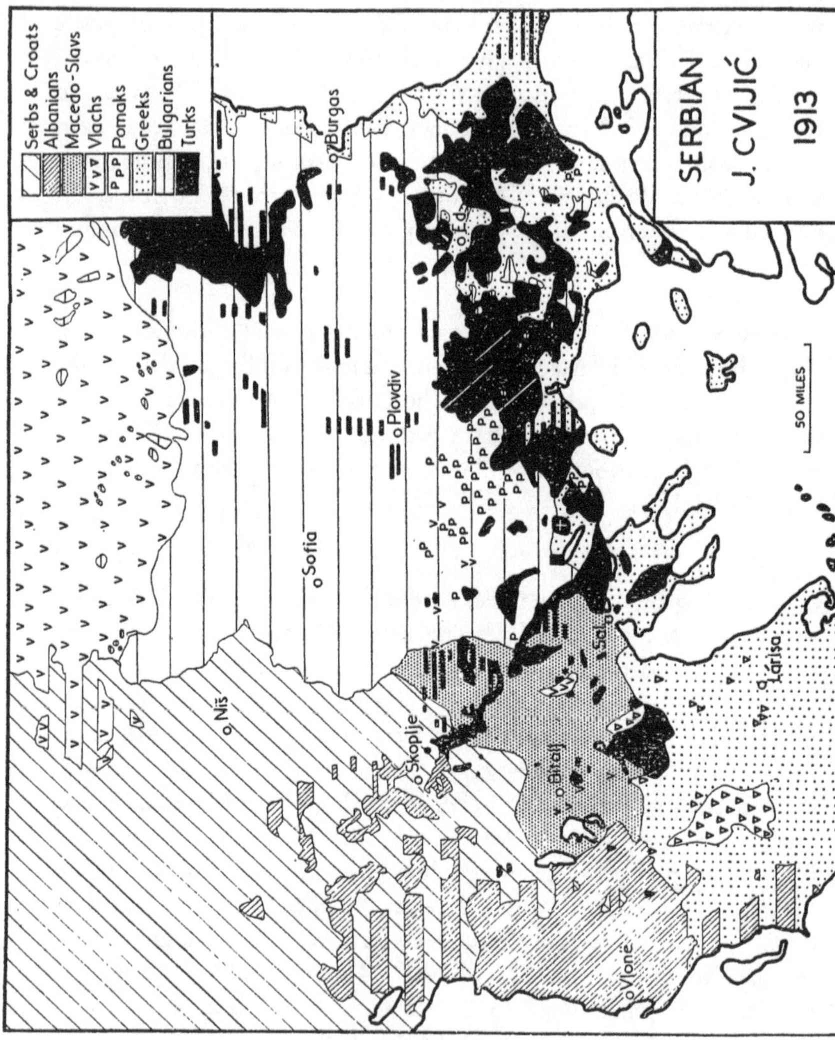
Serbs & Croats
Albanians
Macedo-Slavs
Vlachs
Pomaks
Greeks
Bulgarians
Turks
Burgas
Plovdiv
Sofia
Niš
Skoplje
Bitolj
Vlonë
Lárisa
50 MILES
SERBIAN
J. CVIJIĆ
1913

The *rapprochement* between the Serbians and Bulgarians brought about by the Serbo-Bulgarian treaty was the reason for Cvijić's rejection of Hellenism. In 1903 he had referred to nearly the whole of Macedonian culture as 'Byzantine', and in 1909 he had made a point of recognizing Greek cultural influence in southern Macedonia. He modified his views because the recognition of the Bulgarian sphere of influence in southern Macedonia appeared to be a corollary of the Secret Annex. Whilst the Serbo-Bulgarian treaty stood, Cvijić rejected Greek claims (*cf.* Figs. 40 & 42).

*The Slavs*

Cvijić distinguished three branches of the Slavs in Macedonia and Old Serbia—Serbo-Croats, *Macedo-Slavs* and Bulgarians. He further sub-divided the Serbo-Croats into five—Orthodox, Catholic, Moslem, *albanianized Serbs* or *Arnauts* and *Orthodox Serbs of Albanian speech*. The last two categories of Serbo-Croats differed from all the other sub-groups insomuch as they were not based on religious differences. The *albanianized Serbs* had appeared on his map of 1909. The term was originally used by Gočpević. Their inclusion as Serbo-Croats enabled Cvijić to depict half of northern Albania, as well as a great deal of Old Serbia, as 'Serbo-Croat.' The *albanianized Serbs* were, however, linguistically Albanians, and all other maps except Serbian maps had classified them as such. The *Orthodox Serbs of Albanian speech* formed an entirely new sub-group originated by Cvijić himself. This group was represented in the mountains of north-western Macedonia, hitherto regarded as Albanian territory.

---

FIG. 42

It is to be noted that in the original : (*a*) The **Serbo-Croat** group comprises the following sub-groups which are not shown here for the sake of clarity—Orthodox Serbs, Roman Catholic Serbs, Moslem Serbs, Albanianized Moslem Serbs (*Arnauts*) and Orthodox Serbs of Albanian speech ; (*b*) The **Albanian** group comprises both Albanians proper and Serbized Albanians ; (*c*) The **Macedo-Slav** group comprises both Orthodox and Moslem Macedo-Slavs ; (*d*) The **Vlach** group comprises Romanians, Vlachs, Moslem Vlachs and Hellenized Vlachs (the latter differentiated in the accompanying map by a closed **v**); (*e*) **Pomaks** are classed as a sub-group of the Bulgarians ; (*f*) The **Greek** group comprises both Orthodox and Moslem Greeks ; (*g*) The **Turkish** group comprises Turks, Tatars, Gagauzi and Greek Orthodox Turks (the latter differentiated on the accompanying map by a white cross).

Cvijić's justification for including these two sub-groups as Serbo-Croat was that they represented a population which, although sometimes regarded as Albanian, was in reality Serb. It had had no alternative but to become Albanian under the Turkish regime. By using this historical criterion, Cvijić was guilty of gross inconsistency. Such a criterion applied logically to his other ethnic groupings in the Balkans would have reduced all his distributions to absurdity. For example, by this criterion the Slavs of Niš and Leskovac should have been marked as Bulgarian on Cvijić's map. It would appear, therefore, that Cvijić's choice of criteria was motivated by political considerations. At this time the fate of Albania was being decided by international action and the Powers, particularly Austria, would never have allowed Serbia to annex purely Albanian territory (see p. 182).

Altogether, the Serbo-Croats, according to Cvijić, populated western Macedonia as far south as the northern shores of Lake Ohrid, Kruševo and Prilep. They were also found in the Crni Drim valley south of, and including, Debar. Neither Seton-Watson, nor Cvijić in his maps of 1909 and 1911 and in his own definition of 1912, had made such extensive ethnic claims for the Serbs in Macedonia. Only in the discredited Serbian maps of Gopčević and Andonović had such claims before been made. In practically all earlier ethnographic maps the Crni Drim valley had been marked as Albanian or Bulgarian. However, the ethnographic frontier between the Albanians and the Serbs originated, according to Cvijić, in the small peninsula of Lake Ohrid and then passed along the watershed between the Shkumbin river and the Lake. Albanian-speaking Serbs still existed in the Mokra district of central Albania and there were Serbs even on the coast of Albania. Place-name evidence pointed to the fact, asserted Cvijić, that the Serbs once extended as far west as Elbasan. The impression that Cvijić was labouring to invest the Serbian pretensions in the Drin valley and northern and central Albania with some ethnic validity, is hard to resist. That the Serbs had a case, however weak, in Old Serbia, was generally recognized, but that they were ethnically strong in Albania in view of the overwhelming evidence existing to the contrary, was an idea difficult to assimilate.

The second group of Slavs to be depicted on Cvijić's map was the Bulgarian. It included, as on the Serbian map of 1909, both Orthodox Slavs and Pomaks, but Cvijić did not regard all Pomaks as *ipso facto* Bulgarians: he also distinguished *Macedo-Slav* Pomaks and Greek Pomaks. The Bulgarian frontier on his map ran from Lake Akhinoú to the Dojran depression, and then northwards along the Vardar-Struma watershed to the Bulgarian political boundary of 1913. East of this line, he believed the Slavs were all Bulgarians, in spite of the fact that his map of 1909 had shown many *Macedo-Slavs* here as well. This modification was a concession to the Bulgarian point of view as a result of the Serbo-Bulgarian treaty.

The remainder of the Slavs in Macedonia were classed as *Macedo-Slavs*; they included members of both the Orthodox and Moslem Churches. Perhaps because the idea of the *Macedo-Slavs* had not had a good reception earlier, Cvijić elaborated the idea still further. He stated that since the days when it was supposed that the Slavs of Macedonia spoke Bulgarian, research on the part of such celebrated philologists as A. Belić and V. Jagić—both Serbians—had established the fact that the *Macedo-Slav* dialect was transitional between Serb and Bulgarian. (This argument, incidentally, had been used by Gopčević. See p. 100). For example, although the position of the definite article was a Bulgarian trait, the use of 'dj' and 'tj' sound was essentially a Serb trait. He further maintained that the customs and traditions of the Slavs of Macedonia exhibited Serb as well as Bulgarian features. The *zadruga* for instance—the organization of society into large, closely-knit communities based on family ties—was common to both Serb and Macedonian society but was not characteristic of Bulgarian social organization.[1] The traditional festival of the *Slava* was practised in Macedonia and Serbia but not in Bulgaria. Many of the folksongs of Macedonia had their counterpart in Serbia. For these reasons, concluded Cvijić, the Slavs of Macedonia between Skoplje and Salonika were, from many points of view, transitionary between the Serbs and Bulgarians. They exhibited no permanent national consciousness. Such nationality, either Serbian

[1] This claim has been questioned by Emile Sicard, *La Zadruga Sud-Slave dans l'évolution du groupe domestique* (Paris, 1943), in which the author gives evidence of the *zadruga* system existing also in Bulgaria.

or Bulgarian, as had been induced amongst them, was purely superficial, and owed its existence to religious or educational propaganda, or even to terrorization. Sense of nationality was so weak that in the previous ten years (1903-1913) not only individuals, but whole villages, had switched their loyalty from the Serbian to the Bulgarian 'party'. This great mass of Slavs, oscillating between Serbian and Bulgarian nationality, he therefore proposed to designate in a neutral manner as *Macedo-Slavs.* They extended from the Bulgarian ethnographic frontier, already indicated, to the eastern banks of Lake Ohrid and inhabited the whole of the Lake Prespa region, Flórina, Kastoría, Edhessa and Yiannitsá in south-western Macedonia. In southern Macedonia they occupied the greater part of the plain of Kilkís and reached the Aegean coast at Salonika. In the north they gave way to the Serbs in the region of Ohrid, Kruševo, Prilep, Veles and Zletova. Cvijić went on to admit that even north of this line in the districts of Skoplje, Tetovo and Kratovo the Slavs might also be classed as *Macedo-Slav*. He included them with the Serbs because these districts formed part of the kernel of 'Old Serbia.' This being the case it must be pointed out that Cvijić's conception of 'Old Serbia' had varied greatly from time to time.

Although Cvijić postulated that until further research was undertaken the *Macedo-Slavs* would have to remain neutral, he did suggest a tentative division of the *Macedo-Slavs* into Serbs and Bulgarians by marking the limit of certain Serb characteristics in Macedonia, such as the use of the loud 'dj' and 'tj' (or 'g' and 'k') sounds, the *Slava* and Serb folk songs and melodies. This limit was extremely interesting. It indicated a tentative departure from the rather dogmatic statement issued by Cvijić in 1912 insomuch as it created a new line of partition between Serbs and Bulgarians, much farther south than the 1912 line or the line agreed upon in the Serbo-Bulgarian Secret Annex. The new line brought Strumica, Edhessa, Flórina, Bitolj and Veles within the Serb category and left the Bulgarians with Lake Dojran, the lower Vardar, Yiannitsá, Kastoría and the region round Lakes Ohrid and Prespa. There is little doubt that when the Serbians failed to gain possession of northern Albania, they asked Bulgaria for compensation in these areas delimited by

Cvijić (see p. 183). Cvijić's division of the *Macedo-Slavs* might be interpreted as a preparation for the Serbian claim on the Bitolj gap and a common frontier with Greece. Another puzzling feature of Cvijić's distribution arose from his inclusion of Kriva Palanka, Kratovo, Prilep and Kruševo as Serb territory, when in 1912 he had placed these districts in the 'Bulgarian sphere.'

*The Vlachs*

Cvijić divided the Vlachs into three sub-groups :

(1) the Romanians, *Aromunes* or *Kuzowalachen* of Orthodox faith ;
(2) Moslem *Aromunes* ;
(3) hellenized *Aromunes*.

Besides these groups which comprised for the most part nomadic shepherds, there were also Vlachs engaged in various crafts, who were known in Serbia as *Zinzares*. Altogether the Vlachs according to Cvijić scarcely numbered more than 150-160,000 although their numbers had been estimated in the past as high as half a million.

Cvijić based his distribution of Vlachs in south-western Macedonia on Weigand's map. He modified Weigand's classification, however, by referring to many of them as 'hellenized' Vlachs. Other groups of unhellenized Vlachs, not generally marked on ethnographic maps, were acknowledged by Cvijić in the Bitolj-Kruševo area, in the Plačkovica mountains of central Macedonia and in the Rodopi mountains. Cvijić's analysis of the Vlach distributions was of interest in 1913, because the Romanians themselves produced a number of ethnographic maps in that year, purporting to show the distribution of Romanians in Macedonia. Amongst the best known was that of C. Noé published in Bucarest. It made claims for the *Koutso-valaques* out of all proportions to the numbers estimated by Cvijić. The maps of Leon Lamouche and A. Rubin also made extensive claims for the Vlachs (Fig. 86). It would appear that the Romanians were stressing their claims in Macedonia as part of their anti-Bulgarian campaign over the issue of the Dobrudja.

*The Albanians*

Cvijić's Albanian ethnic group included two elements — the Albanians proper (Moslem, Orthodox and Catholic) and the *serbized*

*Albanians.* The latter group was very small indeed and was confined to an insignificant area of southern Montenegro. Cvijić felt bound, no doubt, to distinguish this group as a counterbalance to his *albanianized Serbs.* Apparently the process of 'albanianization' had been much more widespread than that of 'serbization.'

Cvijić maintained that (1) there never had been any Albanians in the Morava valley nor in the Novi Pazar district, although he admitted that there were occasional groups of Albanians in 'Old Serbia', between Pristina and Prizren; (2) that the two important plains of 'Old Serbia'—Metohia and Kosovo—were populated not by Albanians, as the authors of many earlier maps had erroneously supposed, but by a mixture of Serbs, *albanianized Serbs* and Albanians. He did not allow that the Albanians predominated even in northern Albania. Nor, according to his views, did they form any considerable part of the population of western Macedonia, and further, they were unrepresented in the Crni Drim valley south of Debar. In the south, he put the Albano-Greek ethnographic frontier along the watershed between the Devoll river and Lake Kastoría and along the Grámmos mountains. In Ipiros his boundary was not so clear cut, but south of Gjinokastër the Albanians on his map existed only as a minority along the coast. Kónitsa was a mixed area. Cvijić emphasized the difficulty of distinguishing hellenized Tosks from Greeks in this region. In comparison with his map of 1909, Albanian territory was considerably reduced on Cvijić's map, particularly in the north (*cf.* Figs. 40 & 42).

*Conclusion*

Cvijić's ethnographic map of 1913 was by no means the product of well balanced and impartial scholarship. Like many of the other ethnographic maps of the Balkans, its ideas were dictated both by the march of events and by the patriotic outlook of its author. Cvijić's map was designed to support Serbia's plan for a re-organization of the western Balkans, after the Turkish defeat, along the lines of:

(1) a union of Serbia and Montenegro;

(2) the expansion of Serbia into Old Serbia and Macedonia as far south as the Golešnica mountains, and if the possibility were to arise, as far south as Bitolj;

(3) the incorporation of northern Albania into the Serbian state, including Shkodër, Lesh, Tiranë, Durrës, Elbasen and the whole of the Drin valley and its tributaries.

Nicolas Pašić, the Serbian Premier, had already outlined Serbia's territorial aims in the Balkans in 1912 in the following words:

> Servia's minimum request for her national development is economic independence. [ . . . ] and a free and adequate passage to the Adriatic. It is essential that Servia should possess about 50 kilometres from Alessio to Durazzo. This coastline would be joined to what was formerly Old Servia approximately by the territory between a line from Durazzo to Ochrida Lake in the south and one from Alessio to Djakova (Dakovica) in the north.[1]

Plans were already projected for the construction of a railway from the Adriatic to the Danube, which would run through Niš to Durrës During 1913, therefore, Cvijić had modified his ethnographic ideas once again, in order to give support to these grander aims. His idea of the limits of the Serbian distributions had now changed four times. On each of the occasions, when he had been called upon to state the Serbian case—in 1906, in 1909, in 1911, in 1912 and in 1913—his interpretation of ethnographic facts had varied, but always in favour of the Serbs. As the authors of *Enquête dans les Balkans* put it: "Les notions ethnographiques de M. Tsviyets [Cvijić] varient [ . . . ] avec le développement de prétentions politiques serbes." Nor were they any less flattering in their allusions to Cvijić's idea of a *Macedo-Slav* ethnic group, which they referred to as "euphemisme destiné a dissimuler l'existence de Bulgares en Macédoine."[2]

It would be very difficult to say exactly how much influence Cvijić's map might have had on European ethnographic opinion, had not the War of 1914-18 intervened and invested his map with an importance it did not otherwise deserve. Consequent upon the French translation of Cvijić's geographical treatise on the human geography of the Balkans in 1918, and consequent upon the re-publication of modified versions of his ethnographic map in French and American

[1] *Review of Reviews* (London, 1912).

[2] *Enquête dans les Balkans, Rapport présenté aux Directeurs de la Dotation Carnegie* (Paris, 1914).

geographical journals in the same year, the popularity of his ethnographic map knew no bounds. Later, E. Stanford's map, the *Daily Telegraph* maps of A. Gross, the G.S.G.S. maps, the maps of the French War Office and the maps in Philip's atlases, in the Oxford atlases and in a host of others too numerous to mention, were all influenced by Cvijić's version of the ethnography of the Balkans and in particular by his idea of the *Macedo-Slavs*. But in 1913 there had been little to indicate that Cvijić's ideas were ever likely to prevail, any more than, for example, those of V. Kănčev or F. Meinhard or C. Nicolaides. The interesting process by which Cvijić's ideas of Macedonian ethnography became fashionable in western Europe must form the subject of another chapter.

## Consequences of the Creation of Albania

### *The Treaty of London, 1913*

The outcome of the Balkan Wars was further complicated by the intervention of the Great Powers on behalf of Albania. As far back as 1900, Austria and Italy had come to a secret understanding to create an autonomous Albania stretching from Ioánnina to Shkodër. But the Balkan nations by their concerted action early in 1913 threatened to eliminate the Albanians altogether from the political arena. In response to this situation, the interested powers decided to force the issue of an autonomous Albania, even to the extent of sending an international squadron to take over Shkodër, which had been captured by Montenegrin troops.

The Treaty of London, signed in May, 1913, laid down the boundaries of a new Albanian Principality. Its creation was not the work of the Albanians, who, although they had shown a few sparks of nationalism, were not yet capable of sustained and unified national action nor yet capable of ruling themselves. The new State owed its existence principally to Austrian and Italian diplomacy. The establishment of an autonomous Albania, stretching from the Prokletije mountains and Lake Shkodrs in the north, to the Grámmos mountains and the Straits of Kérkira in the south, headed off the Serbians from their ' open door ' and forced the Greeks out of Northern Ipiros (Fig. 44).

*A. Belić's Map of 1913 and the Second Balkan War*

The constitution of an autonomous Albania by international action sabotaged the Serbian plan for a partition of the western Balkans, as envisaged by J. Cvijić. As Austrian statesmen had perhaps foreseen, the miscarriage of Serbian plans in Albania had repercussions on Serbia's Macedonian policy. Serbian troops still occupied most of Macedonia, including Bitolj. If Serbia could consolidate her claim to Macedonia and thereby retain possession of a common boundary with Greece, her position as an 'imprisoned nation' would not be so desperate. A map reflecting Serbia's new designs compiled by A. Belić, a Serbian philologist, was published in 1913 on the eve of the Second Balkan War (Fig. 43).[1]

The map showed the dialects of Macedonia. A. Belić was then professor of Slavonic languages and he maintained that the influence and extent of the Serbian language in the western Balkans had hitherto been grossly underestimated[2]. He gave his support to the idea that there were two Slav linguistic provinces in Macedonia. 'Serbo-Macedonian' and 'Bulgaro-Macedonian.' The former stretched as far south as Edhessa and divided the Bulgaro-Macedonian province into two parts, one centred around Debar, Ohrid and Kastoría in the west and the other around the Struma valley and the lower Vardar in the east. The inclusion of Debar and Struga in the Bulgarian sphere was surprising for Cvijić had shown these towns to be in Serb ethnic territory. In order to drive a wedge through the Bulgarians towards Salonika it would appear that the Serbians were willing to make concessions here now that Albania was lost.

Bulgaria, however, refused to allow any modifications of the terms of the Secret Annex of the Serbo-Bulgarian treaty, even when Serbia protested that Serbian troops had been used to capture Edirne, that the Bulgarian gains in Thrace were more than adequate compensation for any adjustment that might take place in Macedonia, that the formation of an independent Albania had robbed Serbia of a port and that therefore compensation in Macedonia was in order. But Bulgaria demurred. The Macedonian settlement was further compli-

[1] See evidence in *Enquête dans les Balkans.*

[2] Dr. Belić is now President of the Serbian Academy of Science.

cated by the Greek refusal to allow Bulgarian claims in Greek-occupied Macedonia. Joint Serbo-Bulgarian action might have forced Greece to yield, but Bulgarian and Serbian claims in Macedonia overlapped to such an extent that no agreement could be reached. Thus whereas Bulgarian and Greek claims were in conflict, Serbian and Greek

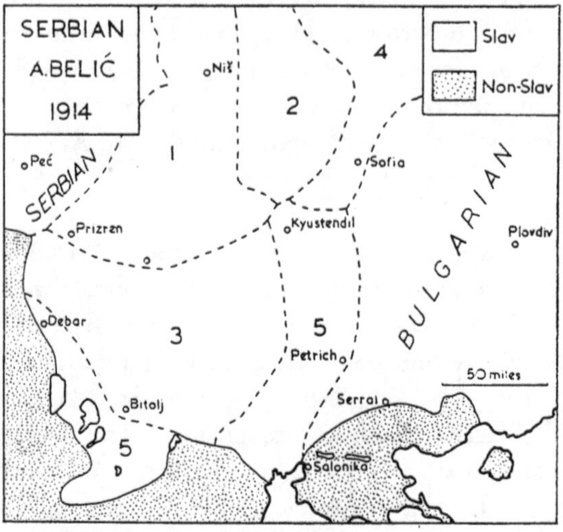


FIG. 43. DIALECTS OF MACEDONIA IN 1913

The numbers on the map refer to the following dialect-zones: **1.** Prizren; **2.** Timok; **3.** Serbo-Macedonian; **4.** Sphere of influence of Serb over Bulgarian; **5.** Sphere of influence of Serb over Bulgaro-Macedonian.

claims were complementary and consequently Greece and Serbia tended to combine against Bulgaria.

The Bulgarian military clique eventually took matters in their own hands and declared war on Serbia and Greece. The result of the Second Balkan War or the 'War of Partition' settled the new political boundaries. Bulgaria was crushed by a combination of powers, including not only Serbia and Greece but also Romania. The Turks also seized the opportunity of re-opening the war in Thrace.

## THE TREATY OF BUCAREST, 1913

By engaging in the Second Balkan War, Bulgaria lost all prospect of acquiring territory in central Macedonia and also the chance of a port at Kaválla (Fig. 44). The Treaty of Bucarest restricted Bulgaria's total gains to the Aegean coastal districts of western Thrace including Alexandroúpolis (Dedeagatch) and in Macedonia, to the Strumitsa valley and the districts of Petrich, Melnik and Nevrokop. To Romania, Bulgaria lost a strip of Dobrudja, and Edirne, occupied in the First Balkan War, was lost again to Turkey. Serbia emerged from the wars with immense prestige and considerable territorial gains including Novi Pazar, Old Serbia and central Macedonia. Montenegro obtained part of the Novi Pazar corridor and a common boundary with Serbia. Greece benefited by southern Ipiros and southern Macedonia, including Salonika and Kaválla.

### *Consequences of the Treaty*

The new boundaries created numerous minorities which played their part in provoking friction between the Balkan nations for the next thirty-five years. In spite of their gains in Thrace and eastern Macedonia the Bulgarians felt cheated of their rights in central and southern Macedonia. They determined on a revisionist policy, which ultimately threw them into the camp of the Central Powers. From the time they occupied the area, the Bulgarians utilized the Petrich district as a forward base for the re-conquest of Macedonia. A state of undeclared war existed there until 1915 and again after 1918, whilst between 1915 and 1918 the Bulgarians directed their main war effort in this theatre. The Serbians, who had hardly figured on ethnographic maps of Macedonia until after 1885, incorporated into their then small state the large alien populations of Old Serbia which had always been regarded by everyone except the Serbians as mainly Albanian, and of northern and central Macedonia which for so long had been regarded as Bulgarian. The possession of these territories gave Serbia the chance of closing all Bulgarian schools and inaugurating a policy of serbization amongst the *Macedo-Slavs*, a policy which proved a signal failure. The presence on Serbian soil of a population which refused to be absorbed, and which looked across the border for help against Serbian centralization, poisoned Serbo-Bulgarian relations and was a perpetual source of embarrassment to the Serbian Government.

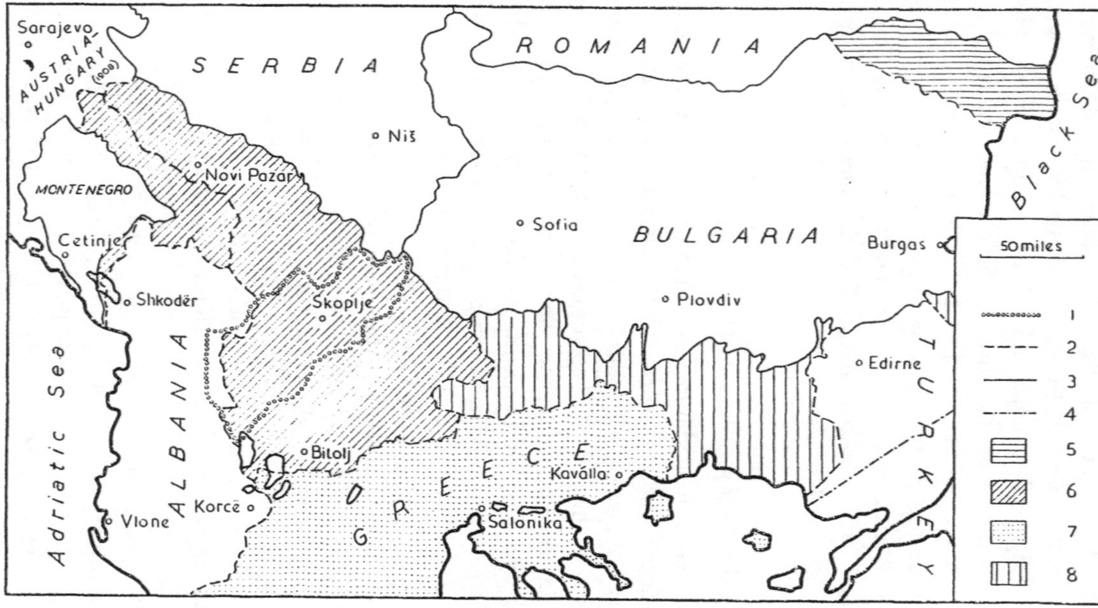


FIG. 44. BOUNDARY CHANGES, 1912–13

The references in the key are as follows : **1.** Limits of territory disputed by Serbia and Bulgaria in the Secret Annex of their Treaty of 1912 ; **2.** New boundaries under the terms of the Treaty of Bucarest (1913) and of the Conference of London (1913) ; **3.** Boundaries in 1912 ; **4.** Bulgaro-Turkish boundary under the terms of the Treaty of London (1913) ; **5.** Romanian gains as a result of the Balkan Wars ; **6.** Serbian and Montenegrin gains ; **7.** Greek gains ; **8.** Bulgarian gains.

Greece had also accepted the responsibility of a large alien population, which was even more difficult to administer because it was so mixed. Even C. Nicolaides had admitted the presence of large minorities of Turks, Slavs and Vlachs in southern Macedonia. According to all non-Greek maps the Greeks were poorly represented. Their communities were scattered and only occasionally did they constitute the rural population. The mixture of the population in southern Macedonia was itself an indication of the extraordinary function of the region as a corridor route from central Europe to the Straits, as a coastal route from the Adriatic to the Black sea, and as an outlet for the interior of the Balkans to the Mediterranean. By annexing this region, Greece, hitherto exclusively an insular and maritime power, took on the additional role of a continental power. A great many of the problems which were to confront Greece in the future arose from this dual function as a land and a sea power, particularly as her ethnic position on the mainland was not as strong as the Greeks themselves had represented it to be. Her northern boundary passed through territory populated almost entirely by *Macedo-Slavs* with Bulgarian affinities, and by Turks. There appeared to be no justification for the inclusion of the Mesta valley, Kaválla, Dráma and part of the Rodopi range within the Greek boundary in the east. There were only a few Greeks in this area and their incorporation meant robbing Bulgaria of the only reasonable port on this part of the coast, extending the boundary into hill country controlled by the Bulgarians and Pomaks, and precluding any possibility of peaceful relations with a disgruntled Bulgaria.

The boundaries laid down by the Treaty of Bucarest were largely dictated by Greece and Serbia. The Great Powers, for the first time since the Eastern Question began to dominate European diplomacy, had been onlookers instead of active participants in the Balkan drama. Austria and Germany witnessed Serbia's success with feelings of alarm, for Serbia's common boundary with Montenegro meant that the Serbians now controlled the gateway to Salonika. The time had come when Austro-German designs in the Balkans could be realized only by taking direct action against Serbia with the object of nullifying the terms of the Treaty of Bucarest.

## Chapter IX

# MAPS OF THE WAR YEARS, 1914-1918

Although the causes of the outbreak of war in 1914 were extraordinarily complex, the struggle had initially a purely local significance for the peoples of the Balkans. For them, the conflict was the inevitable reaction of Austria to the aggrandizement of Serbia under the provisions of the Treaty of Bucarest of 1913. The union of Serbia and Montenegro, which had been achieved, had dealt a death blow to the Austrian *Drang nach Osten*, directed towards Salonika. Moreover the very success of Serbia had infected the Slavs of the Habsburg-controlled Slovenia, Croatia, Bosnia, and Hercegovina, with a restless urge for national freedom which constituted a grave threat to the realization of Germanic supremacy in the western Balkans.

The war between Austria and Serbia could not remain localized. The Balkan theatre became but a part of the general world conflagration. Inevitably the enormous strategic importance of the peninsula led to the diplomatic and military intervention of the Great Powers and to the ultimate clash of German, Russian, Turkish, British and French forces within the region. With the whole of the Balkan balance of power upset, the temptation to fish in troubled waters eventually proved too great for the remaining neutral states. Bulgarian politicians discerned in the conflict an opening for achieving a settlement of outstanding claims in Macedonia and Thrace. In Greece, the school of thought which had for so long cultivated the 'Grand Idea,' the revival of the Byzantine Empire, saw in the war the means of implementing its cherished scheme. Moreover the Serbians, although their country had been overrun, seized the opportunity offered by their alliance with the West, of working for a 'Jugoslavia' to be erected on the basis of a 'Greater Serbia.' Italian statesmen also endeavoured, under cover of the confusion of war, to widen the Italian sphere of influence in Albania. It is only against this background of conflicting aspirations, that the ethnographic maps which appeared between 1914 and 1918 may be fully appreciated. During this period ethnographic ideas were changing constantly. They varied frequently with the fortunes of war, and ethnographic distributions put forward by one authority early in the war, were often as not, modified, if not repudiated altogether, by the same authority before the fighting was done.

It would be impossible to discuss every ethnographic map that appeared during these years, but the general trends of ethnographic thought may be followed by dealing with representative examples of ethnographic maps produced both by the Great Powers themselves and by the Balkan nations. Accordingly, samples of Greek, Bulgarian, Serbian, Swiss, American, Italian, French and British maps are considered in turn.

## Greek Maps

### *C. Nicolaides' Map of 1914*

C. Nicolaides, who had already produced an ethnographic map of Macedonia in 1899, undertook in 1914 to justify the Greek annexation of southern Macedonia which had taken place the previous year. His published work, describing the part played by Greece in the Balkan wars, included an ethnographic map which incorporated several modifications of his earlier work. The map was not confined to Macedonia; significantly, it also covered Thrace and part of Asia Minor. An important modification contained in Nicolaides' later map concerned the Greek ethnographic frontier in the vicinity of Lakes Ohrid and Prespa. On his map of 1899 this frontier had been shown along the Devoll river and thence along the water-shed between the two lakes. On his map of 1913, however, it was pushed north, to include the whole of the upper Devoll valley, as well as all the territory between the lakes which had been occupied by the Greeks during the Balkan wars. The whole of Thrace, not included in his earlier map, he marked

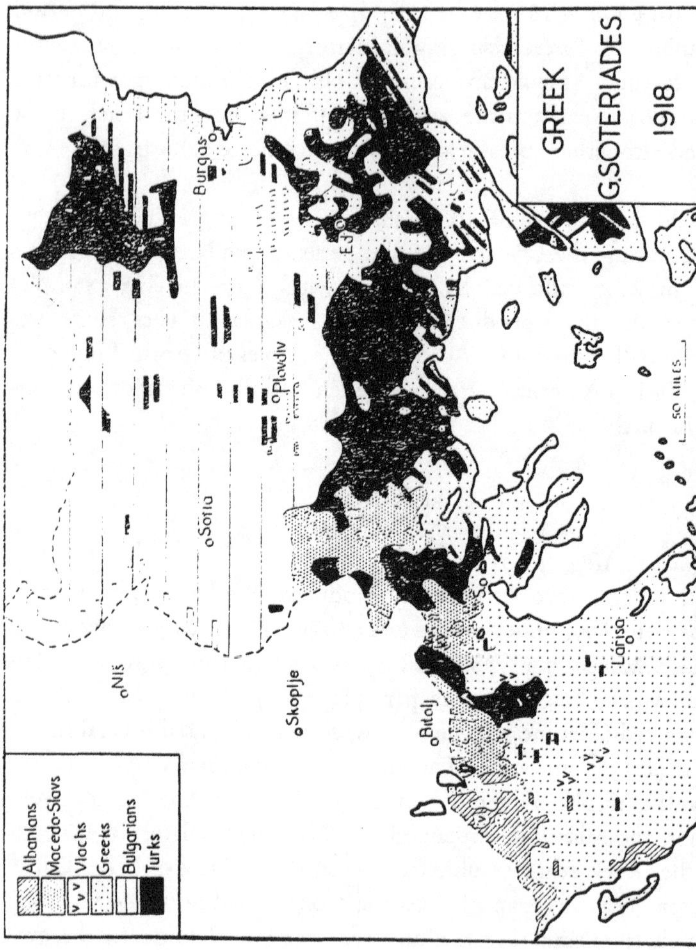


FIG. 45

In the original, the Turkish category is referred to as *Mohomedan* and includes Pomaks. The Vlach category is referred to as Romanian.

as Graeco-Turkish. He estimated the population of the *Vilayet* of Adrianople (Edirne), which covered the greater part of Thrace, to be 1,031,123, including 503,311 Turks, 396,963 Greeks but only 94,843 Bulgarians. Northern Ipiros, also included in his map of 1913, was depicted as Greek. Nicholaides gave his figures for the *Vilayet* of Jannina (Ioánnina), which incorporated most of Ipiros, as 316,651 Greeks, 154,413, Turks and 759 Vlachs. It must be borne in mind that when Nicolaides' map appeared, a commission appointed by the London Ambassadorial Conference was still debating the Graeco-Albanian boundary in its delimitation from Korcë to the coast of the Adriatic.

As was his earlier map, Nicolaides' map of 1914 was little more than a production coloured to suit the mood and purpose of the Greek expansionists. Southern Macedonia had been acquired, and thus a way cleared for the building up of Greek claims in Northern Ipiros and Thrace. The acquisition of those regions was the logical sequence to the Greek successes in 1912-13 and the next step in the realization of the 'Grand Idea' (Fig. 88).

*G. Soteriades' Map of 1918*

On the outbreak of war in 1914, the Greeks were divided in their attitude towards the respective belligerents. As early as December, 1914, the Entente offered southern Albania to Greece, for active participation in the Allied cause. In the January they increased their offer by including Smyrna—the important port on the Anatolian coast largely inhabited by Greeks; later the whole *Vilayet* of Aidin was included in their offers to Greece. The *Vilayet* incorporated most of the western coast of Anatolia. But all these promises failed to induce the Greeks unanimously to enter the war. Allied failure was partly due to confused diplomacy because compromising offers were also being made at the same time, of southern Albania to Italy, and of Thrace to Bulgaria. Moreover, the early successes of the Central Powers in the northern Balkans, combined with the disastrous end to the Gallipoli expedition, predisposed one section of the Greeks, the 'King's party,' towards co-operation with the Central Powers. On the other hand, Allied Mediterranean maritime supremacy and the presence of Allied troops in Salonika, encouraged the hopes of those

Greeks who looked to the Allies to promote Greek schemes on the mainland. This party, led by Venizelos, maintained that Greece could hope for no sympathy from Austria or Turkey and that the future of Hellenism lay in entry into the war on the side of the Allies. For a time the government of Greece was actually divided between Venizelos, in control of Greek Macedonia, and the King, in control of Athens and the archipelago, but ultimately Venizelos' party triumphed, and a Greek army fought with the French and British in the final Macedonian campaign. Greek views at the end of the war were well illustrated by Professor G. Soteriades' map (Fig. 45). He was a professor of history in the University of Athens and his map appeared in a pamphlet published by E. Stanford. It was written, stated the professor, to offset the 'inaccuracies' contained in the map issued in 1918 under the auspices of *The Daily Telegraph* (see p. 226). The work was limited to a consideration of ethnographic distributions in Greece, Ipiros, Bulgaria and Turkey. Serbian Macedonia was excluded from consideration. Soteriades recognized six nationalities—*Mohomedans* (Turks and Pomaks), Greeks, Bulgarians, *Macedo-Slavs*, Albanians and Romanians.

*The Mohomedans.* The most noticeable feature of this Greek map was the radical departure from the distributions favoured by C. Nicolaides. Southern Albania, southern Macedonia and Thrace were not shown as predominantly Greek but Slavs and Turks were acknowledged to be the chief inhabitants. The Turks in particular were given a very wide distribution. Their importance was accentuated by the inclusion of Slav Moslems in the Turkish national or *Mohomedan* group. The portrayal of the whole of the Mesta valley, the Rodopi mountains and the upper Maritsa as *Mohomedan* or Turk, rather than Bulgarian, gave these regions an overwhelmingly Turkish character. To this extent the map was reminiscent of that of G. Amadori-Virgilj.

*The Greeks.* Soteriades gave the distribution of the Greeks in a fair amount of detail. He did not depict them as a majority in Macedonia, except in Khalkidhikí, but he indicated important Greek exclaves around Korcë, Edhessa, Sérrai and Dráma. The three following categories he regarded as Greek nationals :

(1) the Vlachs of Thessalía—because " they expressed a preference for the Greek nationality " (only the *Koutzo-Vlachs* who associated themselves with the Romanians were coloured separately. They formed, stated Soteriades, only a very minor element in the population of Macedonia because most Vlachs looked on themselves as Greek) ;

(2) the Orthodox Slavs, still faithful to the Patriarch—he maintained that only the Exarchist Slavs looked upon themselves as belonging to the various Slav national groups ;

(3) the Albanians of Evvoia, Atticí, Kórinthos, Argolís and Lakonía—because they had always looked upon themselves as Greek.

*The Slavs.* Soteriades referred to the Slavs west of the Mesta river as *Macedo-Slavs*, and he depicted them in three groups as follows, each group separated by zones of Moslem territory :

(1) north of Kastoría extending to the Jugoslav border ;

(2) north of Edhessa and Salonika to the Jugoslav border ;

(3) north of Sérrai and Dráma to the Bulgarian border.

The Bulgarians he limited to Bulgarian territory north of the Rodopi mountains and to the upper Maritsa, with the exception of a few exclaves in Thrace.

*The Albanians.* He confined the Albanians in Northern Ipiros to the coast of the Kérkira (Corfu) channel. Otherwise, he maintained they were only to be found north of a line from Vlonë to Tepelenë and Korcë (for Greek aspirations in Albania at this time see Fig. 46).

*Conclusion.* Soteriades' map may be said to be almost an official Greek view of Balkan ethnography, put forward at a time when the war was going well for the Allies (Bulgaria had just asked for an armistice), and it was intended to support Greek claims in Albania and Thrace. Unlike the Jugoslavs, the Greeks had not been engaging in propaganda in European capitals throughout the war, and, until Greece definitely entered the war in 1917, Greek policy in the Balkans had been suspect by the Allies. In the interim, the Jugoslav interpretation of the ethnography of Macedonia had become fairly well established, and so Soteriades had to recognize *Macedo-Slavs* where Nicolaides had marked only Greeks. Viewing the map as a whole its Graeco-Turkish character was rather remarkable. The impression is gained from this map that Thrace, the Black sea coast, and the

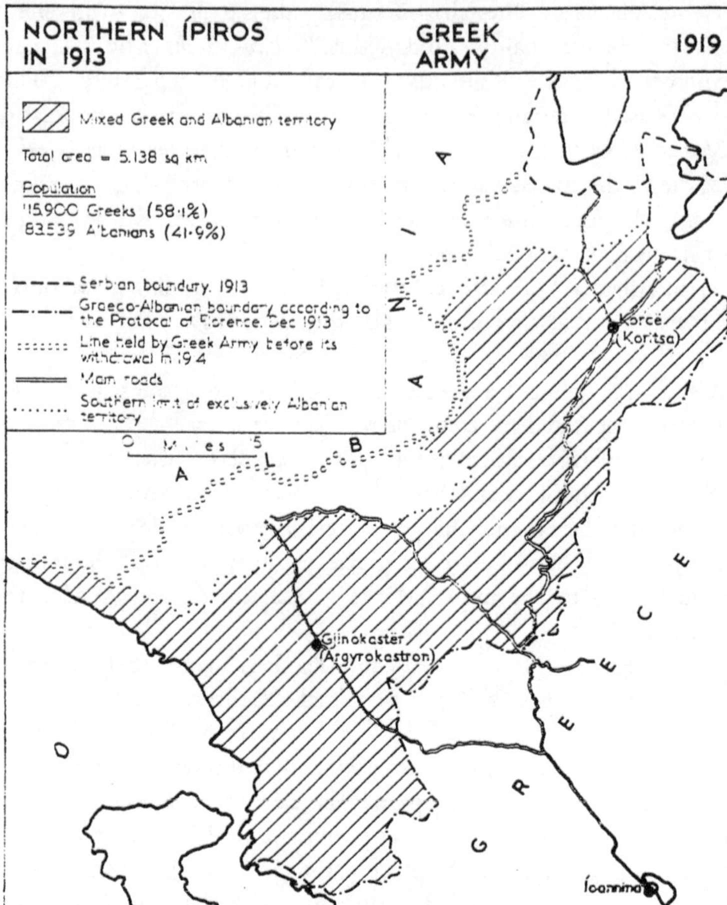


FIG. 46

In addition to the above information, the original incorporated a detailed representation of Albanian and Greek elements in each town and village. The representation was by means of symbols but the scale of the map did not allow of reproduction.

coasts of Marmara and Anatolia were peopled by a mixed population, predominantly Turkish and Greek. One was obviously expected to infer that the Greeks were in a position to supersede the Turks as rulers of the southern Balkans, were the latter compelled to forfeit their control of the Straits as the penalty for aiding the Central Powers.

## Bulgarian Maps

*D. A. Iširkov's Map of 1915*

In 1914 there was a school of thought in Britain—supported by Lloyd George and Winston Churchill—which believed that Bulgaria could be won over to the cause of the Entente in the Balkans. The difficulties met by Allied diplomacy in this theatre arose from the fact that promises of territorial concessions to Bulgaria could be made only at the expense of Greece, Serbia or Turkey. Venizelos stated early in 1915 that he was ready to sacrifice Kavälla to Bulgaria in order to arrive at a Graeco-Bulgarian understanding, and in Thrace, the Allies offered to the Bulgarians the 'Enos-Midia line' (Enez-Midye) in return for their military aid. Macedonia, however, remained the crux of the problem, because the Serbians were not ready to make wholesale concessions there of the territory they had so recently acquired. Russia urged Pašić, the Serbian foreign minister in 1915, to make over Macedonia to Bulgaria, and furthermore, as a result of the British mission to Bulgaria in 1915 headed by Noel Buxton, the Allies themselves brought pressure upon the Serbians, to come to an understanding with Bulgaria over Macedonia. Turkey's entry into the war precipitated the Allied offer of the '1912 line' to Bulgaria, an offer approved, when it was too late by the Serbians (for details of the '1912 line' see p. 170). The eclipse of Russia in Galicia in the summer of 1915, and British failures against Turkey in the same year, undermined Allied influence with King Ferdinand of Bulgaria. Finally, the more tempting offer of the whole of Serbian Macedonia by the Central Powers, induced the King of Bulgaria to declare war against Serbia in September, 1915.

Professor D. A. Iširkov's map appeared in *Petermann's Mittheilungen* late in 1915, when Bulgaria had already decided on war with Serbia.[1] The Macedonian section of the map was prepared by the professor of geography in the University of Sofia—J. Ivanov. Many other

[1] 'Ischirkoff' is the German transliteration of his name.

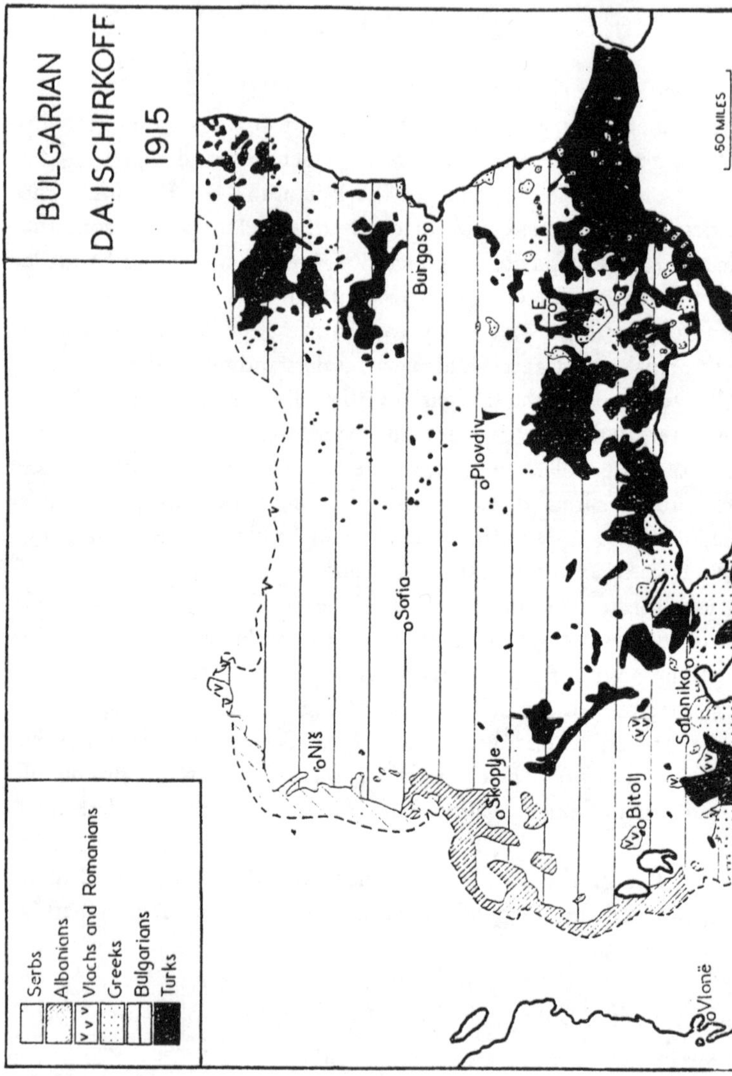


FIG. 47

Tatars, included in the Turkish category above, are separately distinguished in the original.

Bulgarian professors, including philologists and historians, were engaged in the compilation of the map. It purported to show the extent of *Das Bulgarentum*, i.e. Bulgarian ethnographic territory, in the Balkan peninsula in 1912 (Fig. 47). The map was based upon a large-scale survey in which 1,200,000 plans had been used, and it incorporated the results of statistical surveys made by the Bulgarians in 1905 and 1910. Other sources to which reference was made, were the manuscript evidence of the monk *Païssi* (see p. 199), and the descriptions of travellers in the Balkans dating from the fifteenth century and including those of Pouqueville, Grisebach, Boué, Wikenty, Makuscheff and von Hahn. Russian official publications inspired by the Russo-Turkish wars and the personal surveys of the Russians, Eneholm, Heine and Obrutscheff, were also utilized. Furthermore, the evidence in the maps of Boué, Lejean, the Misses Mackenzie and Irby, Mirković and Petermann was stressed. Iširkov claimed that these maps were particularly valuable because their authors had been completely disinterested in politics, and no reference was found in their maps to the *Bulgarophones* and the *Kulturlos* Slavs (according to Iširkov, inventions of the Greeks and Serbians respectively.) He was chiefly concerned in showing the distribution of Bulgarians, but he also recognized Turks, Greeks, Serbs, Albanians and *Kutzo-Vlachs* as national groups.

*The Turks.* When considering the distribution of the Turks it must be remembered that Turkey was operating with the Central Powers of which Bulgaria was a potential ally. On Iširkov's map the Turks were represented very strongly in Thrace, largely at the expense of the Greeks. It is interesting to contrast the distribution of Turks on this map with that on the map of Soteriades. Iširkov classed the Pomaks as Bulgarian. The Turks were not given such a widespread distribution in Macedonia as they had been given by the Bulgarian, Kănčev, in 1900.

*The Greeks.* Iširkov depicted the Greeks in Thrace as a very scattered population indeed. Dráma, Sérria, Salonika and Kastoría, he marked, as the extreme northern limits of the Greeks in Macedonia.

*The Slavs.* Iširkov distinguished neither Serbs nor *Macedo-Slavs* in Macedonia. He represented the Bulgarians as forming the bulk of the population, both of the Niš region and of Macedonia, including the Crni Drim valley and the coast of the Gulf of Salonika. The claims he made for the Bulgarians had hardly been exceeded on any one map before ; even the Russian map of 1867 had portrayed the Crni Drim as Albanian, G. Lejean's map had acknowledged the existence of Serbs around Lake Ohrid and H. Kiepert's map had shown Serbs in the Niš area. Indeed few maps since 1878, not even Russian, had depicted the Niš area as Bulgarian, although that area had been acknowledged as Bulgarian on earlier maps (see p. 103). However, Professor B. Zonev, who had co-operated in the production of Iširkov's map, insisted that the *Torlak* dialect, spoken throughout the upper Morava valley in Niš, Leskovac, Prisren, Vranje and Pristina, was a branch of Bulgarian. Seven out of ten of its distinctive features were Bulgarian and only three were Serb. Bulgarian characteristics included the use of the post-positive article, the use of certain declensions and forms of infinitive, the lack of quantitatives, etc. Pristina and Prizren, however, whilst they belonged to the *Torlak* dialectal province, Iširkov excluded from *das Bulgarentum* because of the overriding considerations of physical geography ; he believed the Sar mountains and the Crna hills ought to be respected as ethno-geographical boundaries.

*The Vlachs and the Albanians.* Iširkov was not primarily concerned with depicting the distribution of Vlachs and Albanians, but it is of interest to note that he acknowledged no Vlach minorities at all in Bulgaria. Also, he excluded the Albanians from the Crni Drim valley between Lake Ohrid and Debar, and from the districts immediately west of Lake Ohrid. But he did show large exclaves of Albanians in north-western Macedonia.

*Conclusion.* As in the case of J. Cvijić and G. Soteriades, Iširkov's views on the ethnography of the Balkan peninsula were put forward not with the idea of ultimately achieving a balanced view of the situation but with the motive of bolstering the plans entertained by Bulgarian politicians for the future of their state. The appearance of Iširkov's map made the gap between Serbian and Bulgarian interpretations wider than ever before. Moreover, the rival map-compilers were already working, in 1915, in closed academic fields, since Serbian propaganda was confined to America and western Europe, whilst Bulgarian ethnographers were working for publication in Germany and central Europe. When the time came to apply the principles of self-determination to the solution of boundary problems, the cleavage between Serbian, Greek and Bulgarian ideas had grown wider than ever, and the possibility of solving the controversy correspondingly more difficult.

*J. Ivanov's Maps*

Of the many Bulgarians who interested themselves in ethnic distributions, by far the most prolific in his output was J. Ivanov, professor

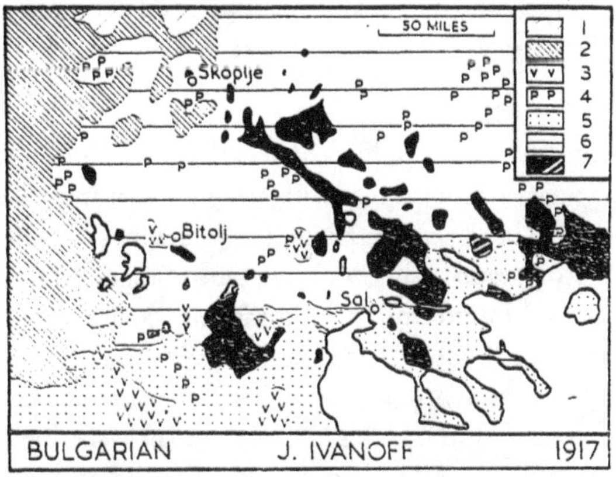


FIG. 48

The references in the key are as follows: **1. Serbs**; **2. Albanians**; **3. Vlachs**; **4. Pomaks**; **5. Greeks**; **6. Bulgarians**; **7. Turks** and **Gagauzi** (Christian Turks, indicated above by white diagonals).

in the University of Sofia, geographer, historian and philologist, and as indefatigable a worker for the Bulgarian cause as was J. Cvijić for the Serbian. One of his first published works on Macedonia had appeared in 1906, when the Macedonian problem was uppermost in everybody's mind. From that time on he produced about one work a year on some aspect of Bulgarian philology, history, antiquity or ethnography. His most celebrated publication was the history of the Slavs according to the monk Païssi, written in 1762 and discovered by Ivanov himself in one of the monasteries of Mount Athos in 1914. Païssi's many references to the Bulgarian character of Macedonia came too late to have very much influence on the solution of the

Macedonian problem, as the Balkan Wars had already settled the boundaries. Ivanov's major work on the Bulgarians in Macedonia appeared in Sofia in 1917, but it was not translated into French until 1919. His thesis was elaborately documented and contained an ethnographic map of Macedonia (Fig. 48).

His definition of Macedonia was purposely left vague but the rough limits to be inferred from his map were the Sar mountains and the Crna hills, the Mesta river, the Drin valley and the Lakes Ohrid and Prespa. His distributions were the same as those he had used in 1915 for Iširkov's map, but he further distinguished religious communities such as the Pomaks and Christian Turks.

### *D. Rizov's Atlas, 1917*

In 1917, the Bulgarians published an atlas in Berlin, usually referred to as Rizov's atlas, although that Bulgarian foreign minister only wrote the foreword. The atlas contained forty maps illustrating aspects of Bulgarian history and ethnography, more especially maps of the various mediaeval kingdoms of Bulgaria and facsimile copies of many ethnographic maps favourable to the Bulgarian cause. These included most of the pre-1878 maps, and the later Russian and Bulgarian maps. In 1917, Bulgaria was in occupation of most of Macedonia and part of eastern Serbia, and the Allies had not experienced any military success in that theatre. Rizov's atlas was part of the preparation made by the Bulgarians for the resettlement of the Balkans along lines to be dictated by the Kaiser. Naturally the Bulgarian point of view did not receive any publicity in the Allied press at this time, but I. Bowman later drew attention to the claims of the Bulgarian revisionists in *The New World* (1921). The contents of the atlas made it clear that Bulgaria hoped to gain, not only Macedonia, but also the Niš-Leskovac region from Serbia, and Kavália and Salonika from Greece, were she successful in the war.

## Serbian Maps

### *The Map of St. Stanojević and D. J. Derocco, 1915*

From the time that Austria declared war on Serbia, the Serbians perceived that the future of their State in the Balkans depended not only on an Allied victory but on the subsequent formation of a 'Greater Serbia' when the war was won. Hence the Serbians' anxiety, not only to retain their Macedonian possessions, but if possible, to achieve unification of all the Serbo-Croat peoples of the Balkans into a single federation which would be powerful enough to preserve its independence by virtue of its own strength and extent. Two Serbians, St. Stanojević and D. J. Derocco, inspired by this idea, produced a map showing 'l'extension ethnique de la nation Serbo-Croat-Slovene,' in 1915. A French version was published in Belgrade and Niš, and it indicated a Serb population in the greater part of Macedonia, as far south indeed as Yiannitsa, including as Serbs many Macedonians formerly acknowledged by J. Cvijić to have Bulgarian affinities.

### *N. Zupanić's Map of 1915*

Consequent upon the occupation of Serbia, propaganda for a 'Greater Serbia' was carried on by exiled Serbians in the capitals of the Allied states. The Croats and the Slovenes, of course, were technically enemies of the Allies at that time, but numbers of them were working for the downfall of Austria, either in their homeland or with the Allies. The dissension which marred the relationships of the Serbo-Croat groups in their homeland was overcome by these exiles, who formed the 'Jugoslav Committee' to work for the union of the Serbs, Croats and Slovenes. This Committee sponsored a map, compiled by Zupanić and published in London in 1915, and in Paris in 1916. Later it was incorporated in A. H. E. Taylor's book *The Future of the Southern Slavs,* 1917. The map outlined the extent of Jugoslav territory in the Balkans. It depicted all Serbian Macedonia, within its boundaries of 1913, as inhabited completely by Serbs. Prominently marked on the map were many places famous for their associations with the Serbs in the mediaeval period. No reference was made to any minorities in this region. Taylor explained in a note on the map, that apart from the omission of the Albanian element in Old Serbia, and a *Macedo-Slav* element in the south, the map was 'substantially accurate."

*J. Cvijić's Map of 1918*

Undoubtedly by far the most important map which the Serbians produced during the war was J. Cvijić's revised ethnographic map in 1918. This map, published originally in Cvijić's *La Péninsule Balkanique : Géographie Humaine*, contained significant modifications of the distributions shown on his map of 1913. These changes were as follows :

(1) the Vidin area, depicted as Bulgarian in 1913, was indicated as Serb in 1918 ;

(2) the whole of western Bulgaria from Vratsa to Kyustendil had been indicated as Bulgarian in 1913, but was shown as mixed Serbo-Bulgarian in 1918 (Fig. 82) ;

(3) in 1913, the eastern limit of the *Macedo-Slav* group had been fixed on the Vardar-Struma watershed, but in 1918 this limit was moved eastwards to the Struma river itself (Fig. 84).

Civijić never adequately explained these modifications. It would seem that in 1913 he had regarded the Bulgarian boundary as inviolable, but by 1918 expansion of Serbia at the cost of Bulgaria had, in his estimation, become feasible. Cvijić himself had always been an advocate of ethno-political boundaries, so he appears to have modified his ethnic dispositions, in an endeavour to prepare the way for Serbian claims on Bulgaria. By such a manoeuvre, the Serbian demands for a strategic Serbo-Bulgarian boundary and, in particular, for the incorporation of the Strumica salient within Serbia, were given support.

By 1918, Cvijić's influence in western Europe and the United States was enormous. His work on the human geography of the Balkans, closely modelled on the French school of possibilist thought, was hailed as a masterpiece, and the fact that his ethnographic map formed part of the classic, lent it an air of infallibility.[1] It was introduced into *Les Annales de Géographie* in 1918, accompanied by an article by L. Gallois, who reproduced Cvijić's views practically verbatim, and it was also reproduced in the American *Geographical Review* of the same year, together with an account of the distribution of the peoples of the Balkans written by Cvijić himself. Cvijić's map must go down

[1] There is a critical review of the book, by Miss M. I. Newbigin, in *The Geographical Journal* (1919).

in history as a powerful factor in the emergence, after the war, of a Serbo-Croat-Slovene national state which included also the greater part of Macedonia. In 1913, the partition of Macedonia had been regarded as having been brought about by a *tour de force*, and the expansion of the Serbians and Greeks had been condemned even in western Europe as the usurpation of the moral rights of the Bulgarians in this part of the Balkans. But by 1918 public opinion in western Europe had been moulded by Cvijić to such an extent that the right of the Serbians to retain Macedonia was taken for granted. Indeed Cvijić's impartiality was hardly questioned inside Allied circles and even the Greeks, faced with the irrefutability of his thesis, themselves modified their own ethnic ideas. Compare, for example, the maps of Nicolaides and Soteriades (Figs. 29 & 45). Cvijić was awarded the *Patron's Gold Medal of the Royal Geographical Society* largely in recognition of his work on the Balkans. A well-known English political geographer has referred to him recently, in 1945, as " the most learned and enlightened, not only of Serbian, but of all the Balkan geographical experts."[1] In the words used in another recent article on Jugoslav geographers : " He [Cvijić] played an important part at the Peace Conference and his wide scientific knowledge was the determining factor in solving many problems related to state frontiers."[2]

Cvijić's map influenced practically all the British maps published in 1918 (see p. 226). Ultimately his concept of the *Macedo-Slavs*, repudiated by the Balkan committee of experts in 1914, was almost universally accepted and became a feature of nearly all ethnographic maps of the 1914-18 post-war years.

## The Map of the Union of Nationalities, 1918

The Union of Nationalities was an organization founded by a Lithuanian, J. Gabrys and a Frenchman, J. Pelissier, in 1911. Its object was to ventilate national minority grievances and to secure by united action the rights of minorities all over Europe. The first

[1] A. E. Moodie, *The Italo-Jugoslav Boundary* (1945).

[2] " The contribution of the Yugoslavs to Geography and Ethnography " by Mark Clement, *Scottish Geographical Magazine*, Vol. 58, No. 3 (1942).

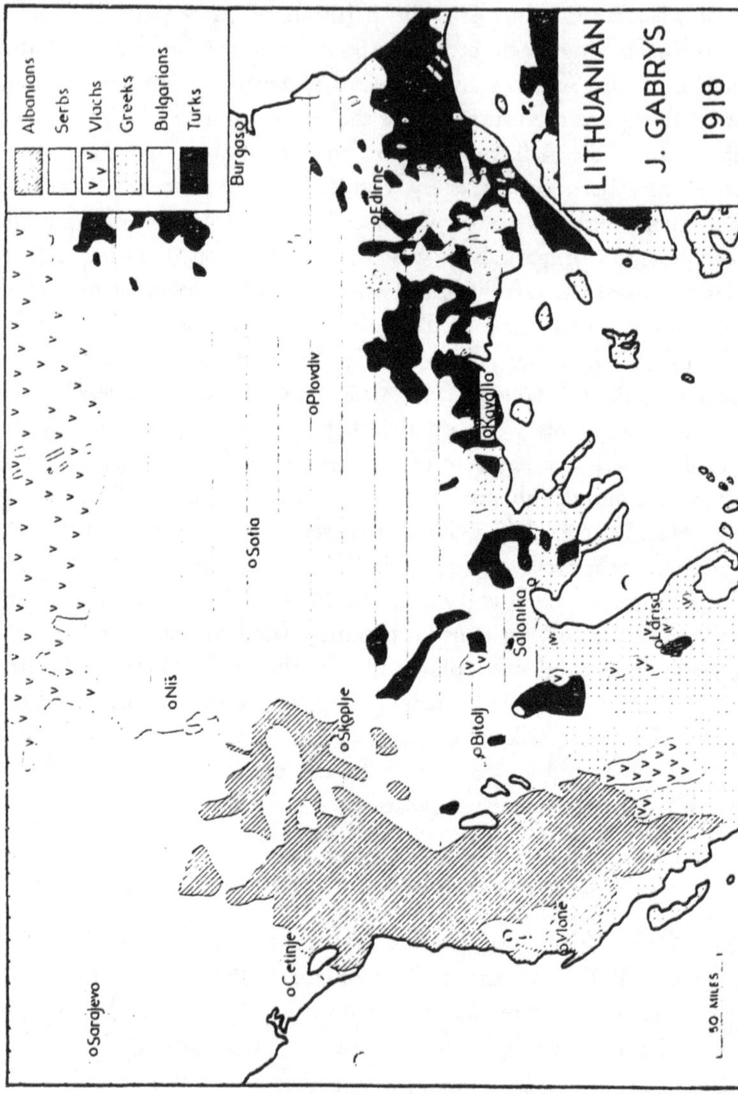
Albanians
Serbs
Vlachs
Greeks
Bulgarians
Turks
LITHUANIAN
J. GABRYS
1918
Burgas
Edirne
Plovdiv
Sofia
Kavalla
Salonika
Larisa
Niš
Skoplje
Bitolj
Sarajevo
Cetinje
Vlone
50 MILES

FIG. 49

meetings of the organization were held in Paris but when the war began in 1914 the headquarters were moved to Lausanne so that representatives from diverse national groups could continue to meet on neutral territory. The activities of the organization were suspect both by the Allies and by the Central Powers. Under the auspices of the *Union of Nationalities* a great deal of literature was published concerning the claims of various minority groups throughout Europe. The Macedonian Bulgarians were amongst minorities represented at the various conferences held in Switzerland during the war. In 1918 *La Librairie Central de Nationalités* published an ethnographic map of Europe compiled by the secretary-general of the Union, J. Gabrys (Fig. 49). The work was dedicated to President Woodrow Wilson. The sources used by Gabrys made up a useful bibliography running to twenty-one pages. He was certainly well informed on European ethnography.

*Gabrys' Distributions*

His distribution of Turks was taken from Iširkov's map, as were also his distribution of Greeks along the Aegean littoral. In Northern Ipiros he represented the Greeks as the majority population along the coasts of the Kérkira (Corfu) channel but he showed few Greeks in the interior. The Slavs of Macedonia were, according to Gabrys, all Bulgarians, and in his distribution of Bulgarians he followed Ivanov and Iširkov. The Vlachs, according to his representation, formed an important element in the population of Thessalía. Both in Old Serbia and in Ipiros, Gabrys indicated large tracts of Albanian territory. For example, the districts of Gjinokastër, Kónitsa and Ioánnina are all marked on his map as Albanian. On the whole, therefore, this Swiss map was decidedly pro-Bulgarian. It was largely based on Bulgarian sources, particularly on the maps of Ivanov and Iširkov (Figs. 47 & 48). Gabrys recognized none of the Serbian claims in northern Macedonia nor did he mention the *Macedo-Slavs*. All Greek claims were likewise disregarded.

The view may be taken that Gabrys rejected Greek and Serbian representations because they rested on insecure evidence. On the other hand his very full bibliography did not even mention Cvijić's map of 1913. His insistence on Bulgarian sources seemed to point to the fact that the map was largely Bulgarian inspired.

### American: Leon Dominian's Maps of 1915 and 1917

During the war years, the chief exponent of European ethnography in the United States was Leon Dominian. Born in Constantinople, he was a geographer of international repute and one who possessed a knowledge of most of the languages of eastern Europe and the Near East. H. W. V. Temperley sought his advice on ethnography when writing his history of the peace conference. In 1915, the *American Geographical Society* published a series of ethnographic maps of Europe compiled by L. Dominian. These were later reproduced in his *Frontiers of Language and Nationality in Europe,* which was published in 1917. This work received very favourable reviews. It was intended to be an impartial examination of the problem of linguistic frontiers in Europe, and it was Dominian's ambition that his researches might be applied to the settlement of European boundary conflicts. Madison Grant, in his introduction to Dominian's book, wrote :

> In the Balkan States the difficulty of finding any political boundaries that in any way correspond to race or language has hitherto been insuperable, but when the Congress of Nations convenes every member of it [ . . . ] should be familiar with all the facts that bear on the case and there exists no book which covers these questions so fully, so accurately and so impartially as Mr. Leon Dominian's "Frontiers of Language and Nationality."

Amongst the sources used for the compilation of the map was *Debes' Handatlas*, 1911.[1] Included in Dominian's very useful bibliography were the works of J. Cvijić, D. M. Brancov and A. Boué. Dominian regarded language as by far the most important criterion of nationality in Europe : his national groups were very largely linguistic groups.

According to Dominian, Macedonia comprised the basins of the

[1] Sheet 12c. The original map had appeared in the first edition of the Atlas in 1895 and had purported to show the situation in 1880 !

Vardar and Struma. It was isolated from the rest of the peninsula by a practically continuous line of mountains including the Píndhos, Sar, Osogovska, Rilo and Rodopi. Within these limits, stated Dominian, the Bulgarian, Serb, Romanian, Albanian and Greek tongues were heard. Many of the inhabitants spoke two vernaculars and Turkish was generally understood throughout the region.

*The Turks*

However, the Turks were not a vitally important element in the population from the point of view of nationality. Dominian observed, " A tradition flourishes to this day among the Turks that their occupation of European territory could not be permanent." For example, in spite of their four centuries' long sojourn, they did not collect furniture in their homes and they expressed unwillingness to be buried on the European side of the Bosporus or Dardanelles. " The state they founded in Europe had a weak head and no heart at all." Hence, the exodus of the Turks from Europe was only a matter of time. After the wars of 1912-13, 125,000 Turks had left the Balkans and a further 50,000 had left Crete. As a national group the Turks were important only in south-eastern Thrace. It must be remembered that these opinions of Dominian were expressed before the great exodus of Turks from Greece in the years 1923-1926.

*The Greeks*

Dominian believed that the Greeks of Macedonia were as mixed a population as could be found anywhere on the surface of the earth. Racially they included strains of Albanian, Slav and Tatar, and it was only on approaching Thessalía that the 'mediterranean type' became more pronounced. Dominian drew the Greek ethnographic frontier along the lower slopes of the eastern Píndhos, along the valley of the Aliákmon river, across the head of the Khalkidhikí peninsula, thence discontinuously along the coast of the Aegean, the Bosporus and the Black sea, but nowhere extending far inland.

He fixed the Greek ethnographic frontier in Ipiros to include Tepelenë, Klisoura, Gjinokastër and Himarë, and pointed out that the non-inclusion of this area within the boundaries of Greece, in 1913, had led to the 'Epirote insurrection' of 1914. Before the Conference of London in 1913, Greece had aimed at securing a line from 'Gramala bay' to the centre of the western shore of Lake Ohrid. This would have meant the inclusion of Himarë, Gjinokastër, Premetë and Korcë in Greece. According to Turkish statistics of 1908, wrote Dominian, there were 340,000 Greeks and 149,000 Moslems in these districts. In his opinion the language divide coincided with the upper course of the Voyussa (Vijosë), and with the road from Delvino (Delvinë) to Ostanitza, passing by Doliano (Dholianá). He maintained that history, legend and myth, as well as language, testified to the Hellenic character of the Epirote land and added: "Every step in the rugged country raises the dust of Hellenic antiquity."

*The Slavs*

Dominian made several rather conflicting statements about the Macedonian Slavs. He stated first of all that they were transitional between the Serbs and Bulgarians, but then went on to refute this idea by demonstrating the essentially Bulgarian character of the Slavs south of the Sar mountains. He quoted the following figures for the population of Macedonia, taken from D. M. Brancov:

*Christian Population of Macedonia*

| | *Total* | *Percentage of total Christian Population* |
|---|---|---|
| Bulgarians .. .. | 1,172,136 | 81·5 |
| Greeks .. .. .. | 190,047 | 13·22 |
| Romanians .. .. | 63,895 | 4·44 |
| Albanians .. .. | 12,006 | 0·84 |

Dominian apparently believed that the Serbs were so few as not to deserve mention, and that the Bulgarians formed a compact mass containing only a slight admixture of alien elements. He maintained that although the Macedonian speech might be regarded as transitional between Serbian and Bulgarian, its affinity with the latter was close enough for the two to be mutually intelligible. Serbian was, on the other hand, not readily intelligible to the natives. There were other reasons why Dominian classed the Macedonian Slavs as Bulgarians. He stated that Turkish historians, such as Evlia Tchelebi and Sa 'aeddin, had constantly referred to them as Bulgarians. It was to the diocese of Skoplje that the first Bulgarian bishop had been appointed as a result of the Turkish census of 1872. He believed that the temperament of the Macedonians was closer to that of the Bulgarians than to that of the Serbians. He declared that in the districts of Edhessa, Yiannitsá and Salonika, there had been preserved forms of the old Bulgarian language which had disappeared even in Bulgaria itself. Bulgarian place-names, too, abounded in south-western Macedonia. Serbian and Slav scholars, such as Rajić, Solarić and Vuk Karadžić, formerly had all concurred in setting the Serb southern frontiers on the Sar mountains. On Dominian's evidence therefore, the Bulgarian territory included Kastoría, Konia, Edhessa, the Drin valley, the coast of Dhitikí Thráki, the Maritsa valley and the interior of Thrace, but not Salonika which was a polyglot town. The Serbo-Bulgarian frontier he placed on the Sar mountains, whilst the Krajste and Vlasina valleys of eastern Serbia he regarded as a transitional zone between the two. The region between Pirot and Vranje was another zone where the Slav dialect departed equally from both Serbian and Bulgarian. In his sketch-map of 1917, however, Dominian placed the limit of Serb speech farther south.

*The Vlachs*

Dominian stressed the importance of the Romanian enclave around Métsovon in the Píndhos; it had been estimated to include half a million *Kutzo-Vlachs*. But there were also other Vlachs in the valleys of the Semen (Semeni) and Devoll rivers, and in the Olimbos (Olympus) mountains—*Vlakko-Livadi*. The *Frasheri*, or Vlachs of southern Albania, numbered many thousands and there were 10,000 around Berat alone; nearly all the towns of Macedonia and Thessalía boasted Vlach colonists. There were 14,000 in the Vardar valley, and many parts of the coasts of Kérkira also had Vlach settlements.

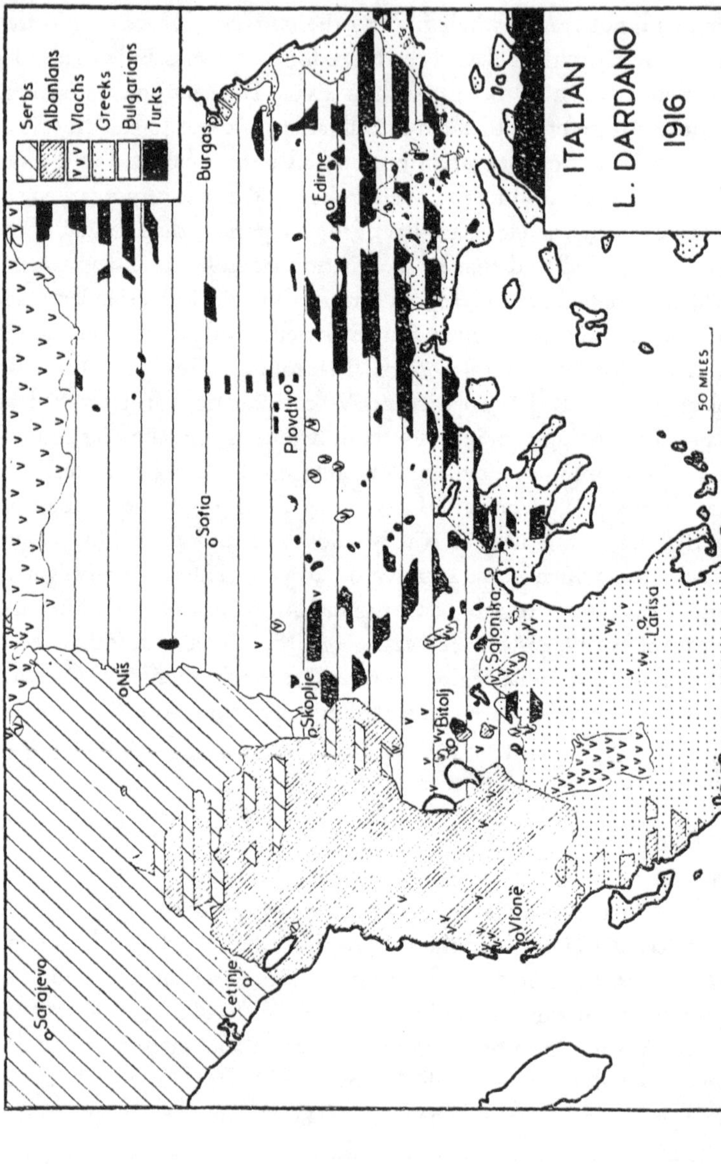
ITALIAN
L. DARDANO
1916
Serbs
Albanians
Vlachs
Greeks
Bulgarians
Turks
Burgas
Edirne
Plovdivo
Sofia
Niš
Skoplje
Bitolj
Salonika
Larisa
Vlonë
Cetinje
Sarajevo
50 MILES

FIG. 50

*The Albanians*

Dominian gave some pertinent facts about the Albanians. Etymological explanations of the word 'Albanian,' he pointed out, were numerous, but it appeared to be related to the Celtic form *alb* or *alp*, meaning mountain.[1] Dominian maintained that all the inhabitants of this part of the peninsula who spoke *Skip* (Albanian) should be regarded as Albanian nationals. He declared that the language was exclusively Aryan in form, but noted that of the 5,140 entries in G. Meyer's *Etymological Dictionary of Albanian* "only four hundred could be listed as unalloyed Indo-European." Tatar-Turkish accounted for 1,180, Romanic for 1,420, Greek for 840 and Slav for 540 words.

He described the Roman Catholic *Gheks* (*Gegs*) in the valleys of the Drin and Mat as largely under Italian influence. The Christian *Tosks* of the south were mostly Orthodox but many were Moslems. The Moslem Albanians were often referred to as *Arnauts*. In his opinion, Albanians in 1913 had been totally devoid of national feeling. Only the rivalry of Italy, Austria, Serbia and Greece, each competing for the use of Albanian ports, had resulted in independence. Italian influence in Vlonë was maintained through the Roman Catholic Albanians.

*Conclusion*

Dominian's views were very influential in the United States, and a map of the ethnographic distributions of Europe produced later (1919) by *The National Geographic Society* adopted similar distributions to those of Dominian—as far as the Balkans were concerned. In his opinion, the resumption of hostilities in this part of Europe in 1915 had been due to the pressing need for boundary revision in the light of ethnographic distributions. He was prepared to see the Bulgarians remain in possession of the part of Serbian Macedonia which they had occupied. He advocated a line in Thrace, very favourable to their claims, which would have included the Maritsa valley in Bulgaria. He emphasized too the necessity for allowing Bulgaro-Macedonian territory an outlet through Kaválla and Salonika. America, of course, had not entered the war when Dominian made these observations.

[1] 'Albanach' is still used to-day to refer to certain of the highlands of Scotland.

They were thus made by a citizen of a neutral state, and he was better able, perhaps, to see the Bulgarian point of view. But, on the other hand, he placed too much reliance on Brancov's evidence in his final summing up of the case for the Bulgarians in Macedonia.

## ITALIAN: A. DARDANO'S MAP OF 1916

Early in the war, *L'Istituto Geografico de Agostini* of Novara prepared a number of ethnographic maps of Europe. They were first published in Italy in 1916, and in 1917 a French version of the maps appeared. The eastern European map in this series was representative of the Italian view of ethnographic distributions in the Balkans. It might be remembered that pre-war Italian maps had been compiled by Amadori-Virgilj and Barbarich. A. Dardano prepared this map and his ideas were often diametrically opposed to those of the Serbian ethnographers, nor did he see eye to eye with the Greeks on the question of the presence of Greeks in Northern Ipiros (Fig. 50). He was not so favourably inclined towards the Greek cause as Amadori-Virgilj had been in 1908 (*cf.* Fig. 39).

### *The Distributions*

The ethnographic distributions on this Italian map showed quite a close relationship to those on H. Kiepert's map of 1877 (*cf.* Figs. 15 & 50). The distribution of the Turks, for example, was strongly reminiscent of Kiepert's map. But whereas more Greeks were shown along the Aegean coast than had been indicated by Kiepert, less were indicated in the vital region of the coastal districts of Northern Ipiros. Graeco-Italian relations were far from cordial at this time. The Allies had, by the Secret Treaty of London in 1915, promised Vlonë and its hinterland to Italy. Italy had also been given concessions in the Greek-speaking Dodecanese Islands. The Greeks, on the other hand, had occupied most of Northern Ipiros and the Allies were seeking to bring Greece into the war by the promise of even more of the southern Albanian territory. Italy, however, was firmly entrenched in the Albanian ports and was determined to prevent any further advance of the Greek frontier along the coast.

Dardano's Slav distributions were rather interesting as Italy and Serbia were at this time allied against Bulgaria ; yet the Italian view favoured the Bulgarians with an ethnographic frontier along the Timok and Morava rivers, extending as far west, in Macedonia, as Lakes Ohrid and Kastoría, thus including Skoplje and Bitolj in the Bulgarian sphere. In the south, the Bulgarian boundary was set as far south as Salonika, Sérrai and Xánthi. Only a very small part of northern Macedonia was designated as Serb, including Tetovo and Kumanovo. The Vlachs were depicted after J. Cvijić, but the Albanians were shown everywhere as solidly inhabiting territory which stretched well beyond the boundaries offered to them in 1913. Thus Dardano indicated them as a strong element in the population of southern Montenegro as far north as Cetinje. He insisted that the whole of Old Serbia west of a line, Novi Pazar-Mitrovica-Pristina, was Albanian. Macedonia, north and west of Kruševo, Prilep and Veles, was also depicted for the most part as Albanian. The fact that the Serbians were Allies had not modified the Italian ideas in their favour. In fact, Italian politicians hoped that the war would be the means of consolidating Italian influence in Albania and they envisaged a ' Greater Albania ' under Italian tutelage.

## French Maps

### *Vte. De la Jonquière's Map of 1914*

Vte. D. la Jonquière was one of the foremost French orientalists. In 1914 his *L'Empire Ottoman* was published and it remains a classical work of reference on Turkey. It included an ethnographic map of Turkey-in-Europe which would appear to have been based fairly faithfully on H. Kiepert's map of 1876. The only concession made to the Serbian claims was in the Niš-Lescovac area, where the population was shown, not as pure Bulgarian as on Kiepert's map, but as mixed Serbo-Bulgarian. Thus the Serbian views of Macedonian ethnography put forward twenty years earlier, and partly conceded by such scholars as F. Meinhard and T. D. Florinski, were apparently rejected by one of France's premier authorities on Turkish affairs.

### *Cte. A. D. Arlincourt's Map of 1914*

Another authoritative French view, concerned more particularly with the limits of Hellenism in the Balkans, emanated from the ethnographer, A. D. Arlincourt. An ethnographic sketch-map of his appeared in the periodical *L'Ethnographie*, in 1914. Arlincourt, after discussing the contributions of René Pinon and Victor Berard, appraised the population figures given by J. Cvijić for Macedonia (those of S. Gopčević, V. Kănčev, C. Nicolaides and K. Oestreich) and came to the following conclusions: firstly, that the northern limit of Hellenism might be drawn from Himarë, on the Adriatic, to Ayastafanos on the Black sea, passing through Gjinokastër, Korcë, Bitolj, Sérrai, Dráma and Edirne (Fig. 80). North of this line lay the domain of the Slavs, and south of it existed a basic mixture of Turks and Greeks containing exclaves of Bulgarians. Secondly, he maintained that certain localities had a distinctive ethnic character. He distinguished, for example, a purely Bulgarian district around Lake Ohrid and another between Kastoría and Flórina. There was, he asserted, no disputing the Turkish and Bulgarian character of the peoples living immediately north of Salonika, and the Pomaks of the Mesta valley were also a distinctive group. So too were the Greeks of Melnik. Arlincourt made no mention of any areas which might definitely be called Serb.

### *The Map of the French Ministry of War, 1915*

This map, compiled by the French Intelligence in 1915, illustrated the French view of the distribution, more especially, of the Albanians. It completely ignored J. Cvijić's claims for the Serbs in northern Albania and Old Serbia but accepted the Serbian conception of the *Macedo-Slavs*. It also ignored C. Nicolaides' claims for the Greeks in southern Albania and went so far as to depict all Ipiros, as far as the Píndhos, as Albanian (Fig. 51).

Typical of the impartial French attitude at the beginning of the war was a series of articles published in *La Revue Hebdomadaire* for 1915. They dealt with the views of the Greek, Bulgarian and Serbian schools of thought, and incorporated maps by the Bulgarian, Ivanov, and by the Italian, Amadori-Virgilj.

*French Views in 1918*

By the end of the war, however, French views on the ethnography of Macedonia had hardened in favour of Serbian interpretations and the maps of both J. Cvijić and N. Zupanić received widespread publicity. One of the best known of the contemporary French geographers, L. Gallois, supported Cvijić's ideas in an article written for *Les Annales de Géographie* in 1918. He dismissed the Bulgarian linguistic claims, maintaining that it was going too far to classify people by syntax and grammar. As for ecclesiastical claims, nationality was not to be confused with religion. In any case Niš had been part of the Exarchate, and yet the inhabitants of that area had not protested at their inclusion in Serbia in 1870. It is rather interesting to note that Gallois then proceeded to argue that from a 'geographical' point of view Macedonia was separated from Bulgaria by relief features of considerable altitude. He protested that to install the Bulgarians in the middle Vardar valley would be a crime against nature, which would cut off Serbia's access to the Mediterranean.

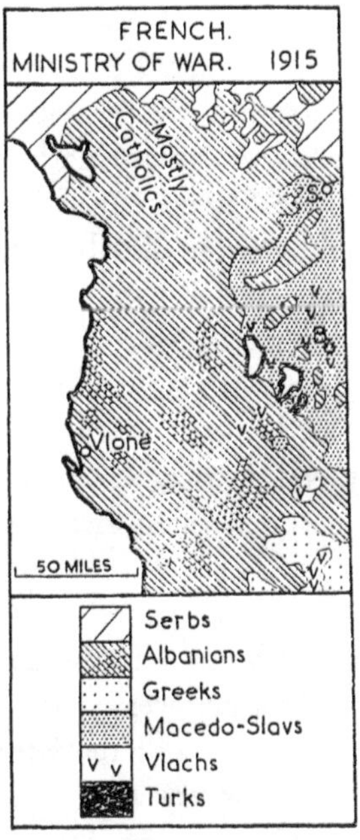


FIG. 51

In the original, the following religious sub-divisions of Albanians are shown: 1. Roman Catholics—in the north-west; 2. Mixed—between Tiranë and Shkodër; 3. Greek Orthodox—as shown above by stipple; 4. Moslems—in remaining areas.

A. Meillet, the French philologist, also lent his support to the Serbian thesis in his book, *Les Langues dans L'Europe Nouvelle*, published in 1918. Writing of the Serbian and Bulgarian affinities of the Macedonian dialect he wrote : " En réalité ces parlers n' appartiennent en propre ni à l'un ni à l'autre des deux groupes qui se les disputent ; c'est la politique qui décidera de l'avenir linguistique de la Macédoine." The use of the future tense here was significant.

Thus French thought on the ethnography of Macedonia inclined more and more towards Cvijić's viewpoint as the war progressed. There was a startling contrast between the maps being published in France in 1914 and those in 1918. Ethnographic thought in the British Isles developed on a similar pattern and underwent a parallel revolution.

## British Maps

### *A. J. Toynbee's Sketch-map of 1915*

A typical British view of Balkan ethnography at the beginning of the war was that held by A. J. Toynbee, the historian. His *Nationality and the War* contained a sketch-map of the Balkans, which incorporated ethnographic distributions and a proposed revision of political boundaries. His distributions were reminiscent of H. Kiepert's map of 1878. He depicted Macedonia as, in the main, Bulgarian, with the exception of the north-western districts of Tetovo and Gostivar which he marked as Serb overlain with Albanian, and the extreme south-west and the coastal districts which he indicated as Graeco-Turkish (see also Fig. 90). He stated categorically :

> There is no truth in the Serbian contention that the Slavonic dialect spoken in Central Macedonia is a variety of " South Slavonic " in the narrower sense. It is not even an intermediate link between South-Slavonic and Bulgar. The two languages are sharply differentiated from one another, and there can be no ambiguity in the classification of the Macedonian patois under one head or the other. Linguistically, the Macedonian Slavs are as unmistakably Bulgars as the Slavs of Sofia or Plevna, and the Bulgarian propaganda of the last twenty years has roused in them a keen sense of national brotherhood with the speakers of their tongue who live beyond the Bulgarian frontier.

### *M. I. Newbigin's Map of 1915*

Miss M. I. Newbigin, editor of *The Scottish Geographical Magazine*, was one of the few British geographers to write about the Balkan region. In her book *The Geographical Aspects of Balkan Problems*, 1915, she discussed the distribution of the peoples of the peninsula. Amongst her sources appeared J. Cvijić's map of 1913, H. N. Brailsford's *Macedonia*, S. P. Tučić's *The Slav Nations* and articles from *Questions Diplomatiques et Coloniales*. Her book included a small sketch-map of ethnographic distributions in which the Turks and the Greeks were shown roughly according to Cvijić's map. All the Slavs, however, were indicated in the same fashion—as 'South Slavs'—but this group was sub-divided into Serbs and Bulgarians. Her comments on their respective distributions were as follows:

> In the Sanjak of Novi Pazar the people are partly Orthodox Serbs, partly Moslem Serbs and partly Albanians. Farther south, in what was once Old Serbia, but is now New Serbia, the percentage of Albanians greatly increases—of this there can be no doubt. The region has been one of constant ethnographical change, and while, according to one view, the Albanians have actually pushed the Serbs back, according to another many of the inhabitants are "Albanised Serbs," i.e. Serbs in race who found it an advantage under Turkish rule to become Albanians. There is no doubt that in this region Serbia has difficulties before her in the future. Still further south, i.e. beyond Uskub (Skoplje) we come to a region which was recognised by Serbia in her secret treaty with Bulgaria, of March, 1912, as falling into a Bulgarian zone, but which is nevertheless now partly Serbian and partly Greek, Bulgaria receiving in 1913 an insignificant part of what was once Macedonia. Not unnaturally, Serbian authorities now find that "Macedonian Slavs" is a much more appropriate name for these peoples than "Bulgarians," whilst the Greeks have suggested that Bulgarian is not a race name at all but merely means a "countryman" as contrasted with "town dweller!" The point, at least, is that from a short distance south of Uskub to the northern shore of the Gulf of Salonika the land is chiefly inhabited by persons, who have hitherto been attracted to Bulgarian propaganda, with whom are mingled many Turks now, as usual in such circumstances, tending to emigrate, and with not a few Vlachs.

Miss Newbigin came to some conclusions which are perhaps worth mention here. She pointed out that Albania had never existed except on a scrap of paper and was hardly likely to survive the war, and that Serbia's outlet to the sea should be gratified at the expense of Albania. Greece's occupation of southern Albania could also be

FIG. 52

Albanian-speaking Greeks of Ipiros, and Pomaks of the Rodopi district are roughly indicated in the original, the former being classed as Greeks and the latter as Bulgarians.

justified. Satisfaction granted to Serbia in northern Albania, and to Greece in southern Albania, would encourage those states to consent to territorial changes in Macedonia in favour of the Bulgarians. Bulgaria's situation called for control of the Struma and Maritsa and she should have had Kaválla as her chief outlet to the south.

*The Map of Nevill Forbes, 1915*

A map incorporating similar distributions to those in the maps of Toynbee and Miss Newbigin was included in a book on the Balkans, written by Nevill Forbes and other well-known historians and students of eastern Europe. The work was a symposium and included short histories of Greece, Serbia and Bulgaria. The map was only a simplified sketch but it serves to illustrate the opinion of the time (Fig. 52). The views that Nevill Forbes himself expressed are contained in the following extracts from his work :

> It is the Slav population of Macedonia that has engendered so much heat and caused so much blood to be spilt. The dispute as to whether it is rather Serb or Bulgar has caused interminable and most bitter controversy. The truth is that it " was " neither the one nor the other, but that, the ethnological and linguistic missionaries of Bulgaria having been first in the field, a majority of the Macedonian Slavs had been so long and so persistently told that they were Bulgar, that after a few years the Bulgars could, with some truth, claim that this fact was so [ . . . ] If the question [of control of Macedonia] were to be settled purely on ethnical considerations, Bulgaria would acquire the greater part of the interior of Macedonia, the most numerous of the dozen nationalities of which is Bulgarian in sentiment if not in origin.

The depiction of Albanian territory in Forbes' map was of interest. The whole of the population living in the region between the 1913 political boundary of Albania and the Morava valley was indicated as a mixture of Albanians and Serbs. The population of Northern Ipiros was composed, he believed, of ' Albanian-speaking Greeks.'

J. A. R. Marriott used Forbes' map to illustrate his *Eastern Question*

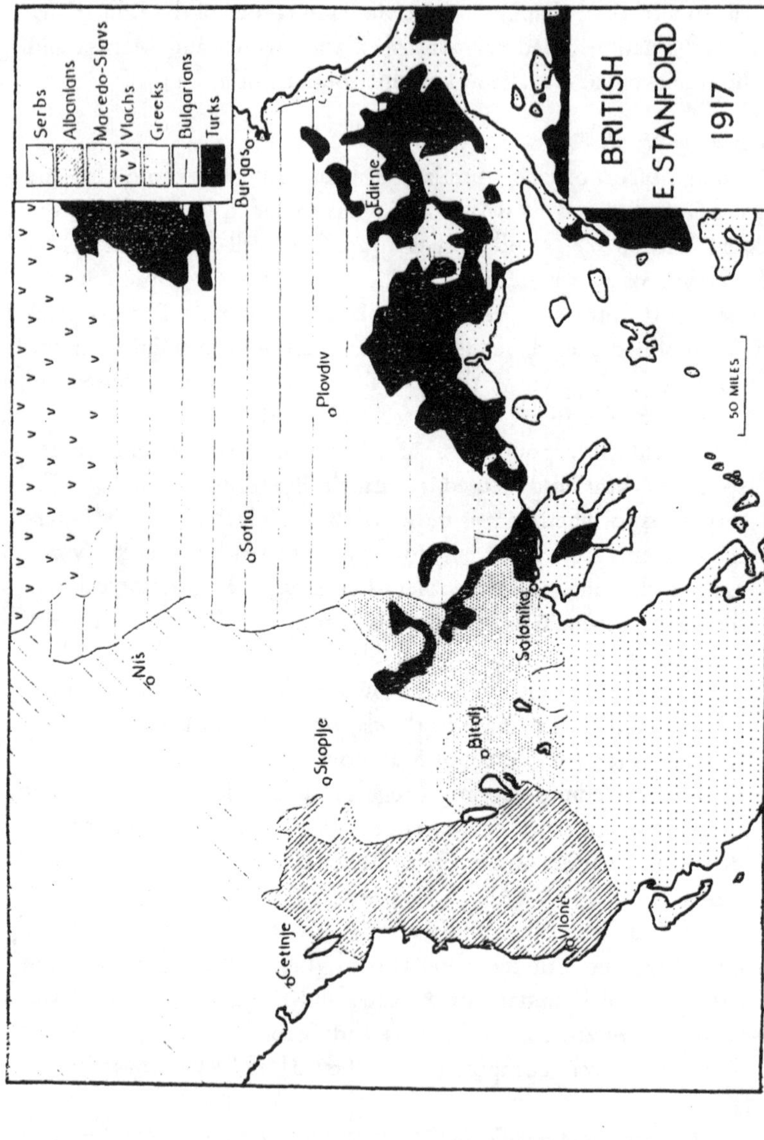
Serbs
Albanians
Macedo-Slavs
Vlachs
Greeks
Bulgarians
Turks
BRITISH
E. STANFORD
1917
Burgas
Edirne
Plovdiv
Sofia
Nis
Skoplje
Solonika
Bitolj
Cetinje
Vlonë
50 MILES

Fig. 53

which was first published in 1917. In his preface, he warned his readers not to accept the map as portraying anything but a " rough indication of the distribution of races."

In the light of the evidence afforded by the maps of A. J. Toynbee, Miss Newbigin and N. Forbes, British ideas at the beginning of the war appeared to be extremely conservative. The most these scholars conceded to the Serbian thesis, was to acknowledge a small portion of Macedonia, north of Skoplje, as Serb. The Slavs of Macedonia were generally classed as Bulgarians.

*The Map of A. J. B. Wace and M. S. Thompson, 1914*

Of a more specific nature, insomuch as it dealt only with the distribution of the Vlachs, was a map produced by two Englishmen—A. J. B. Wace and M. S. Thompson—whose archaeological duties had taken them to Greece on many occasions (Fig. 86). They cultivated an interest in the Vlachs, learned something of their language and recorded impressions of their folklore, distribution and numbers, and finally published their findings in 1914. They maintained that the numbers and importance of the Vlachs in Albania, Greece and Bulgaria had been sadly underestimated in the past, even by G. Weigand. Wace and Thompson extended Weigand's distribution to show considerable enclaves of Vlachs in the valleys of the upper Mesta and upper Struma, the existence of which had been ignored hitherto except on Romanian maps. Whereas Cvijić had estimated the number of Vlachs to be not more than 200,000, Wace and Thompson put the figure as high as 500,000. They stated : " Prevailing beliefs that the numbers are less are due to the reluctance of the Vlachs, in the face of the policy of the Greeks, to confess their nationality. Weigand's estimate of 373,520 was too moderate as it was based on a calculation of five to a family. The Authors' experience is that this is far too low."

The work of Wace and Thompson stimulated interest in this minority of whom so little was known in the British Isles—an interest which increased when Romania became an ally.

*The G.S.G.S. Map of 1916*

After the Bulgarians had decided to give their allegiance to the Central Powers, and the *Jugoslav Committee* had been operating for

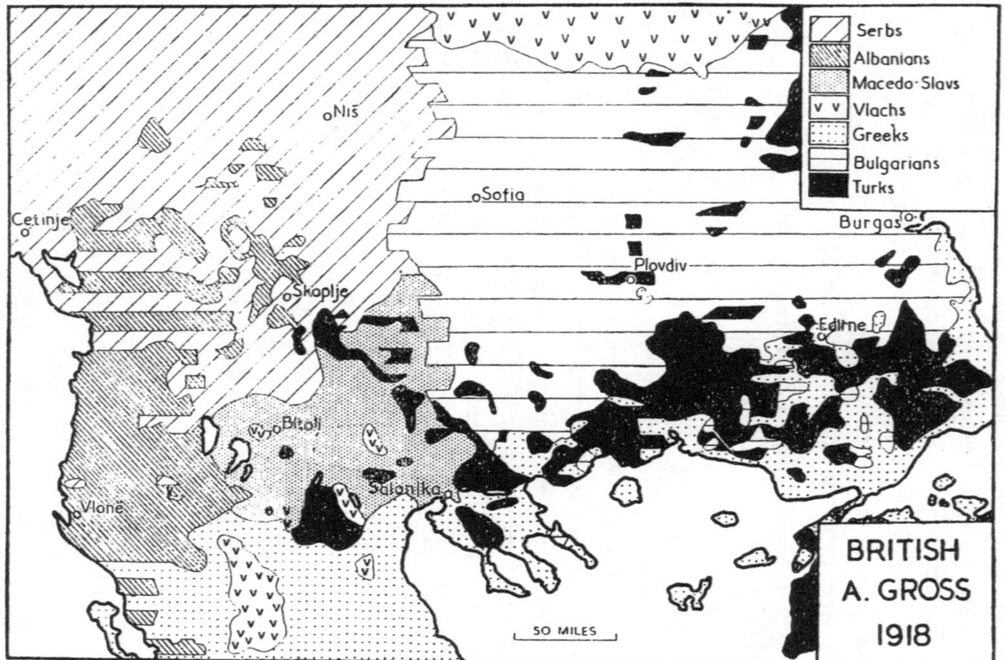


FIG. 54

*Note.* The distributions shown above were incorporated in two maps published under the auspices of the *Daily Telegraph* in 1918. In one map the distributions were referred to as 'races,' and in the other as 'languages'.

many months with success in London and in Paris, British opinion on the ethnography of the Macedonian region began to undergo just such a change as had characterized French opinion. The War Office produced a map in 1916 which attempted to simplify ethnographic distributions. On this map the distribution of Greeks was extremely limited. The Graeco-Albanian frontier was made to correspond to the 1913 political boundary. The most noticeable feature of the map was the depiction of the *Macedo-Slavs* in an area between Lake Ohrid, the Struma valley, Kastoría and Skoplje (Fig. 84). The Morava valley as far south as Kumanovo was marked as Serb. The distribution of the Bulgarians was limited, because of recognition of the *Macedo-Slavs*, to the region east of the Struma, but Sérrai and Dráma were shown as Bulgarian. The Albanians were depicted in the Debar region as well as in the Novi Pazar corridor.

### *A. H. E. Taylor's Opinion in 1917*

Another well-known student of Balkan affairs, A. H. E. Taylor, included in his book, *The Future of the Southern Slavs*, N. Zupanić's map of Jugoslav territory which had indicated all Jugoslav Macedonia as Serb (see p. 201). He stated :

> As for the ethnology of the Macedonian Slavs the best opinion is that it is not unlike what we might expect from this previous history [which dealt with Serb and Bulgarian Empires in Macedonia]. The original Slavs must have been of the same general stock as their Serb neighbours and the original Slav inhabitants of Bulgaria and that original stock has at different times received an infiltration of Bulgarians. East of the middle Vardar valley they may be described as Bulgarians and north of the line Stip-Gostivar as Serbs, the remainder living in the Slav portions of the former Vilayet of Monastir are neither pure Serb nor pure Bulgar.

### *R. W. Seton-Watson's Map of 1917*

R. W. Seton-Watson's publication, *The Rise of Nationality in the Balkans* (1917), also contained an ethnographic map, one somewhat

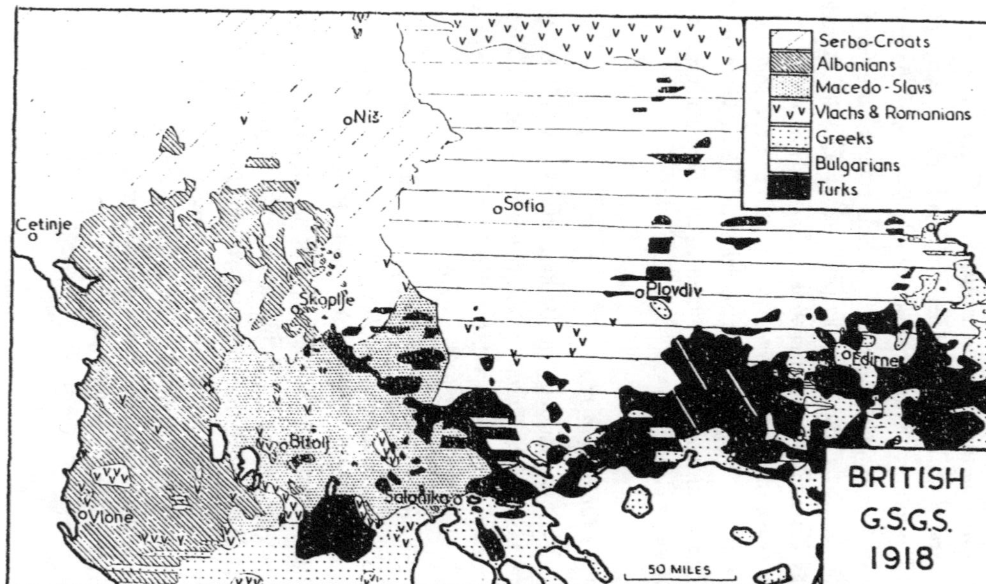


FIG. 55

*Note*. The Geographical Section of the General Staff published an ethnographic map in 1916 which showed simplified distributions different from those adopted in 1918.

unfortunately styled 'The Races of the Balkan Peninsula.' Seton-Watson was obviously using the term 'race' to mean 'ethnic group' and not physical type. He had already published one ethnographic map showing the distribution of the Serbs in 1911 (see p. 166). He had then been of the opinion that most of the Macedonians were Bulgarians. His new map was based on that of J. Cvijić but he somewhat modified Cvijić's distribution of *Macedo-Slavs* (Fig. 84). He had therefore changed his views in favour of the Serbian thesis. For example, he wrote in 1917 : "Peopled by a fluid population of Turks, Albanians, Jews, Greeks, Bulgars, Serbs and Vlachs, Macedonia has been the home of ceaseless and varied racial animosities, of rival racial and ecclesiastical propaganda, each backed as the Christian states of the Peninsula grew stronger, by its particular racial affinity beyond the Turkish frontier."

This picture of Macedonia as 'ethnically neutral' territory was precisely the point of view which Cvijić had laboured so assiduously to establish. The acceptance of the concept of the *Macedo-Slavs* by such an authority as Seton-Watson was, for the Serbians, a triumph which helped to justify their political control over northern Macedonia. About this time (1917) an interesting sketch-map appeared in the March issue of *The National Review*, suggesting new political boundaries for Europe on a national basis. It proposed to retain the 1913 boundaries in Macedonia and showed its inhabitants as *Macedo-Slavs*. There was no question of boundary revision in Macedonia, such as had been suggested in A. J. Toynbee's map of 1915.

### *Edward Stanford's Map of 1917*

The growing adherence to the views of Cvijić began to reach its height with the issue of Stanford's *A sketch map of the linguistic areas of Europe* in 1917 (Fig. 53). As far as the western Balkans were concerned, the ethnic distributions were almost entirely those of Cvijić. The map was simplified. It ignored Vlach distributions altogether and showed virtually no Albanians in Old Serbia. The *Macedo-Slavs* were depicted as on the version of Cvijić's map which had been published in 1913 (see p. 177).

*The Daily Telegraph Maps of 1918*

Under the auspices of *The Daily Telegraph*, a number of maps of Europe were issued throughout the period of the war. In 1918 there were two of these maps published—one a language map of eastern Europe, the other a map of the 'races' of eastern Europe. Both were compiled by Alexander Gross. As far as Macedonia was concerned the two maps showed identical distributions. 'Race' was here again being used to mean 'ethnic group.' The maps were designed to provide background information for the public, so that the coming Peace Conference proceedings might be followed with understanding. Gross's distributions of Turks, Greeks and Albanians were based entirely on those of Cvijić, but he made an interesting attempt to reconcile Cvijić's conflicting Slav distributions. The *Macedo-Slavs* he gave according to Cvijić's map of 1913, but such mixed Serbo-Bulgarian districts as appeared on the map he took from Cvijić's map of 1918 (Fig. 54).

*The G.S.G.S. Map of 1918*

The War Office issued another ethnographic map in 1918, showing more details of distributions than had been portrayed on the 1916 map (Fig. 55). If the two maps are compared, considerable differences in the various distributions emerge (Fig. 84). The influence of Cvijić's map was once more apparent. All the detail in eastern Bulgaria and Thrace had been taken from Cvijić's map, as had also the distribution of Greeks and Turks in Macedonia. Cvijić's distribution of *Macedo-Slavs* for 1913 was adopted but the northern limit of the group was extended. The Serbs were excluded from the lower Crni Drim valley around Debar. The distribution of the Vlachs showed differences from Cvijić's map, no doubt due to Wace and Thompson's findings. The Albanian distributions had evidently been influenced by Italian views, for Cvijić's claims for the Serbs in Old Serbia and northern Albania were disallowed. Albanians were also marked around Debar.

*J. S. Barnes' Map of 1918*

A map dealing specifically with the distribution of the Albanians appeared in the April issue of *The Geographical Journal* for 1918 (Fig. 56). Its author, Captain Barnes, had been engaged in intelligence work in Albania and held a brief for the Albanians. He supplemented his experiences by consulting the works of A. Baldacci and E. Barbarich, and also the articles of Bouchier, *The Times*' correspondent. He suggested that the 1913 boundary of Albania should be considerably modified in favour of the Albanians. He was adamant on the Albanian nature of Old Serbia and the Crni Drim valley below Debar; also he maintained that part of Montenegro was overwhelmingly Albanian in character. He agreed, however, that the boundary between Greece and Albania, as fixed in 1913, was a just compromise from an ethnic point of view. Finally he suggested that an enlarged Albania should be constituted and placed under the care of Italy (Fig. 56). Miss E. Durham, who might have been regarded, in 1918, as the foremost British authority on Albania, agreed with Barnes' remarks in a discussion on his paper, and the then President of *The Royal Geographical Society*, Sir Thomas Holdich, himself a political geographer, expressed a hope that an Italian protectorate might bring Albania peace.

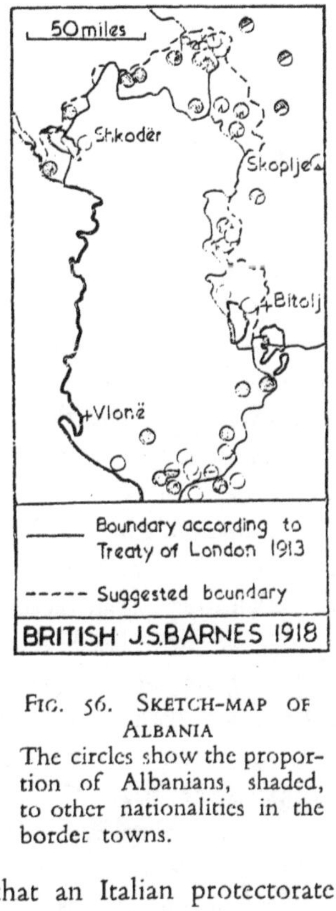


FIG. 56. SKETCH-MAP OF ALBANIA
The circles show the proportion of Albanians, shaded, to other nationalities in the border towns.

There were thus two schools of British thought on the distributions of the Albanians in 1918. One supported Cvijić's ideas and minimized Albanian superiority in Old Serbia, northern Albania and the Drin valley. The other supported what might be called the Italian view—that the Albanians were the most numerous element in the population of the territory which both Serbia and Greece hoped to incorporate in their respective states on the conclusion of the war.

## Chapter X

# ETHNOGRAPHIC MAPS AND THE PEACE-MAKING 1919–1923

### Bulgarian Maps

The five years following upon the end of the War of 1914–18 witnessed important changes in the political geography of eastern Europe. Many traditional boundaries were scrapped and many new ones created, and such was the expectant atmosphere engendered by President Woodrow Wilson's enunciation of the principle of 'self-determination,' that the irredentist minorities of the whole of Europe looked forward to the fulfilment of their desires for political independence and territorial recognition. Not since 1878, when the Tsar of Russia had formulated ethno-political boundaries for a Greater Bulgaria, had the principle of drawing up political, to coincide with ethnographic boundaries, excited so much attention. Nor was this sense of expectancy confined to the minority populations in the Allied camp. It animated the hopes also of those minorities to be found amongst the defeated peoples.

The representatives at the Peace Conference in Paris of the five Balkan nations, Bulgaria, Albania, Greece, Jugoslavia and Romania, were all careful to lay stress on the 'ethnic rights' of their respective populations, when asking for rectifications of political boundaries in Macedonia and the surrounding territories. But of all of them the Bulgarians placed the greatest reliance on this argument, since as the one-time ally of the Central Powers, they were desperately short of other legitimate means of sustaining their claims for revision of boundaries in their favour.

*Bulgarian Claims in* 1919

The unofficial memorandum which the Bulgarians laid before the delegates at the Peace Conference—they had not been asked to submit one—pleaded Bulgaria's historic rights in Macedonia and also her need for an outlet to the Aegean, and emphasized the existence of a million Bulgarians in the territory annexed by Greece and Serbia in 1913. The evidence they put forward to support their contentions seems to have been assembled by J. Ivanov, to whom reference was made in the last chapter. The French versions of two books written by Ivanov and explicitly devoted to a statement of Bulgarian claims, were published in Berne in 1919. Together they composed a symposium in which were expressed all the pro-Bulgarian ideas which had been in circulation since 1847. In addition to these books mentioned above, a French translation of Ivanov's book on Macedonia was prepared and issued in 1920 (see p. 199).

All these works of Ivanov contained ethnographic maps of various kinds. The view of the distribution of the Bulgarians which had been expressed in Iširkov's map of 1915 was reproduced, as also was Ivanov's ethnographic map of Macedonia, which had first appeared in 1917, but with the addition of French place-names. The map purporting to show the extent of the Bulgarian Exarchate between 1870 and 1912, from Rizov's atlas of 1917, was another of the maps to be found in these books. In his choice of maps Ivanov remained constant to his portrayal of Macedonia as Bulgarian territory.

Two novel maps utilized by Ivanov in his vindication of the Bulgarian cause summarized in a striking manner the evidence of certain older ethnographic maps bearing on the western and south-western limits of Bulgarian territory in the Balkans (Fig. 57). One map incorporated ideas held previous to 1878 ; on it were traced the limits imposed by A. Boué, G. Lejean, the Misses Mackenzie and Irby, and others who had shown the whole of the Niš-Leskovac area as well as Macedonia, to be within the Bulgarian sphere. The super-imposition of these boundaries created an impressive composite in favour of the Bulgarian claims. The other map treated ideas held since 1878. The composite effect in this case excluded Niš and Leskovac but still included most of Macedonia. In this instance,

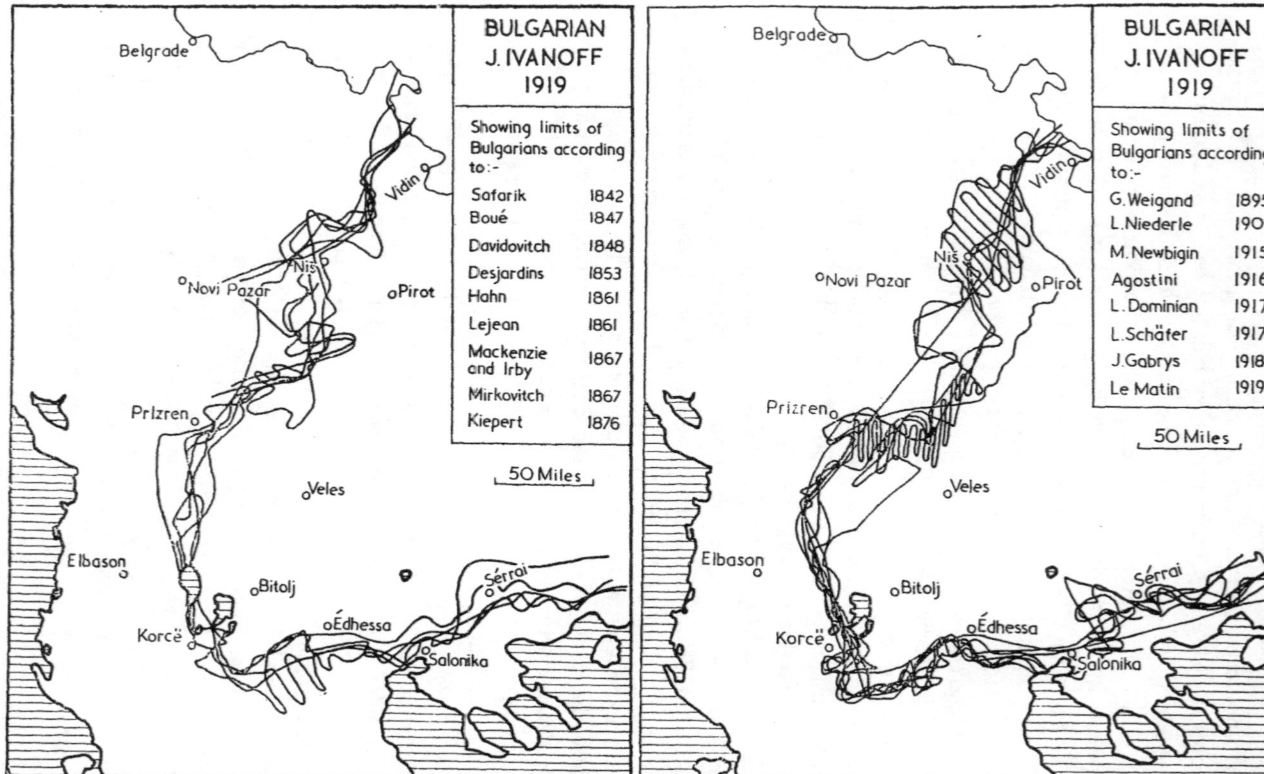


FIG. 57. A BULGARIAN VIEW OF THE COMPOSITE ETHNOGRAPHIC FRONTIER OF THE BULGARIANS IN MACEDONIA
The left-hand map refers to ideas held before 1878, the right-hand to those held after 1878.

however, Ivanov's choice of maps was not a genuine random selection, for after 1878, as it has been amply demonstrated, ethnographers were no longer unanimous about the Bulgarian character of Macedonia.

Ivanov's summing up of the situation in 1919 was as follows: The western limit of the Bulgarian nation in Macedonia was formed by the Sar mountains and by the line of the mountains of Jablanica on the western side of the Crni Drim valley. In the north-west the limit had formerly been fixed by the valleys of the Timok and Morava rivers, but since 1878 'serbization' had occurred in these areas and the boundary was no longer so definite as it had been. In the south the boundary was an imaginary line drawn from the Grámmos mountains to Salonika, thence to Sérrai and Dráma. In the east, there were mixed agglomerations of Turks, Greeks and Bulgarians and the frontier of the Bulgarians was no longer clear-cut, but it touched the Aegean at Pórto-Lágo, Alexandroúpolis and Enez, and thence followed the river Ergene. Finally it reached the Black sea at Midye.

*The Rejection of Bulgarian Claims*

But it was in vain that Ivanov produced his maps; nor did his quotations, often out of context, from the works of L. Niederle, C. Jireček, V. Oblak, L. Lamouche and a host of other well-known authorities on Balkan philology, history and ethnography, appear to have had much effect on the views generally held by the delegates at the Peace Conference. H. W. V. Temperley, for example, noted concerning the Bulgarian claims that, "in no area was the racial predominance (of the Bulgarians) clear cut and free from arguable uncertainties." Yet it is noteworthy that in the Allied reply to the Bulgarian memorandum the Bulgarian claims were traversed with the single exception of their claim in Macedonia, and the reply ended with the significant remark that "if all the questions raised by the Bulgarian Delegation had not been answered it was because, after studying them, the Allied Powers had not deemed it possible to accede to the requests made."[1]

There was a number of reasons why Bulgarian claims in Macedonia were ignored, or deliberately overlooked, at this time. First, the

[1] *History of Peace Conference of Paris*, vol. II (p. 450).

Bulgarian army had disintegrated and the Serbians and Greeks had re-occupied the territory they had seized in 1913. Their armies could only have been removed by force. Second, in the atmosphere prevailing in 1919 the Allies could not have ignored the guilt incurred by a defeated power which had broken the line of communication between Russia and the Allies and had prolonged the war by linking Turkey and Germany. It was impossible for the Allies to have accepted the explanation that Bulgaria had entered the war to right

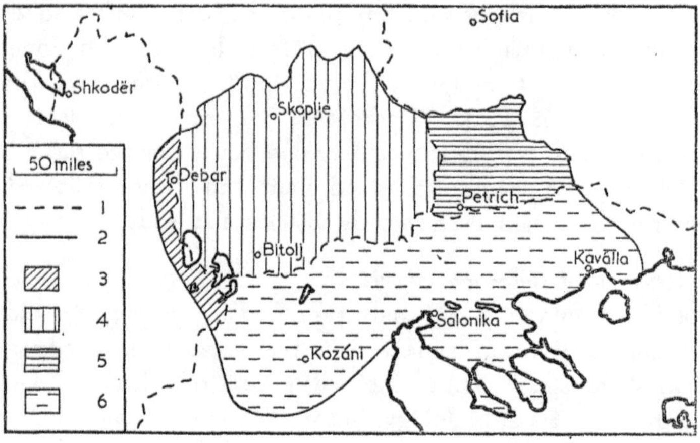


FIG. 58. MACEDONIA ACCORDING TO I.M.R.O.

The references in the key are as follows : **1.** Present-day political boundaries (1949) ; **2.** Limits of Macedonia ; **3.** Albanian Macedonia ; **4.** Jugoslav Macedonia ; **5.** Bulgarian Macedonia ; **6.** Greek Macedonia. Based on a map in R. Busch-Zantner, *Bulgarien. Geschichte und Politik im Zentrum des Balkans* (Leipzig, 1941).

the wrongs perpetrated against her by Serbia in 1913, and thus to have regarded Bulgarian relations with the Central Powers as incidental. (The Bulgarians had undoubtedly concentrated their war effort on the conquest and occupation of Macedonia and the Morava valley, even at times to the detriment of the other fronts of the Central Powers ; indeed, there were indications that the Central Powers had been unable to receive much help from Bulgaria, apart from that indirectly achieved by the occupation by Bulgarian troops of the vital Macedonian theatre). Third, the Allies wished to conserve the Jugoslav and Greek hold on the Vardar-Morava route from Central Europe to the Aegean. In the command of this strategic highway lay the key to the political control of the whole of the Balkan peninsula. Fourth, in 1919, eastern Macedonia together with Dhitikí Thráki (Western Thrace) had to accrue to Greece in order to give that country access to Turkish Thrace and Anatolia. The vision of a friendly, strong Jugoslav state dominating the Balkan passes, and of a rejuvenated Greek Empire in the eastern Mediterranean guarding the exit from the Black sea, appears to have been far brighter in the minds of influential delegates at the Peace Conference, than any of the colours on ethnographic maps produced by Ivanov and his contemporaries. Fifth, President Wilson himself made it clear, during the course of the Peace Conference, that his principle of 'self-determination' applied only to the territory of the defeated powers and that it was not the business of the Peace Conference to "inquire into ancient wrongs."[1]

Nevertheless the delegates of the Allied Powers did consider the possibility of creating an autonomous Macedonia. The Italian delegate put forward an elaborate scheme for an independent state similar in extent to that defined by I.M.R.O., the limits of which were to be settled by the Powers themselves (*cf.* Fig. 58).

The British and French delegates opposed the proposition on the grounds that it impugned Jugoslav sovereignty and that the creation of an autonomous Macedonia would, in any case, only provide fresh grounds for intrigue and thus increase rather than reduce political instability in the Balkans.[2] From the discussions which took place, it appears that the possibility was strong in the minds of the British and French delegates that an independent Macedonia would inevitably gravitate towards Bulgaria.

*The Final Settlement*

The final settlement in Macedonia modified the Serbo-Bulgarian boundary in favour not of Bulgaria but of Jugoslavia (Fig. 59). The Jugoslav memorandum of 1919, which had born every trace of the pen of J. Cvijić, had claimed a strip of Bulgarian territory, on ethnic and

[1] H. W. V. Temperley, *op. cit.*, (p. 433).
[2] M. Lorković, *Das Recht der Makedonier auf Minderheitensschutz* (Berlin, 1934).

FIG. 59. BOUNDARY CHANGES, 1915–1923

The references in the key are as follows: **1.** Boundaries in 1914; **2.** Modification in the Turko-Bulgarian boundary, 1915; **3.** New boundaries under the terms of (*a*) the Treaty of Neuilly, 1919, (*b*) the Treaty of Lausanne, 1923, and (*c*) agreements reached by the Great Powers and Albania, Greece and Jugoslavia; **4.** Jugoslav gains as a result of the Peace Treaties; **5.** Greek gains; **6.** Albanian gains.

strategic grounds. These claims were indeed only partially satisfied, but even so the result was a transfer to Jugoslavia of the passes of Bosiligrad and of the Strumica salient. Only on such maps as that of Gopčević, 1889, and on that of Cvijić of 1918 had these areas been shown as Serb or *Macedo-Slav*.

In addition to the material changes of boundary effected under the terms of the Treaty of Neuilly, there were other aspects of the settlement that deserve mention here. No minority rights were conceded to the Slavs of Macedonia under the terms of the Treaty, nor even to the few thousand Bulgarians living in the newly transferred territory. Since the *Macedo-Slavs* had been incorporated into Serbia in 1913 their cultural status was not considered at the Conference. This fact, together with the idea which had gained currency in the west in the latter part of the war, that the *Macedo-Slavs* were in any case not Bulgarians, stiffened the Serbians in their resolve to absorb these Slavs into their own Serbo-Croat group. Ultimately the Slavs of Serbian Macedonia were denied any freedom of self-expression, and for all practical purposes were held to be Serbians in culture and in national outlook.

## Albanian Maps

The Albanian state had been created by international action in 1913, but during the years of war which followed its existence had been academic rather than real. No central Albanian government had ever contrived to exist and although boundaries had been drawn on the map, they had been fixed only partially on the ground. The outbreak of war in 1914 witnessed the occupation of Ipiros by the Greeks, of northern Albania by the Austro-Hungarian forces, and of Vlonë by the Italians. The status of Albania was obscured still further in 1915 by the provisions of the Secret Treaty of London, whereby the Allied Powers agreed to a virtual partition of the state between Jugoslavia, Italy and Greece, thus vitiating the international agreement of 1913. Nor did subsequent events clarify the situation. In 1917, for example, Italy declared Albania to be under her protection and the French, then in Korcë, proclaimed an independent republic there.

Towards the end of the war the Albanians themselves had taken a hand in their own affairs. Fighting broke out in the north between Albanians and Serbians. In the south, Albanian guerilla action eventually forced the Italians to withdraw their 'protection,' and the Albanians also succeeded in re-occupying northern Ipiros as the French forces withdrew. Under the stimulus of war and occupation, a central provisional Albanian Government was established which, for the first time, spoke with some authority for the whole of Albania. Due to its existence, Albania received semi-official recognition at the Peace Conference.

*The Albanian Claims in 1919*

The Provisional Albanian Government, in 1919, presented a memorandum to the Allied Powers in which were proclaimed not only the right of Albania to the boundaries laid down in 1913, but also the right to expand beyond those boundaries into so-called Albanian territory in Old Serbia, Montenegro, Macedonia and Ipiros. The memorandum put forward proposals for a new boundary based on ethnic limits. A map showing the proposed boundary has been published in *A History of the Peace Conference of Paris* (Fig. 60).[1] It included Bar, Podgarica, Peć and Pristina in the north and extended within ten miles of Skoplje in the east. The towns of Gostivar, Debar and Ohrid were also included, and in the south much of the coast of Ipiros.

In support of these claims, the powerful and wealthy colony of Albanians in Turkey also produced a detailed ethnographic map of Albanian territory, which was published in Paris in 1920. This was the first map which the Albanians had compiled for themselves. Albanian territory on this map exceeded that claimed by the Provisional Albanian Govenment, for it included Mitrovica in the north, Skoplje in the east and the whole of 'Tchamuria' in the south (Fig. 61). This distribution was reminiscent of those on earlier nineteenth-century maps—for example that of G. Lejean on which the Albanians were shown to inhabit the whole of Ipiros and Old Serbia.

[1] See also, E. P. Stichey, *Southern Albania or Northern Epirus in European International Affairs, 1912-23* (Standford, Ca., 1926).

It might be recalled that in H. Kiepert's maps of 1876 and 1878, the view had been expressed that much of what had passed for Albanian in southern Ipiros was really Greek territory. G. Weigand had agreed with this conclusion in his map of 1895. On the other hand, Kiepert had increased Albanian territory in western Macedonia and Old Serbia; the Austrian, K. Sax, in 1877, and subsequently Italian ethnographers, had concurred with this interpretation. Only in the Serbian maps, more particularly those of S. Gopčević and J. Cvijić, had Kiepert's conclusions about Albanians in Old Serbia been refuted. British and French maps, appearing in the early part of the war, had at first favoured the idea that the Albanians formed the predominant element in both Northern Ipiros and in Old Serbia. The map issued by the French Ministry of War, in 1915, was typical of ideas then held (Fig. 51). As the war had proceeded, however, Allied opinion had become more and more susceptible to Greek and Serbian influences, the more so, as a partition of Albanian territory would have solved many of the problems facing the Allies during the war and during the peace-making. Indeed, even after the Joint Franco-British-American Memorandum of December, 1919, in which the northern and eastern Albanian boundaries of 1913 were recognized and the Albanian claim to Gjinokastër conceded, Clemenceau and Lloyd George continued to agitate for part of north Albania to be made

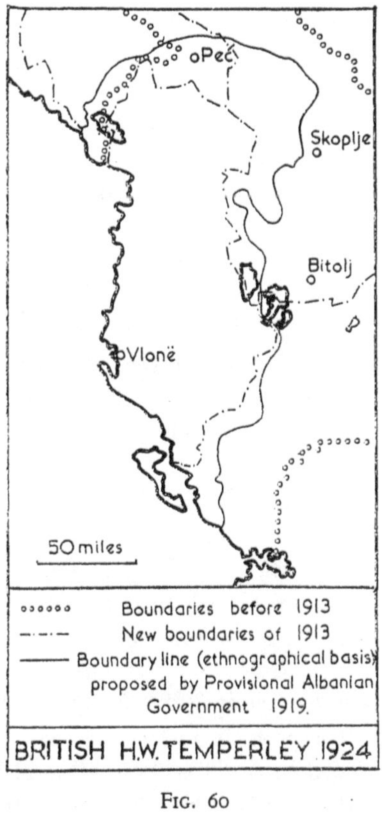


FIG. 60

into an autonomous province under Jugoslavia, and for Korcë and Gjinokastër to go to Greece.

*The Settlement in Northern Albania*

A proposal for new Albanian boundaries had been expressed by J. S. Barnes in 1918 (see p. 227). This had been an indication that informed opinion considered the boundaries of 1913 unsuitable. The boundary of 1913 with Jugoslavia had been the result of a compromise between the Serbian desire to control the coast and the Italo-Austrian desire to prevent such a contingency. It ran through 'impossible' country: Prizren, Debar and other big towns on the Serbian side were seriously exposed to Albanian tribal raids. Added to this danger was the possibility of unrest amongst the Albanian tribesmen in Serbian territory. The Jugoslavs themselves were eager to extend their control over the whole of northern Albania, and they made clear this aim in the memorandum they presented to the Peace Conference in 1919. Clemenceau and Lloyd George were inclined to compensate Jugoslavia for the loss of Trieste, but President Wilson himself insisted on the maintenance of the Albanian northern and eastern boundaries of 1913. The problem was not settled at the Conference and the Jugoslavs never managed to gain more than minor modifications on the line of 1913, nor, in the face of Jugoslav opposition, did the Albanian request for a boundary to include Old Serbia ever receive serious consideration at the Peace Conference.

*The Settlement in Southern Albania*

The problem in the south was equally as acute as that in the north. The occupation of part of southern Albania by the Greeks in 1914 had made possible the administration of the territory acquired by Greece in Ipiros in 1913. So difficult are the passes over the Píndhos and associated ranges, that communications by land between Ipiros and Macedonia may only be successfully maintained through Korcë, which had fallen on the Albanian side of the boundary in 1913. The Greeks, therefore, wished to retain this town, manifestly Albanian even according to the views expressed by ethnographers sympathetic to the Greek claims. They were also eager to incorporate the

Gjinokastër district into Greece. Gjinokastër in addition to being an important centre commanding the route from Vlonë to Ioánnina, was alleged to be pro-Greek in sentiment. Moreover, its incorporation into Greece would have shortened the tortuous Albano-Greek boundary and given Greece complete control of the Straits of Corfu (see also Fig. 46).

Thus the Greeks were able to offset any claims which the Albanians made on Southern Ipiros by counter-claims on Northern Ipiros. It has been pointed out in the last chapter, that ethnographic ideas in vogue in the west towards the end of the war tended to support Greek rather than Albanian claims, but that the Albanians were not without support even in Britain. The conflicting claims were so difficult to assess that the Peace Conference delegated its authority to a special commission, the members of which were instructed to seek further evidence and finally to decide on the course of the Albano-Greek boundary. The issue remained in doubt for some years and the modifications of the boundary finally agreed upon amounted to little more than variations on the line suggested in 1913.

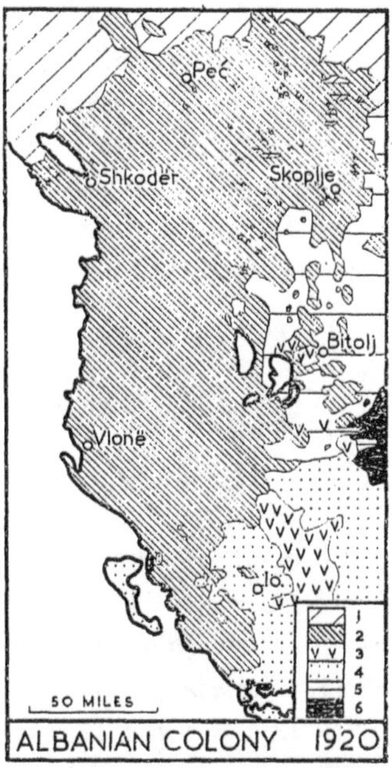


FIG. 61

The references in the key are as follows: **1. Serbs; 2. Albanians; 3. Vlachs; 4. Greeks; 5. Bulgarians; 6. Turks.**

*Conclusion*

The Albanian state thus emerged from the Peace Conference and the subsequent negotiations with substantially the same boundaries as those framed in 1913. The two biggest changes were first, the reduction in favour of Jugoslavia of the Albanian salient south of Prizren and second, the rectification in favour of Albania of the boundary in the neck of land between Lakes Ohrid and Prespa (Fig. 59). That the Albanian boundary did not coincide with the limits of the Albanians and that it was highly disruptive of the natural economy of the whole of the western Balkans can hardly be disputed. The final boundaries bore relation neither to ethnography nor to economy for the simple reason that they were the result of a number of forces, each of which was exerting its influence in a different direction. The solution was different from anything desired by the interested parties and gave satisfaction to no one. And so the boundary remained, a distasteful compromise until the equilibrium of forces was once more upset in 1940.

## Greek Maps

Although the Greeks had taken possession of southern Macedonia in 1913, their hold on that territory was never secure during the war. This insecurity was due first, to the threat of Bulgarian aggression and second, to the possibility of the re-adjustment of the Serbo-Greek boundary in favour of Serbia. Control of the whole *Macedo-Slav* sphere would have given the Serbs an outlet to the Aegean. The entry of Greece into the Allied bloc in 1917 had put an end to the latter possibility, but the Greeks felt constrained to restate their rights in Greek Macedonia in the face of the claims made by the Serbians, Bulgarians and Albanians when the peace was being made. They were anxious also to support their own claims on the neighbouring territories of Thrace and northern Ipiros.

### *An Official Map of Ipiros*

An original map showing ethnographic distributions in Ipiros was published in 1919 by a department of the Greek War Office. The main criterion on which its distributions were based appears to have been religion. Language was virtually disregarded. On this map, therefore, much of southern Albania including the district of Gjinikastër was shown by symbols as Greek territory (Fig. 46). Greek territory exceeded that shown on the earlier pro-Greek maps of G. Weigand and H. Kiepert and was, of course, totally at variance with that attributed to the Greeks by the Albanians themselves.

*V. Colocotronis's Book on Macedonia, 1919*

The Greek hold on Macedonian territory was justified by the work of V. Colocotronis published in 1919. He vindicated Hellenism in Macedonia by reference to both history and ethnography. Although he included no original ethnographic map in his book he reproduced those of E. Stanford (1877), H. Kiepert (*Carte Ethnocratique*, 1878), F. Bianconi (1877) and G. Amadori-Virgilj (1908), all pro-Greek maps. In declaring that many ethnographic maps of a kind unfavourable to the Greek cause in Macedonia were prejudiced, he asserted the maps of G. Lejean, M. F. Mirković, the Misses G. M. Mackenzie and A. P. Irby, and J. Erben to have been biased in favour of Bulgaria because of the sympathy felt by their authors for the members of the Exarchate.

The principal arguments put forward by Colocotronis were of an historical nature and only a small part of the book was devoted to ethnography. Many of his maps, designed to stress the historical role of the Greeks in Macedonia, were facsimile reproductions from classical atlases. The theme of his book was the traditional and spiritual leadership of the Greeks to whom looked for guidance " les populations slavophones, qui, en outre, ont conservé une conscience nationale purement grèc."

*Dhitikí Thráki (Western Thrace)*

The Greek interpretation of the ethnography of the lands adjoining the Aegean undoubtedly exerted a strong influence on the terms of the Treaty of Neuilly, under which Greece received Dhitikí Thráki from Bulgaria (Fig. 59). The transfer of this coastal territory was of considerable significance for the future political geography of the peninsula. This strip of territory served the dual purpose of linking Greek Macedonia with Turkish Thrace, and of providing Bulgaria with an outlet to the Aegean. The change of boundary therefore had the following consequences : first, it isolated Bulgaria from the Aegean and on that account gave further cause for resentment; second, it burdened Greece with a much extended boundary on the mainland and one most difficult to defend; third, it increased the polyglot population in the northern provinces of Greece; fourth, it provided Greece with a land corridor to Turkish Thrace and beyond, and thereby encouraged Greek designs on Constantinople and Anatolia.

In 1922, all hope of a new Greek empire based on Thrace and Anatolia had to be abandoned consequent upon the Smyrna disaster. But Greece contrived to retain Dhitikí Thráki, in spite of the fact that the circumstances under which it had passed to Greek control had materially changed, and retention of the region could at the best be only a strain on the small Greek state, and a constant source of friction with Bulgaria.

## Romanian Maps

On every occasion when changes of boundary in the Balkans had been imminent, Romanian interest in the fate of the Romanian-speaking peoples of Macedonia had revived and maps depicting the Romanian view of their ethnographic distribution had been produced. The year 1919 was no exception to the rule and A. D. Atanasiu, a Romanian professor, compiled two maps which emphasized the importance of the Romanian minorities, in the Timok area and in Macedonia respectively. The relationship of the Vlachs of Macedonia to their Romanian cousins north of the Danube has been touched upon in previous chapters. The Romanians of the Timok area were, of course, much closer in language and were to be regarded merely as an outlier of the main body. Those in Macedonia were, however, far removed from the Danube, and only very loosely connected with the Romanians proper. During the latter part of the nineteenth century, the Romanians had provided schools and churches for their 'Macedonian cousins' and had otherwise subjected them to political propaganda designed to stress the relationship between the Vlachs and themselves, but any possibility of the extension of the Romanian political boundaries into Macedonia had been precluded by the scattered nature of Vlach settlement. There may have been considerable truth in the theory propounded by L. Niederle and G. Weigand, and reiterated by Romanian professors, that the whole of the Niš-Leskovac area had once contained a large Latin-speaking element in the population, but many centuries had elapsed since this condition had existed, and the Macedo-Romanians in the interim had become well isolated from the main body (see p. 156).

### *A. D. Atanasiu's Map of 1919*

A. D. Atanasiu's map was published in Paris in 1919. It was described as an ethnographic map of the Macedo-Romanians or Kutzo-Vlachs, but it also showed interesting distributions of Turks, Greeks, Serbs, Bulgarians and Albanians (Fig. 62). The map was based, according to Atanasiu, on the work of J. Lejean, H. Kiepert and G. Weigand and on the work of the Romanians, C. Noé, L. T. Boga and D. Abeleanu.

Atanasiu credited the Turks with extensive territory in the Aliákmon valley, in central Macedonia, in the Vardar valley and in eastern Macedonia. To the Greeks he gave a very limited distribution, confining them to the coast in the east and to the line of the Aliákmon valley and the Píndhos in the south. He represented Ipiros as being Greek only in the south, between Ioánnina and the Gulf of Arta. The Serbs, he practically excluded from Macedonia altogether except for an exclave around Skoplje ; indeed, he showed them only as a minority in Old Serbia. He depicted the Slavs of Macedonia as Bulgarians, this in spite of the fact that Bulgaria and Romania had fought as enemies in the war, and that they had a quarrel of long standing over the southern Dobrudja. He marked the Bulgarian Moslems or Pomaks, however, as Turks. The Vlachs, he portrayed as occupying a number of large enclaves in Macedonia and the surrounding territory. The particularly big enclave in the Píndhos region, he suggested should be the nucleus of an independent Vlach state, and he marked on his map the limits of claims put forward by the Kutzo-Vlachs concerning the extent of 'Pindus,' the proposed independent state. A very large part of western Macedonia, Old Serbia and Ipiros, he portrayed as Albanian territory.

All these distributions were radically different from those appearing on Serbian and Greek maps, and they differed in a large degree also from those on Bulgarian maps. These differences stress once again that kaleidoscopic property of ethnographic distributions of changing in character and size when viewed from different angles. Although Atanasiu's map was supposed to have been based upon that of G. Lejean, G. Weigand and H. Kiepert, a comparison of the distribution of the Vlachs, on these three maps with that on Atanasiu's map reveals

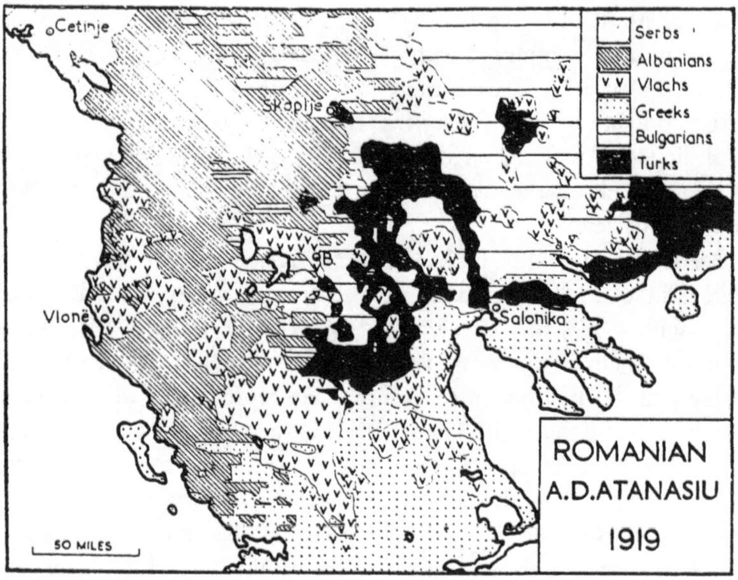


FIG. 62

In the original, numerous references are made to towns which contained a Vlach element in their populations.

considerable differences. Nor did his map agree with the opinions recorded by A. J. B. Wace, G. Weigand, and other authorities on the distribution of the Vlachs. In fact, it is necessary to refer back to F. O'Etzel's map of 1821 to find a cartographer, other than Romanian, so favourably disposed to the importance of the Vlach community in this region (Fig. 86).

The suggestion put forward by Atanasiu, that the Píndhos area should be made into an independent state, was an indication of the length to which Balkan politicians were prepared to stretch the principle of 'self-determination.' It must be remembered that Romania, as one of the Allied and Associated Powers, received enormous transfers of territory partly as a result of the application of this principle. However, these transfers were made at the expense of defeated powers. Romanian minorities in Greek and Serbian Macedonia, since their fate had been determined in 1913, were not entitled to consideration at the Peace Conference, nor, for the same reason, were Macedo-Vlachs entitled to any special treatment as minorities. Even if the estimates of their territory submitted by the Romanians were accepted, there could have been no question of political independence, apart from that to be gained within the framework of the states of Jugoslavia and Greece. But as it happened, the refusal of the Allied Powers to consider a settlement of the Macedonian problem in its historical perspective deprived the Vlachs of such protection as they might have received as a legitimate minority, and without which their communities within Greece and Jugoslavia found it increasingly difficult in subsequent years to maintain their cultural identity.

## British Maps and American Maps

### *E. Stanford's Map of 1919*

The firm of Edward Stanford produced a Peace Conference Atlas in 1919, which contained a map of Jugoslavia showing the distribution of Serb speech. Its limits were given in the south as an imaginary line joining the towns of Ohrid, Kruševo, Prilep and Strumica, and thus much of Macedonia denoted as *Macedo-Slav* on J. Cvijić's map was marked as Serb on this map (Fig. 81).

### *H. W. V. Temperley's Map*

H. W. V. Temperley's work on the Peace Conference of Paris did not include a map specifically covering Macedonian ethnography. According to the preface, L. Dominian, the American geographer, contributed some valuable advice on the ethnographic side. In the chapter on the New Bulgaria in volume IV there was a map showing ethnographic distributions within the new boundaries. The distributions were not those of Dominian, however, but those of Cvijić. They may have been taken from the British G.S.G.S. map of 1918.

### *A Map in the Times Atlas, 1920*

British opinion in 1919, therefore, was still largely influenced by Serbian maps but not exclusively so. In the edition of 1920 of *The Times Atlas*, for example, there was an ethnographic map of the Balkans, based on G. Gerland's map, which had appeared in Berghaus's atlas in 1892. This map portrayed the Serbs only as a small enclave in northern Macedonia, and the Bulgarians as the predominant Slav group in the central and south-western districts. It also favoured the Greeks in northern Ipiros, to the disadvantage of the Albanians, but the latter were given a distribution in the east which stretched well beyond the Drin river, into Macedonia and Old Serbia.

### *The Views Expressed in the Admiralty Handbook of 1921*

The British Naval Intelligence Division had engaged during the war in the production of a number of Geographical Handbooks. Although Macedonia was not a political unit, it was regarded as such for the purpose of this series, and a whole Handbook was devoted exclusively to its treatment. That the Director of Naval Intelligence considered Macedonia worthy of individual treatment was in itself a significant fact. From this decision it might be inferred that the political future of this territory was doubtful at the time when the Handbook was undertaken, and that its unique character warranted treatment apart from the political units of Greece, Serbia and Bulgaria.

The Handbook on Macedonia referred the reader to the G.S.G.S. map of 1918 for the details of Macedonian ethnography, but its authors expressed themselves thus on the Macedonian Slav controversy : " For ethnological purposes it is generally agreed that racially the Macedonian is somewhat distinct from both Serb and Bulgar. For political purposes it is agreed that the index of nationality is the inclination of the people and that this is mainly Bulgarian." The distinction made here between ethnological and national group is far from clear. The authors then stated : " The language of the Macedonian Slavs has affinity both to Serb and Bulgarian : the distinguishing characteristic of the latter, the use of the definite article at the termination of the noun, prevails in Macedonia." Something about the inclination of the Slavs may be gathered from the report that " the French forces on the Crna River, relying on Serbian maps, were surprised to find the villagers eager to help the Bulgarian invaders." These statements cannot be regarded as pro-Bulgarian but they do indicate that in the opinion of the authors of the book on Macedonia, the term *Macedo-Slav* was but a thinly disguised connotation for Bulgarian.

*H. A. Gibbons' Map of 1921*

There were also maps published in Britain which favoured the Greek cause. A symposium on Venizelos, the Greek statesman, to which H. A. Gibbons, the historian, and other writers contributed, was published in 1921. It contained an ethnographic map based on that of G. Amadori-Virgilj. H. A. Gibbons had already written on the Armenian question and on the problem of nationalities in the Near East. He supported the aspirations of the Greek expansionists—a political reunion of all the Greeks in Thrace and Asia Minor. It might be remembered that religion had been used as a criterion of nationality by Amadori-Virgilj and he had taken care to indicate this on his map. On Gibbons' map, however, all Slavs of the Greek Orthodox religion were described as Greeks, Albanians of Moslem faith as Turks, Pomaks as Turks, etc. The resultant ethnographic pattern was distinctly favourable to the Greeks. No better map could have been chosen to sustain the policy then being pursued by the Greeks in Thrace and Asia Minor.

*Facsimile Reproductions by Saxon Mills, etc., 1919*

Another British publication favoured the Greeks by utilizing ethnographic maps based not on religion but on language. The maps, those of Boué, Lejean, etc., were facsimiles drawn directly from Rizov's Atlas and incorporated into *The Question of Thrace* by Saxon Mills and M. G. Crussachi. These maps, it might be remembered, has shown considerable Greek territory in Thrace. Because they had shown Bulgarians in Macedonia, they had been condemned by V. Colocotronis as unreliable, but this did not prevent their being used by Saxon Mills and Crussachi to bolster Greek claims in Thrace.

*An American View, 1919*

In the United States, the ethnographic problems of Europe were summarized in map-form in a publication of *The National Geographic Society* in 1919. Because of the scale adopted, distributions were not given in detail, but generally, Greek Macedonia was depicted as mixed Graeco-Bulgarian, Jugoslav Macedonia as mostly Bulgarian, Old Serbia as Albanian, and Northern Ipiros as Greek.

## German and Austrian Maps

*H. Herzberg's Map of 1919*

Under the editorship of Hermann Haacke a number of maps were published during and immediately after the war in the *Physikalische Wandatlas* series used for teaching in German schools and universities. One of these maps, an ethnographic map of Europe, compiled by H. Herzberg, appeared in 1919. German geographers had had the opportunity of studying the ethnic situation in Macedonia during the war, because Germany had been in partial occupation of the territory.[1] At the same time, German interest in, and sympathy with, Bulgarian expansion, might be expected to be reflected in their ethnographic maps. Herzberg's map, since it was designed as a wall map, was simplified, but the ideas he held concerning broad distributions are worth noting. There was nothing extraordinary about his Turkish distribution, except perhaps the tendency to minimize its importance in eastern Thrace. Greek territory, he showed as restricted to the coast in eastern Macedonia and as bounded by the Aliákmon river in the south-west, but he did favour the Greeks with considerable territory in Ipiros. He regarded the Bulgarians to be the most important Slav group in Macedonia. All that area marked on Cvijić's map of 1918 as *Macedo-Slav*, he marked as Bulgarian. He did make concessions, however, to the Serbs in the north, crediting them with territory

[1] This research had been directed by G. Weigand (see Appendix A).

in the Drin valley of northern Albania, and with the Skoplje region. This distribution was very similar to but not so extensive as that given by Cvijić himself. Herzberg followed Weigand's ideas on the distribution of the Vlachs. He did not indicate such a widespread distribution of Albanians, as did Bulgarian and Italian sources.

### *J. I. Kettler's Map of 1919*

Perhaps the most interesting German map appearing in 1919, was one in the series produced by Karl Fleming. His well-known geographical institute commenced to prepare ethnographic maps of Europe and Asia during the war, and it produced, amongst others, an ethnographic map of the Balkan peninsula accompanied by a statistical abstract dealing with the population. It was compiled by J. I. Kettler, who appears to have specialized in this branch of cartography. He approached the problem of distributions historically but used mainly German sources for his information. It would be difficult to say how far personal observation influenced his opinions.

His distribution of the Turks was detailed and contrasted greatly in the region of eastern Thrace with that of Herzberg. The Greeks fared ill. He depicted them as a minority in Thrace, and as occupying only small stretches of coast in eastern Macedonia. In the north-west he confined Greek territory to a zone south of the Aliákmon valley, and showed only a mixed Greek population on the western side of the Píndhos. He was very liberal with his Bulgarian dispositions. Not only did he allow them to be the predominant group in the Niš-Leskovac area but he also showed all the Slavs of Macedonia as Bulgarian, and represented them strongly in Dhitikí Thráki. The Serbs he excluded from Macedonia altogether and showed them as forming exclaves only amongst the population of Old Serbia. The Vlachs he depicted after G. Weigand. The Albanians he credited with considerable territory beyond the boundaries laid down in 1913. On few modern maps had so much Albanian territory been shown. He extended it well into Montenegro, and to Novi Pazar, Mitrovica, Vranje, Kumanova and Skoplje in Serbia. Northern Ipiros was shown as Albanian and Southern Ipiros as partially Albanian.

## Chapter XI

### Maps of the Inter-war Years, 1924–1938

By 1924, most of the post-war boundary changes had been effected and the mass-movement of population which took place as a result of the War of 1914–18 and of the Graeco-Turkish War had subsided. The year 1939 marked the beginning of a new phase in the Eastern Question with the opening of the War of 1939–45. The years between had their quota of ethnographic maps. They varied in character and significance. Some of them were retrospective. In others an attempt was made to produce new maps based on the official censuses of population taken by Greece and Jugoslavia, and to incorporate modifications in distributions arising out of the profound effect of war and post-war migrations.

### L. Schultze-Jena's Map, 1927

A most useful account of the geography of Macedonia was published in Jena in 1927, written by a German geographer, L. Schultze-Jena. He limited his study to Macedonia proper (Fig. 2). He was primarily concerned with the evolution of Macedonian landscape and with the classification of settlement patterns, but his book also included references to Macedonian ethnography. Schultze-Jena's bibliography made mention of A. Boué, J. Cvijić, K. Oestreich, O. Maull, G. Weigand, J. Ivanov and E. Stanford amongst other authorities, so that he was well aware of the many diverse points of view existing on ethnographic distributions in his region. Although the book contained no complete

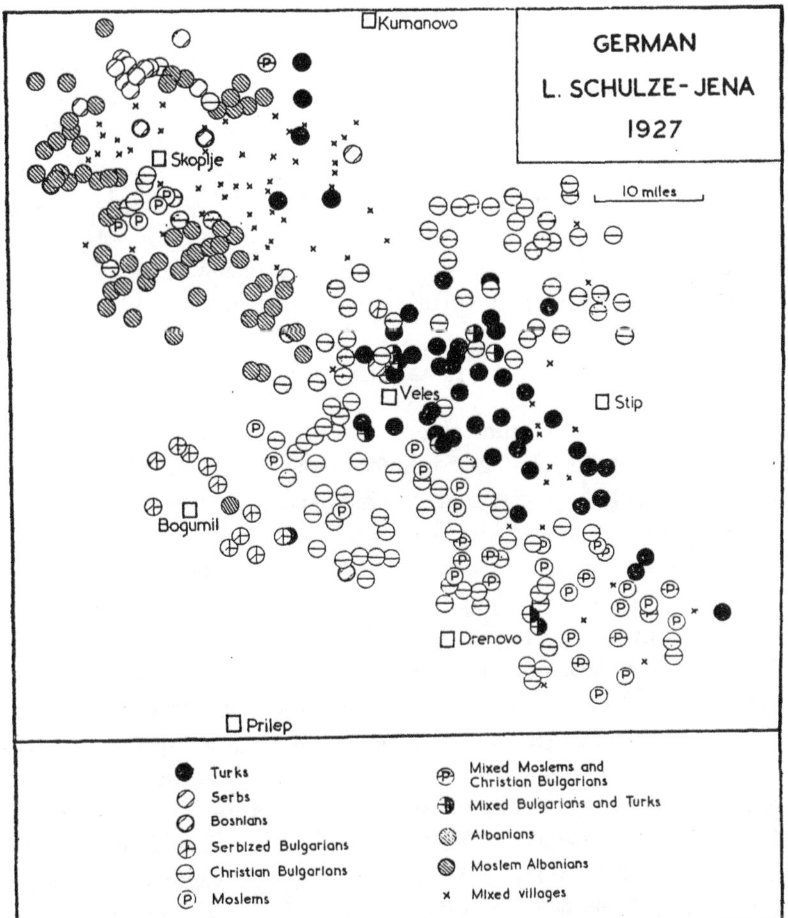


FIG. 63. ETHNIC AFFINITIES OF VILLAGES IN THE PLAIN OF SKOPLJE DURING THE LAST YEARS OF THE TURKISH EMPIRE

No indication is given in the original of size of villages.

ethnographic map, it included a map of the ethnic affinities of villages and towns in the upper and middle valley of the Vardar, during the last days of Turkish rule (Fig. 63). No source was acknowledged on the map but it appears to be based on the Bulgarian survey of 1901 (see p. 131). On Schultze-Jena's map the affinity of each village was indicated by a symbol. This method enabled the intermixture of Turks, Slavs and Albanians to be appreciated at a glance.

The Turks, according to the map, were well established in the Stip-Veles corridor. Pomaks were separately identified. The Slavs were represented as Bulgarians, only a few Serb villages being marked to the north of Skoplje ; in the region of Bogumil, however, there was a group of ' serbized ' villages which had formerly been Bulgarian. North and south of Skoplje, Albanian villages predominated, together with a number of a mixed character.

## Miss M. M. Hasluck's Map of 1930

A map of a type rather similar to that of L. Schulze-Jena was used by Miss M. M. Hasluck to illustrate the ethnography of South-West Macedonia as it was in 1923 (Fig. 64). She was interested in the ethnographic distributions from a purely anthropological point of view. One of the Karl Pearson school of Cambridge biometricians, she had spent some time compiling mass-measurements of Macedonian population in an effort to establish biometrical differences between Turks, Greeks, Vlachs and Slavs. She remarked on the fact that intermarriage seldom occurred between the various ethnographic groups and that villages retained their identity for long periods. Her map, also, would seem to have been based on the Bulgarian survey of 1901. It showed very clearly the concentration of Turks in the Koniar region, of Greeks in the Aliákmon valley and of Bulgarians in the districts of Kastoría, Flórina and Edhessa. These distributions coincided with those given by the Bulgarians A. Iširkov, J. Ivanov and D. M. Brancov. The map emphasized the stability of the Graeco-Slav frontier in this region because it followed very closely the limits prescribed on the original maps of G. Lejean, G. Weigand and J. Cvijić.

One interesting point in connection with this map was made by Miss Hasluck concerning the size of villages. The average size of villages differed from one ethnic group to another. Her figures are summarized in Fig. 64.

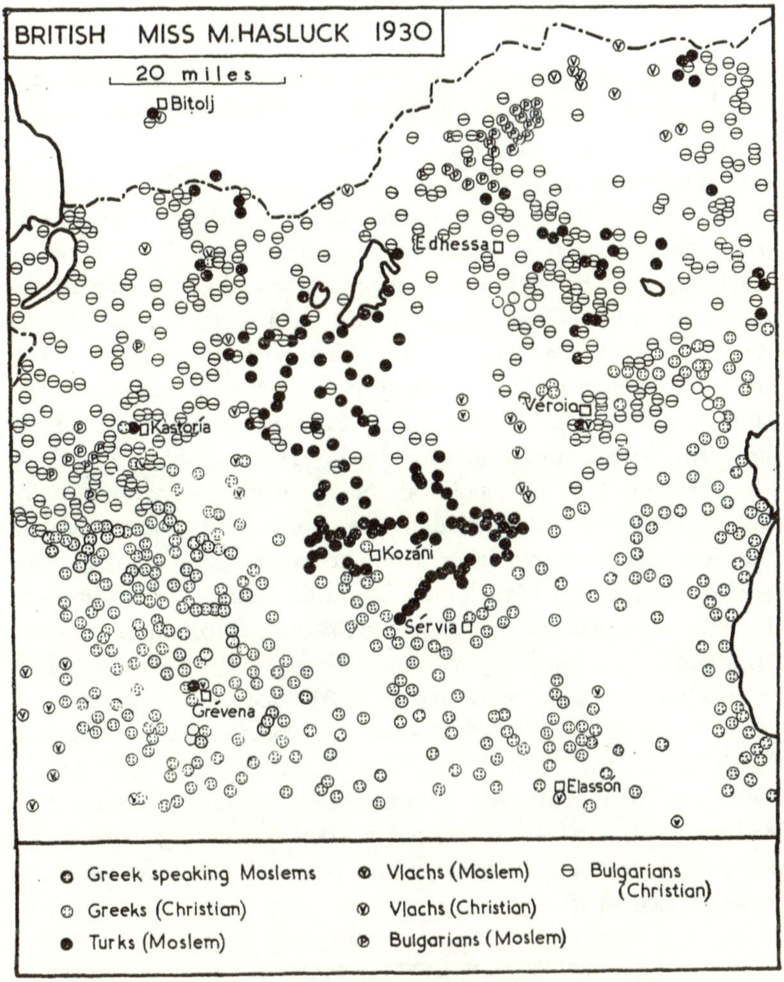


FIG. 64. SOUTH-WEST MACEDONIA IN 1923

Ethnographic affinities of village-communities are shown. The average number of inhabitants in each community was given by Miss Hasluck as follows:

| | | | |
|---|---|---|---|
| Vlach | 1800 | Moslem Greek | 300 |
| Turkish | 500 | Christian Bulgarian | 280 |
| Moslem Bulgarian | 350 | Greek | 270 |

## W. R. Shepherd's Map of 1930

A map showing the peoples of south-eastern Europe and Asia Minor in 1913 formed one of the plates in the edition of 1930 of W. R. Shepherd's well-known *Historical Atlas.* The implication of including such a map was that the ethnic situation in 1913, whatever it may have been subsequently, was an important factor to be considered in the development of the political geography of this region. In Shepherd's view the distribution of the Turks and Greeks was similar to that given in J. Cvijić's map of 1913 but he depicted the Slavs of Macedonia as Bulgarians. In the Niš-Leskovac region he made the Serbo-Bulgarian divide coincident with the political boundary. Thus he conceded the Serbs the Niš-Leskova region but he did not show any Serb territory in Macedonia. He must have considered Vlach minorities of little political consequence in Macedonia for he showed them only in the Píndhos area and around Mount Olimbos. The Albanians, on the other hand, he credited with considerable territory in northern Macedonia and in Old Serbia. Their limits in the north extended well into southern Montenegro, almost to Novi Pazar and Niš, and in the north-east to Skoplje and Veles. The whole of the Drin valley, he represented to be Albanian.

It is rather striking that these three retrospective maps emphasized the Bulgarian character of Macedonia and two of them made extensive claims also for the Albanians. In the case of Schultze-Jena and Miss Hasluck the choice of sources appears to be something more than a coincidence, insomuch as both persons were free to verify their maps by actual field-work.

## The Serbian Census Maps of 1924

In 1921, the Serbs carried out a census of the whole of the Kingdom of the Serbs, Croats and Slovenes and for the first time a reliable count of the population was made in those parts of Macedonia that had fallen to the Serbians in 1913 and 1919. The Kingdom of the Serbs, Croats and Slovenes did not enter into any agreements with Bulgaria or Turkey for an exchange of populations and no considerable changes in the ethnic composition of its southern territories had taken place between 1913 and 1924. The extension of Serbian control to this region had resulted in an influx of officials in 1913 but this trend was reversed in 1917 when the Bulgarians occupied southern Serbia. After the Bulgarians had been driven out in 1918, Serbs once more began to enter Macedonia but not in sufficient numbers to change the composition of the population. J. Ancel estimated that as late as 1929 barely 15,000 newcomers had settled in Serbian Macedonia.[1] In view of the relative stability of ethnographic distributions in this part of

*Kingdom of the Serbs, Croats and Slovenes*
*Nationalities,* 1921

| | |
|---|---|
| Serbo-Croats | 8,911,509 |
| Slovenes | 1,019,997 |
| Czechoslovaks | 115,532 |
| Ruthenes, etc. | 25,615 |
| Poles | 14,764 |
| Russians | 20,568 |
| Magyars | 467,658 |
| Germans | 505,790 |
| Albanians | 439,657 |
| Turks | 150,322 |
| Romanians and Vlachs | 231,068 |
| Italians | 12,553 |
| French | 1,163 |
| English | 453 |
| Others | 68,262 |

Macedonia, the results of the census of 1921 were all the more interesting.

To distinguish nationalities in the census returns the criterion of 'mother-language' was adopted, in contradistinction to the criterion of 'religion' adopted in connection with the exchange of populations in Greek Macedonia (see p. 263). The accompanying table gives the nationalities returned in 1921 for the country as a whole. No specific reference was made to Bulgarians or *Macedo-Slavs*. For Southern Serbia, i.e. Old Serbia and Macedonia, the relative proportions

[1] *La Macedoine* (Paris, 1930).

of various nationalities were summarized as below. *Macedo-Slavs* were officially regarded as Serbo-Croats. The explanatory text made no reference to them. It merely stated : " Il est vrai qu' en Serbie du Sud la population est des langues serbe (65·4 per cent), mais plus d'un tiers de la population serbe habite aussi les arrondissements dont la majorité est de langue albanaise et turque." Apart from the failure

*Southern Serbia: Relative Strengths of Nationalities,* 1921

| *Nationality* | *Percentage of Total Population* |
|---|---|
| Serbo-Croat .. .. .. | 59·7 |
| Slovene .. .. .. .. | — |
| Other Slavs .. .. .. | 1.0 |
| Romanian .. .. .. | 0.6 |
| Italian .. .. .. .. | — |
| German .. .. .. .. | — |
| Magyar .. .. .. .. | — |
| Albanian .. .. .. | 28·4 |
| Others .. .. .. .. | 11·2 |

to give any figures for the *Macedo-Slav* dialect, the census returns must be judged in the light of the following quotation.

> [ . . . ] census figures on ethnic composition are inevitably weighted in favour of the dominant nationality. Questions are customarily phrased so as to favour the dominant group and in their replies many doubtful borderline persons of double language or mixed nationality find it convenient to identify themselves with the dominant element. The political and economic advantages of belonging to the majority group undoubtedly result in an exaggeration of the percentage of that element in the reported census distribution, entirely aside from the manipulations of the central statistical offices.[1]

[1] Dudley Kirk, *Europe's Population in The Inter-War Years,* League of Nations (1946). See also L. Tesnière's analysis in *Les Langues dans L'Europe nouvelle* by A. Meillet (Paris, 1928).

*The Map of Nationalities*

On the basis of the returns of nationalities a number of maps was published in the census reports. There were altogether nine maps dealing with languages and religions. These maps received widespread publicity. I. Bowman made use of them in his political geography, *The New World*. They were also published in *The Geographical Review* during 1925. The most interesting of these maps depicted the general distribution of nationalities or linguistic groups within the kingdom

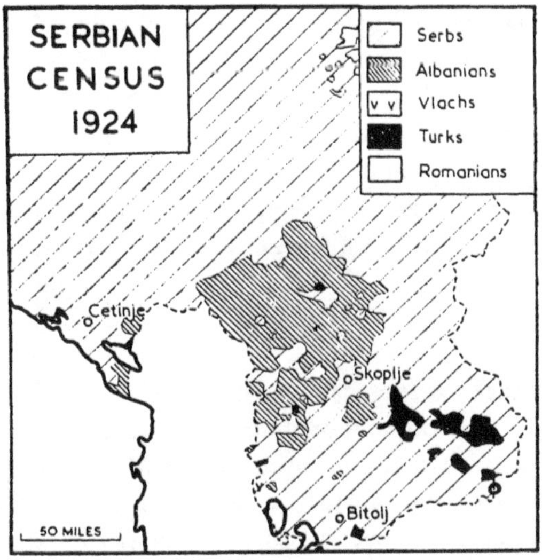


FIG. 65. DISTRIBUTION OF NATIONALITIES IN SOUTHERN JUGOSLAVIA IN 1921

In the original, the distribution is shown by Communes and a distinction is made between 50 and 75 per cent majorities.

(Fig. 65). But since the *Macedo-Slav* dialect was regarded as a branch of Serbo-Croat, the map lost a great deal of its value.

*The Turks.* On this map the Turks emerged as relatively much less important than might have been expected from Cvijić's distribution of 1918. There was no reason to suppose, however, that their strength had been deliberately misrepresented. The Turks of this part of Macedonia had never constituted a solidly settled peasantry as they had done in the Koniar region of Greek Macedonia. The erection of the boundary in 1913 had led to the voluntary emigration of large numbers of Turkish officials and Turkish landlords. After 1918, although the Serbians did not officially press for the transfer of their Turkish-speaking minority to Turkey, many Turks continued to leave voluntarily for Anatolia and Thrace. No official figures of this movement have been published but there seems to be no doubt of the fact that after 1913 the Turkish minority in Jugoslavia languished both in numbers and in territory.

*The Greeks.* No Greeks figured on the map. Many of the Greeks of Serbian Macedonia had crossed the border into Greece between 1913 and 1921 but, in any case, the number of Greek speakers had never been great. Those cartographers who had portrayed extensive Greek territory in the Bitolj region had relied on the criterion of Greek Orthodox religion to justify their claims.

*The Slavs.* All Slavs were depicted as Serb on the map and this representation connoted a homogeneity which of course the Bulgarians disputed. As the *Macedo-Slavs* had no minority rights, the Government of the Kingdom of the Serbs, Croats and Slovenes refused to recognize them as a separate nationality. To have done so would have been to acknowledge the weakness of the Serbian claims on Macedonia. The school of thought of which J. Cvijić had been such a vigorous exponent had always described the *Macedo-Slavs* as incipient Serbs. To depict them officially as Serbs, on the census map, therefore, was the logical conclusion to this line of thought. Indeed, it became the express policy of the Government to absorb these Slavs into the Serbo-Croat group in the hope that the dangers of irredentism in Macedonia might thereby be reduced. The Serbians feared a separatist movement on a large scale were the *Macedo-Slavs* to be granted any separate status or any real freedom of cultural expression. On the occasion of the Jugoslav protest over the Graeco-Bulgarian Protocol of September, 1924 (which placed the 'Bulgarian' minority in Greece under the protection of the League of Nations), the Jugoslav Foreign Minister, M. Maringović, stated categorically to the Greek Minister in Belgrade ; "We do not wish nor can we wish to rely on arms alone for the defence of Serb Macedonia. For us it is essential that no third party should be able to dispute the Serb character of that territory. Therefore, the dogma that the Slav inhabitants of Macedonia are Serbs is a basic principle of our policy. We cannot admit that north of the frontier these Slavs are Serbs while beyond that frontier these same people become Bulgars. To recognize that the Slavs of Edhessa and Flórina are Bulgar would be to destroy the very foundations of our policy in regard to Serb Macedonia. That is the basis of our policy, and should Greece be unwilling to back us up on this question, then we shall regretfully be compelled to change this basis and seek an agreement with Bulgaria by dividing up Greek Macedonia into spheres of influence."[1]

The policy of 'serbization' which had succeeded in the Niš-Leskovac area in the years following 1878 had established a precedent, but events subsequent to 1921 proved the *Macedo-Slavs* of Serbian Macedonia more obdurate in their refusal to accept 'serbization.' The Macedonian revolutionary society (I.M.R.O.) throve on this policy and a state of *komitadji* warfare disturbed the peace of southern Serbia for the next two decades.

*The Vlachs and Albanians.* Only one very small area was attributed to the Vlachs on this map. Presumably this was because their communities in Serbian Macedonia were so scattered that their numbers were concealed amongst those of the Albanians and Slavs. The widespread distribution of the Albanian minority, however, was acknowledged. Since 'mother-tongue' was the criterion used, all Albanian speakers had been counted as of Albanian nationality in the census returns; the 'albanianized Serbs' of J. Cvijić's classification were thus ignored. The map showed a fairly compact distribution of Albanians over the whole of Old Serbia and much of the eastern part of Serbian Macedonia. There were a few enclaves of Slavs within this area, otherwise Albanian speakers predominated, forming over seventy-five per cent of the population of most of the communes. The distribution of Albanians was similar to that which had been portrayed on the map of J. Gabrys. It fell short, however, of those given on the maps of Italian cartographers, and on that of J. I. Kettler.

[1] Quoted from P. Pinenellis, *History of Greek Foreign Policy*, 1923-41 (Athens, 1948) by A. A. Pallis in a typescript historical study, "Macedonia and the Macedonians" (London, April, 1949).

The greatest discrepancies in comparison with the latter maps occurred in the regions of Novi-Pazar and Debar : here the Serbian map showed Moslem Serbs where pro-Albanian maps indicated only Albanians.

*Conclusion*

The distributions given on the Serbian map were accepted subsequently by many ethnographers, historians and geographers, because they emanated from a 'reliable official source.' S. Rundle reproduced these distributions in his sketch-maps with minor modifications (Fig. 76). H. M. Chadwick depicted Jugoslav Macedonia as Serb in his generalized sketch of the nationalities of Europe (Fig. 81). During the years between 1924 and 1939 no other official map of the ethnography of Macedonia was reproduced by the Jugoslav authorities.

The policy of centralization and ' serbization ' brought little peace or development to Macedonia. Down to 1933, I.M.R.O. (see p. 150) practically usurped the authority of the government in this area. After the assassination of King Alexander in 1933, and a threat of war, Jugoslavia and Bulgaria came to some understanding over Macedonia. The Bulgarian government exiled I.M.R.O. which had hitherto had its headquarters in Pirin Macedonia and its leaders fled to Cairo and Vienna. But even after the exclusion of I.M.R.O. agents the Jugoslav hold on Macedonia remained insecure.

## The Maps Published by The League of Nations in 1926

In very few of the maps so far considered may modifications in ethnographic distributions be attributed to movements of population having taken place. Between 1850 and 1912 there was little evidence to suggest that any movements of population had taken place on a scale large enough to affect greatly the relative strengths of the respective ethnic groups in Macedonia and the neighbouring territories. The many variations and modifications to be found on maps of differing origins were to be ascribed for the most part to the contrary ideas held by rival scholars ; the changes in maps from one generation to another were not due to population changes but to the acceptance of new ideas about the character of a relatively immobile population. In some cases changes of nationality were brought about by nationalist propaganda. Admittedly, small shifts of population had taken place before 1912, particularly after the disturbances of 1903, but neither guerilla warfare nor Turkish oppression sufficed to drive away many peasants from the soil. Movement of population was in any case frowned upon by the partisans of the respective 'parties' who did all they could to prevent a single gap occurring in their ranks. Conversely, new colonization was difficult, not only because of the opposition offered by rival *komitadjis* but because of the general unwillingness of aliens to settle in the difficult Macedonian territory.

*Migrations, 1912-26*

During and after the events of 1912-13, however, and subsequently down to 1925, mass-displacements of the Macedonian population took place, resulting in substantial changes in the ethnic structure. First, there were many Turkish refugees who fled before the successful armies of the Balkan League in 1912. Second, when the Balkan allies fought over the spoils, their respective nationals were forced to flee from the battle areas. Third, consequent upon the political partition of Macedonia in 1913, there occurred a semi-voluntary repatriation of many Bulgarians, Greeks and Turks who had found themselves on the wrong side of the newly-erected boundaries. Fourth, fresh flights of refugees once again took place as a result of troop movements between 1915 and 1918. Fifth, consequent upon the modifications in the boundaries of Bulgaria under the terms of the Treaty of Neuilly, provision was made by means of the Graeco-Bulgarian Convention of 1919, for the voluntary emigration of Greek and Bulgarian minorities from both countries. Sixth, after the Graeco-Turkish war of 1922, great migrations of Turks and Greeks took place between the two countries.

*Changes in Ethnic Structure*

The net changes in the ethnic structure of Macedonia produced by the first five movements of population was not as striking as may be expected due to the fact that contrary movements cancelled out each other when a full reckoning was made. The Mixed Commission of the League of Nations published in 1926 the figure of 53,000 Bulgarians emigrating from Greece under the terms of the Graeco-Bulgarian Convention, and 30,000 as the total complement of Greeks leaving Bulgaria, all of whom did not, of course, settle in Macedonia. Considerably fewer Bulgarians took advantage of leaving Greek Macedonia than had been anticipated. Their transference was opposed by the revolutionary society, I.M.R.O., on the grounds that it would be tantamount to a betrayal of their sacred rights to Macedonian territory for the *Macedo-Slavs* to desert to Bulgaria. Even amidst the frightful conditions arising from the wars of 1912-13 and 1914-18, and in spite of insecurity associated with the changes of political boundaries between 1913 and 1919, it is patent that the Macedonian peasantry was not willing to change its abode. It is apparent also that the accuracy of any estimates of the *net* results of such movements as did take place, was difficult to assess, as many of the migrations were only of a temporary character.

Of an altogether different magnitude were the net results of the sixth movement involving Greek and Turkish populations which occurred after the defeat of the Greek Army in Turkey in 1922. The impulse towards the re-establishment of the Greek Empire of Constantinople had gradually gathered strength since the *Grand Idea* had been first conceived (see p. 18 *et seqq.*). A climax was reached in 1919 when what had hitherto been but a dream became a practical possibility. The revival of the Byzantine Empire was within an ace of achievement when Greece declared war on a Turkey already weakened by war and revolution. Then came the phenomenal reconsolidation of the Turkish forces in Anatolia under Mustafa Kemal. The Greek army suffered a disastrous defeat and the Turks exacted fearful vengeance on the Greek community within their power ; many were killed or died of exposure and disease ; others fled in panic from western Anatolia and Pontus to Greek territory. When soon after this disaster events forced the Allied Powers to consign Eastern Thrace to Turkey, Greek refugees from this territory and from Constantinople swelled the total numbers fleeing into Greece.[1]

[1] 914,300, net, is the estimate given by S. P. Ladas, *The Exchange of Minorities, Bulgaria, Greece and Turkey* (New York, 1932).

During the course of the negotiations which followed the Greek defeat, both the Greek and Turkish governments decided to take the opportunity offered by the unsettled conditions to effect a complete mutual exchange of minorities. This decision, whatever its wisdom, implied a compulsory exchange of population, for there were still many Greeks remaining in Turkey and many Turks in Greece, unwilling to quit their homes of their own free will. Under the terms of the Graeco-Turkish Convention of January, 1923, these reluctant minorities were forcibly uprooted and despatched to Greece and Turkey respectively. It is estimated that some 356,000 Turks and 190,000 Greeks were involved in these transfers.

The Graeco-Turkish Convention defined in its first article, the persons subject to this compulsory exchange as follows: "Turkish nationals of Greek Orthodox religion established in Turkish territory and Greek nationals of Moslem religion established in Greek territory." 'Race' and language were ignored because of the difficulty of interpretation of such terms. There were only three general exceptions to this ruling; excluded from compulsion were:

(1) the Greek inhabitants of Constantinople;
(2) the Moslem inhabitants of Dhitikí Thráki;
(3) Albanian-speaking Moslems.

Thus the criterion of religion was preferred to that of language, which meant that many Moslems of Greek, Bulgarian and Romanian speech had to leave Macedonia. It also meant that a large proportion of the immigrants to Greece spoke no other language than Turkish.

The great exodus of Moslems and the great influx of persons of Greek Orthodox faith, after 1922, materially changed the ethnic composition of Macedonia and Dhitikí Thráki. Of the million or so Greeks arriving from Turkey, a fair proportion settled in these territories. It is difficult, however, to estimate the net changes occurring in the total Greek population of Macedonia as a result of the migrations. According to A. A. Pallis, by the end of 1924 some 700,000 Greeks had settled in Macedonia.[1]

Raoul Blanchard writing in *The Geographical Review* in 1925

[1] A. A. Pallis, "Racial Migrations in the Balkans during the years 1912-25," *The Geographical Journal* (1925).

suggested a figure of 660,000 Greeks settled in Macedonia and Thrace before the autumn of 1924, but this figure included all manner of refugees—from Jugoslavia, Bulgaria, the Caucasus and Turkey. According to the official records of the Mixed Settlement Commission of the League of Nations, published in 1926, 430,295 individuals were settled in the rural areas of Macedonia. These figures also included all manner of refugees but they did not incorporate the number of persons settled in towns nor the very large number of persons settling in Macedonia without direction from the Commission, since the number of individuals concerned in independent settlement was not known. It becomes obvious that all attempts to estimate the net growth of the Greek population of Macedonia due to the Turkish debacle were at the most mere hazards. So fluid was the Macedonian demographic situation during the years 1922-24 that it was found impossible to take a census of refugees from Turkey. But however unreliable the estimates were, it cannot be denied that first, the mass-population movements occasioned by the Graeco-Turkish war were far in excess of those arising out of events between 1912-19, second, that Macedonia and Dhitikí Thráki were the main zones of settlement for the immigrants, and third, that the ethnographic character of these territories was profoundly modified in consequence.

*The Ethnographic Map*

In an attempt to summarize these changes, the League of Nations published a report, *Greek Refugee Settlement* (Geneva, 1926), which incorporated an 'Ethnographical map of Greek Macedonia showing the proportion of the different ethnographical elements in 1912 (before the Balkan Wars) and in 1926 (after the settlement of refugees).' A note on the map read as follows: "The result of the settlement in Macedonia of about 500,000 rural and 300,000 urban refugees, followed by the emigration in 1923 of 348,000 Moslems under the Treaty of Lausanne for the Exchange of Greek and Turkish Populations, has been radically to change the ethnical composition of the population of Greek Macedonia and definitely to hellenize that province." The map was compiled by the Refugee Settlement Commission at Athens (Fig. 66). It purported to be based on the

Greek census returns of 1913 and 1920, as well as on statistics supplied by the Greek Ministry of Public Welfare, the Mixed Commission for the Exchange of Greek and Turkish Populations, the Mixed Commission for the Voluntary Emigration of Greek and Bulgarian Populations, and the Refugee Settlement Commission.

The Settlement Commission consisted of four members, two of whom were Greek. In 1925, one of the serving Greeks was A. A. Pallis and the compilation of the map would seem to have been mainly his work. He was well qualified to pronounce on Macedonian population problems. He had already served as Relief Officer in Macedonia in 1913, as Secretary-General of the Refugee Settlement

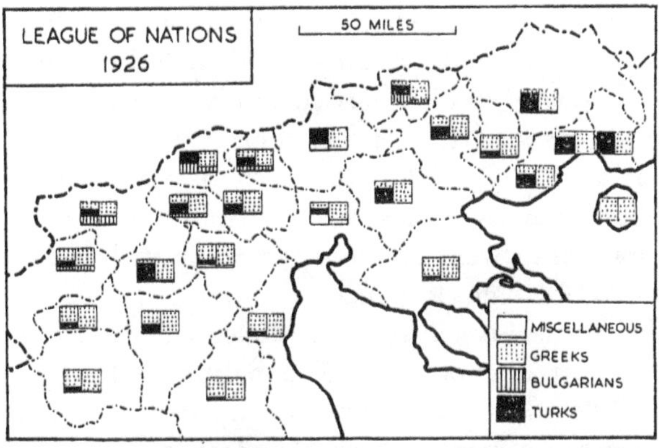


FIG. 66. ETHNOGRAPHICAL MAP OF GREEK MACEDONIA

The proportions of different ethnographic elements are shown in 1912 (before the Balkan Wars) and in 1926 (after the settlement of the refugees).

Commission in Hellenic Macedonia in 1914-15 and as the Deputy-Governor-General of Salonika in 1917-18. He had also been associated with the resettlement of Eastern Thrace in 1919. Some of Pallis's ideas on the migrations have already been referred to. It is to be remarked that the League of Nations' map was based largely on his figures, in particular upon the number of Greeks, Moslems and Bulgarians in Greek Macedonia in 1912, and on estimates partly prepared by him and issued by the *Gouvernement Général de Salonique* in 1925 (see p. 267). The League of Nations' map was, indeed, based on these two sets of statistics because the Greek censuses of 1913 and 1920 did not return nationalities and they could only have provided the map-makers with totals of population. Moreover, as already indicated above, no reliable statistics of refugees had ever been computed. The League of Nations' map, therefore, must be regarded purely as a Greek interpretation of the ethnographic situation, both in 1912 and in 1926.

The criterion determining the character of an ethnic group adopted by the Greeks was religion and not language, and therefore only Orthodox Greeks, Moslems and members of the Schismatic Bulgarian Church were considered as national communities. No attempt was made to distinguish Albanians, Vlachs, Pomaks, Serbs or *Macedo-Slavs*. Thus Bulgarian, Vlach and other non-Greek speakers of Orthodox faith were considered to be Greeks, and Pomaks were considered to be Moslems. A unique method of depicting the ethnography was adopted, which showed relative strengths but gave no indication of absolute size of the respective groups. The distribution was shown by departments which were of a size comparable with English counties. This method renders a comparison with earlier maps, which were concerned primarily with the delimitation of ethnic territory, difficult to make.

According to the map, in 1912 the Moslems numbered 475,000 and formed 39.4 per cent of the total population. They were in the majority in eleven of the twenty-five departments. In 1926, only 2,000 Moslems remained in Greek-Macedonia ; they were Albanians exempted from the transfer.

The Greeks in 1912 numbered 513,000, according to the Greek interpretation, and they had formed 42·6 per cent of the total population. They had a majority in twelve of the twenty-five departments. In 1926, the Greek community, according to the map, had swelled to a total of 1,341,000 and it accounted for 88·8 per cent of the total population. It had a majority in all twenty-five departments.

The Bulgarians were alleged to have numbered 119,000 in 1912 and to have formed only 9.9 per cent of the population. These figures excluded, of course, those Bulgarian speakers who did not belong to the Exarchate, and also it excluded Pomaks. In no departments did the Bulgarians have a majority but they were strongly

*A Greek View of the Ethnic Elements in the Population of Greek Macedonia in 1925*

| Department | Total Population | Greeks | | Slavs | Vlachs | Albanians | Jews |
|---|---|---|---|---|---|---|---|
| | | Indigenous | Refugees | | | | |
| Sidhirókastron | 34,129 | 10,234 | 17,156 | 4,543 | 2,066 | — | — |
| Sérrai | 59,676 | 30,419 | 25,022 | 2,376 | 1,158 | — | — |
| Zíkhna (Fillís) | 37,069 | 22,353 | 12,902 | 606 | 161 | — | — |
| Nigríta (Visaltía) | 26,085 | 17,984 | 8,054 | — | — | — | — |
| Khalkidhikí | 38,354 | 31,552 | 6,741 | — | — | — | — |
| Liarigovi | 17,320 | 12,800 | 4,500 | — | — | — | — |
| Kilkís | 35,655 | 374 | 34,870 | 251 | 141 | — | — |
| Langadhás | 43,407 | 21,275 | 22,024 | — | — | — | 65 |
| Peonia | 37,579 | 16,103 | 10,478 | 7,075 | 3,918 | — | — |
| Thessalonikí | 439,627 | 144,462 | 209,659 | 1,664 | 988 | 1,500 | 60,001 |
| Véroia | 61,124 | 34,050 | 15,660 | 751 | 10,235 | — | 428 |
| Piería | 39,462 | 24,426 | 9,106 | — | 5,823 | 57 | 36 |
| Enotía (Almopía) | 32,306 | 5,059 | 20,659 | 5,057 | 348 | — | — |
| Granítsa | 49,554 | 1,854 | 25,589 | 22,031 | — | — | — |
| Edessa | 27,174 | 6,309 | 10,726 | 8,750 | 940 | 4 | — |
| Eordhaía | 43,596 | 13,092 | 25,668 | 3,449 | 931 | 2 | 3 |
| Kozáni | 47,613 | 30,710 | 16,554 | — | — | — | — |
| Anaselítsa (Vóïon) | 32,264 | 27,947 | 3,918 | — | — | 3 | — |
| Grevená | 49,445 | 32,031 | 5,837 | — | 11,155 | — | — |
| Flórina | 58,899 | 13,155 | 7,449 | 34,244 | 3,590 | 27 | 349 |
| Kastoría | 48,758 | 25,076 | 5,962 | 14,617 | 2,330 | 213 | 525 |

Based on J. Ancel, *La Macédoine* (Paris, 1930).

represented in the northern and western departments—Flórina (35 per cent), Enotía (46 per cent), Sidhirócastron (37 per cent) and Kilkís (29 per cent). In 1926, the Bulgarian strength, according to the map, had fallen to 77,000 and their proportion to 5·1 per cent of the total population of Greek Macedonia. In only three departments did they form any considerable proportion of the population.

On the basis of this map the authors of *Greek Refugee Settlement* felt justified in summing up the situation in 1926 thus :

> At the present time the Greek Republic is united in race, language and religion [ . . . ] including the pro-Bulgarians and pro-Romanians of Macedonia, the Jews of Salonika, the Albanians of Chamuria and the Turks of Western Thrace, alien elements do not number 300,000 out of a total population of six and half millions. The collapse of Hellenism in its wider sphere in the Eastern Meditterranean and the end of the Grand Idea was the price paid for this unity.

A similar conclusion was expressed later by Sir John Hope Simpson in *The Refugee Problem*, R.I.I.A., Oxford, 1939. He wrote :

> The exchange of population, though at the time it caused infinite misery was an element of crisis for Greece, at least resulted in the solution of a difficult political situation. What Lord Curzon once described as the 'unmixing of populations' terminated a period of constant and dangerous international friction and rendered to the Greek and to the other States of the Near East a homogeneity of population which could have been effected in no other way. It unquestionably strengthened the Greek State by the influx of a stable and hard-working element, whose ideas and ideals are Greek ; moreover the disappearance of an irritating minority question has made it possible for Greece to live on good and even friendly terms with her enemies.

But these conclusions depended upon the validity of the League of Nations' map. This map was based on the figures provided by A. A. Pallis and he in turn depended on those provided by G. Amadori-Virgilj in 1908. Amadori-Virgilj, in his turn, had gained his information from the estimates of Graeco-Turkish origin published in the

*Bulletin d'Orient* of 1904. The accuracy of these figures was to be doubted for many reasons. First, because they were based purely on religion, other criteria being entirely disregarded : second, Turkish statistics were notoriously inaccurate and incomplete : third, no information of note existed whereby such figures could have been brought up to date.

The accuracy of the League of Nations' map was also dependent upon the validity of the estimates of the net growth of the Hellenic population in Greek Macedonia made in 1926. It has been pointed out above that such estimates were of necessity purely arbitrary and not

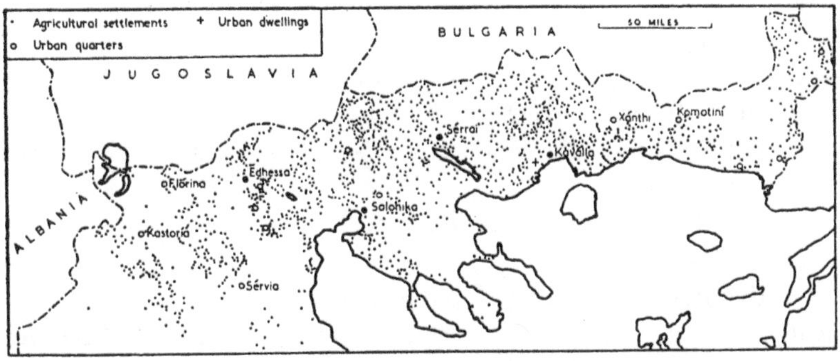


FIG. 67. THE SETTLEMENT OF RURAL AND URBAN REFUGEES IN GREECE, 1923-1926

by any means reliable. All kinds of inconsistencies were apparent in even the official statistics. In the conclusions expressed in *Greek Refugee Settlement*, for example, the figure of 700,000 Greeks was mentioned as the number settling in Macedonia. Yet from the information given on the map, the net growth of Greek population amounted to 828,000.

Finally, the method of depicting the groups on the map did not admit of an analysis of their detailed distribution. The departments were too large for the significant character of frontier districts to be portrayed and thus important frontier minorities remained concealed. The significance of the detailed distribution of Greek settlement may, however, be appreciated by reference to a cognate map also accompanying *Greek Refugee Settlement.*

*The Settlement Map, 1926*

This map showed, by means of symbols, the distribution of rural and urban settlements made under the auspices of the Mixed Commission (Fig. 67). No indication of the size of settlements was given, which meant that a visual impression was likely to concede an equal value to smaller groups in outlying areas and to larger groups in the plains. A glance at this map shows a concentration of settlement in Greek Macedonia and Dhitikí Thráki. But there was no uniformity about the distribution of settlements. They avoided certain difficult frontier regions and were concentrated in the more fertile tracts evacuated by the Turks, or in newly-reclaimed marshlands. The outstanding zones of settlement were to be found in the following areas :

(1) the Aliákmon valley and around Konia ;
(2) the Moglenítsas valley ;
(3) the Strimón valley and on the hill slopes and terraces to the west of it ;
(4) the plains of Sérrai and Dráma ;
(5) the Néstos Valley ;
(6) the coastal plains of Dhitikí Thráki.

It is to be noted that the Rodópi mountains of Dhitikí Thráki and eastern Macedonia, the Píndhos range, the Grámmos mountains, the Kaïmakchalán range and the Páïkon highlands contained hardly a single settlement. These were the regions where alien Slav minorities were entrenched. The reasons for the congestion of Greek settlements within certain limits were not far to seek.

(1) The new settlements had to be made as far as possible in favoured spots evacuated by the Turks, as little time or capital was available for the development of new areas. Many settlers expressed a preference for regions already densely settled by Greeks because of the resemblance of the terrain to the settlers' land of origin. Coastal plains and river valleys were settled in numbers greater than was considered politic by the Commission and minute holdings were the consequence.[1]

[1] J. H. Schultze, " Greek frontier colonization in Thrace and Macedonia " (Translated), *Scottish Geographical Magazine* (March, 1937).

(2) Nearly half of the refugees were artisans, tradespeople and urban dwellers of various types with no interest in agriculture. They flocked into the coastal towns, particularly into Salonika and Kaválla.

(3) Even those refugees who were farmers and who had practised viticulture, tobacco growing, silk-worm rearing, olive growing and cereal production in parts of Thrace and Anatolia, found themselves in 1922 in a land far different in its geographical character from that to which they had been accustomed. Many of these agriculturalists were skilled gardeners who sought favoured sites for the pursuit of their rural occupations: they needed a convenient outlet for their products and therefore avoided difficult and inaccessible country.

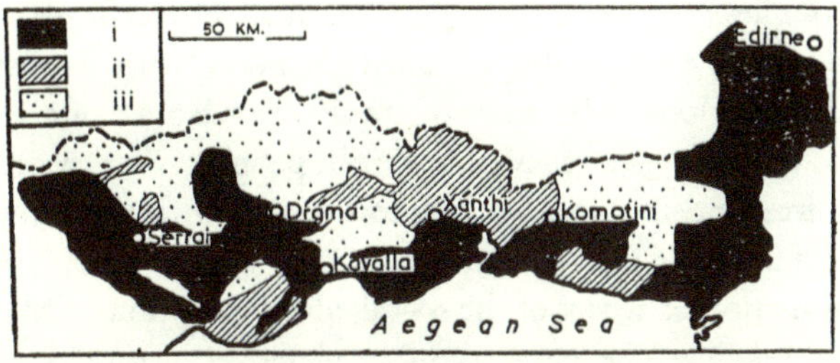


FIG. 68. COLONIZATION IN NEW-GREECE (AFTER J. H. SCHULTZE)

The references in the key are as follows: **i.** Districts with a net population-growth; **ii.** Districts where growth and decline of population may be equated; **iii.** Districts with a net population-decline.

Based on statistics and observations for the years 1913-28 in Eastern Macedonia and 1920-28 in Western Thrace.

(4) *Komitadji* operations, and the hostility of alien minorities in the northern and western districts of Macedonia, prevented isolated settlement, the more so as the Greek newcomers often lacked male representatives to defend pioneer groups opening up new territory and they were unaccustomed to guerilla warfare.

The net result was that in spite of the great influx of Greeks, hellenization of the whole of Macedonia was not achieved. It may be deduced from the map that an intensification of settlement took place in areas which had already been occupied by Greeks together with the hellenization of territory formerly Turkish. It must also be noted that initial settlement in the mountainous areas often resulted in failure. The Greek immigrants were not adapted to the hard environment and declared their abhorrence of the mountains. J. A. Schultze travelling in eastern Macedonia and Dhitikí Thráki in 1937 found many evidences of abandoned settlement, particularly in upland areas (Fig. 68).

*Conclusion*

The frontiers of Greek ethnic territory were not by any means made coincident with the north-western and northern political boundaries of the state in 1926, but they remained for most of their length to the west and to the south of the boundary. There were many dangers inherent in a situation in which a small state had to defend 800 miles of difficult country, with many of the frontier districts populated, however scantily, by alien elements with a pronounced ultramontane outlook. In times of trouble the danger of a rising had to be faced on the part of these non-Greek peoples, who, prompted by a spirit of irredentism, were willing to connive with their neighbours on the far side of the Greek boundary.

There was another aspect of the so-called hellenization of Macedonia, easily overlooked. The new colonists were very conscious of their own diversity. They were not 'Greeks' in the national sense of the word but persons of Greek Orthodox faith drawn from different corners of Europe and the Near East. "Their affections remain true to their land of origin rather than to the country of their adoption [ . . . ]. Nostalgia for the Pontus, Anatolia or the mountains of Caucasus, where their ancestors held the land before them tends to lend enchantment to the old way of life in the old hereditary places. They are suspicious still of their new country and of their near neighbours. The reality of Greek colonization in these lands will not be achieved before a new generation has inherited them as 'home'."[1]

Under the circumstances the authors of the League map were perhaps guilty of over-representation of the Greek elements in the population of Macedonia in 1926. The Greek Government was too optimistic of the ability of the Greek to absorb the Bulgarian minority in Macedonia before the recovery of Bulgaria brought with it bitter

[1]J. A. Schultze, *op. cit.*

recriminations. No official maps of Macedonian ethnography were produced by the Greeks in subsequent years. An attempt was made in the Census of 1928 to count minorities and the figure of 81,844 *Macedo-Slavs* was returned in the published figures, as well as 16,755 Moslem Bulgarians—presumably living in Dhitikí Thráki (see p. 274 & 286).[1] Bulgarian estimates of the number of Bulgarians in these territories were, of course, much higher. For example, V. Batakliev put the figure at 330,000, a threefold increase on the Greek estimates.[2] German authors, in the main, were critical of Greek figures. Most English and American authors, however, accepted them and relied on Pallis's account of the population exchanges. Sir John Hope Simpson's survey of 1939 quoted Pallis's figures and agreed with the conclusions arrived at in *Greek Refugee Settlement*.[3]

## A. Haberlandt's Map of 1927

A German professor of ethnography, A. Haberlandt, was one of the principal contributors to the work on the illustrated ethnography of Europe and its confines, edited by Georg Buschan and published in Stuttgart in 1926.[4] This work, sponsored by the great ethnographic museums of Hamburg, Berlin and Vienna, remains a valuable work of reference. Haberlandt compiled a series of maps for the work including a linguistic map of Europe. This map was revived and re-issued in 1927 on a scale of 1: 3,000,000 by *Freytag und Berndt* of Vienna. It quickly became popular as a wall-map in the inter-war years and was made use of in most of the universities of Europe as a most reliable reference to the ethnographic distributions of the continent. Some idea of Haberlandt's sources may be gained from a perusal of the bibliography accompanying the book mentioned

[1] See also A. A. Pallis, " The Greek Census of 1928 " in *The Geographical Journal* (June, 1929).

[2] "Die Wanderungen der Bulgaren in den letzen dreissig Jahren," *Zeitschrift für Geopolitik* (1941).

[3] Sir John Hope Simpson, *op. cit.*, gave the figure of 121,677 Bulgarians emigrating from Greece between 1913 and 1939.

[4] *Illustrierte Völkerkunde-Europa und seine Randgebiete.*

*Languages and Religions in Greece in 1928*

| Languages | Religions | | | | | | | |
|---|---|---|---|---|---|---|---|---|
| | *Orthodox* | *Catholic* | *Protestant* | *Moslem* | *Jewish* | *Others* | *Without Religion* | *Totals* |
| Greek .. .. .. | 5,716,100 | 27,747 | 3,867 | 2,623 | 9,090 | 15 | 81 | 5,759,523 |
| Turkish .. .. | 103,642 | 327 | 760 | 86,506 | 17 | 1 | 1 | 191,254 |
| Macedo-Slav .. .. | 81,844 | 68 | 11 | 2 | 58 | — | 1 | 81,984 |
| Spanish .. .. | 28 | 58 | 41 | 72 | 62,999 | — | 2 | 63,200 |
| Armenian .. .. | 31,038 | 1,136 | 1,432 | 16 | 10 | 2 | — | 33,634 |
| Kutzo-Vlach .. | 19,679 | 9 | 2 | 3 | 10 | — | — | 19,703 |
| Albanian .. .. | 95 | 59 | 17 | 18,598 | 3 | 1 | — | 18,773 |
| Bulgarian .. .. | 20 | — | — | 16,755 | — | — | — | 16,775 |
| Gypsy .. .. .. | 3,853 | — | 1 | 1,130 | — | 14 | — | 4,998 |
| Russian .. .. | 3,177 | 49 | 14 | 3 | 40 | — | 12 | 3,295 |
| Italian .. .. .. | 98 | 2,878 | 18 | 1 | 203 | — | 1 | 3,199 |
| English .. .. | 201 | 274 | 1,605 | 1 | 15 | — | 2 | 2,098 |
| Others.. .. .. | 1,754 | 2,577 | 1,235 | 307 | 346 | 12 | 17 | 6,248 |
| Totals of Alien Languages .. | 245,429 | 7,435 | 5,136 | 123,394 | 63,701 | 30 | 36 | 445,161 |
| Grand Totals .. | 5,961,529 | 35,182 | 9,003 | 126,017 | 72,791 | 45 | 117 | 6,204,684 |

Based on *Annuaire Statistique de la Grèce* (Athens, 1930.)

above. It ran into forty-three pages and included references to A. Boué, J. Cvijić, and A. Iširkov. Haberlandt's map was, of course, not confined to the Balkan peninsula but he himself had studied the ethnography of the peninsula in 1917 and 1918.[1] G. Buschan, the editor, had also written on the Bulgarians.[2] It might be expected therefore that the part of the map referring to Macedonia was based on first-hand knowledge.

It is interesting to observe that the base-map used for the distributions was a map showing the density of population. This resulted in certain difficulties in interpretation because as many as five different shades of the same colour were used to indicate any one ethnic group. The results of this interesting experiment however were happier than might have been expected. A further refinement introduced by Haberlandt was a set of symbols to represent religions, which could be superimposed upon ethnographic colouring.

A glance at the map shows that Haberlandt had made use of the Serbian census-map of 1924 and that he had made allowances for the movements of populations taking place between 1912-25. He was obviously of the opinion that the Turkish minority in Macedonia had virtually disappeared even in the Serbian part. He showed only small lightly populated regions of Turkish territory in the Vardar valley and in Dhitikí Thráki. The transfers of population induced him to mark as Greek territory the greater part of Greek Macedonia and Thrace with the exception of the frontier districts of the north-west. He also included the Greek Orthodox population of Albania within the Greek group.

In his classification of the Slavs of Macedonia, Haberlandt departed from the views held hitherto by Austrian scholars, in that he represented them to be partially Serb in the north-west, and *Macedo-Slav* south of Skoplje. He depicted *Macedo-Slavs* in the Crni Drim valley (Debar district), and in the northern part of Greek Macedonia, but not

[1] (1) "Kulturwissenschaftliche Beitrage zur Volkskunde von Montenegro, Albanien und Altserbien," *Ergänzungband XII zur Zeitschrift für Osterreich.* (2) *Volkskunde* (Wien, 1917).

[2] *Volkskunde der Balkanländer* (Wien, 1918).

in south-eastern Bulgaria (Fig. 84). He indicated Bulgarians, only in that portion of Macedonia under Bulgarian political control, and none in Serb or Greek Macedonia. Haberlandt showed only small enclaves of Vlachs in south-western Macedonia and Thessalía. In his portrayal of Albanian territory, in the north he accepted the distributions given in the Serbian census-map of 1924, but he indicated no Albanians in Greek Ipiros and he marked the Albanian ethnic limits to fall short of the Graeco-Albanian political boundary.

Haberlandt's map was obviously very much influenced by the ideas on Macedonian ethnography which had been developed during the war-years. He accepted the concepts of the *Macedo-Slavs* and of the Orthodox-Greek Albanians. His views on the hellenization of Greek Macedonia were reasonably in accordance with conclusions to be drawn from the League of Nations' map. He did not accept the Serbian official classification of Macedonian Slavs as Serbo-Croat but he chose to ignore their affinities with the Bulgarians.

## A. Meillet's Map of 1928

A distinguished French philologist, A. Meillet, had written a survey of the languages of Europe and their distribution in 1918. He was of the opinion that language was the principal factor determining national sentiment in Europe. In 1928, he revised his work and a second edition was published which contained a sketch-map showing linguistic distributions, together with a statistical appendix compiled by L. Tesnière. This gave a criticism of European linguistic censuses and useful estimates of the numbers of persons speaking the various languages of Europe.

Unfortunately, the transfers of population between Greece, Turkey and Bulgaria, were not taken into account when the sketch-map was drawn so that the distributions shown in Greek Macedonia were out of date. In spite of this deficiency the map is interesting on account of its portrayal of the distribution of the *Macedo-Slavs* in Jugoslav Macedonia (Fig. 84). Meillet attributed to them a much greater area even than J. Cvijić himself had done. He pointed out that the Serbian statistics had erred in referring to the *Macedo-Slav* dialect as Serbo-

Croat. He believed it deserved to be separately distinguished, and he gave estimates of the number of Macedonian speakers concealed in the Serbo-Croat returns as follows :

*The Macedo-Slav Population of Serbian Macedonia in 1924*

| *District* | *No. of Macedonian speakers* |
|---|---|
| Kumanovo .. | 111,797 |
| Tetovo .. | 53,508 |
| Skoplje .. | 84,107 |
| Stip .. .. | 72,801 |
| Ohrid .. .. | 48,318 |
| Bitolj .. .. | 134,627 |
| Tikves .. | 73,160 |
| Total .. | 578,318 |

Based on statistics given by A. Meillet.

Deserving separate recognition also, in his opinion, were the 68,457 Bulgarians in the frontier regions transferred to Jugoslavia in 1919 by the Treaty of Neuilly. By 1926, he calculated the number of *Macedo-Slavs* in Jugoslavia to be 627,897 and the number of Bulgarians, 74,226. He was careful also to draw attention to the linguistic peculiarities of the dialect spoken in the Niš-Leskovac area, pointing out its affinities with both Bulgarian and Serbian. In this region, however, he believed national sentiment to be Serbo-Croat rather than Bulgarian.

## S. Mladinov's Map of 1929

Stephan Mladinov is probably the best known of Bulgarian philologists. He contributed a history of the Bulgarian language to the series of books on Slav philology and cultural history edited by the German professors, Reinhold Trautmann and Max. Vasmer. The book contained a map showing the distribution of the Bulgarian

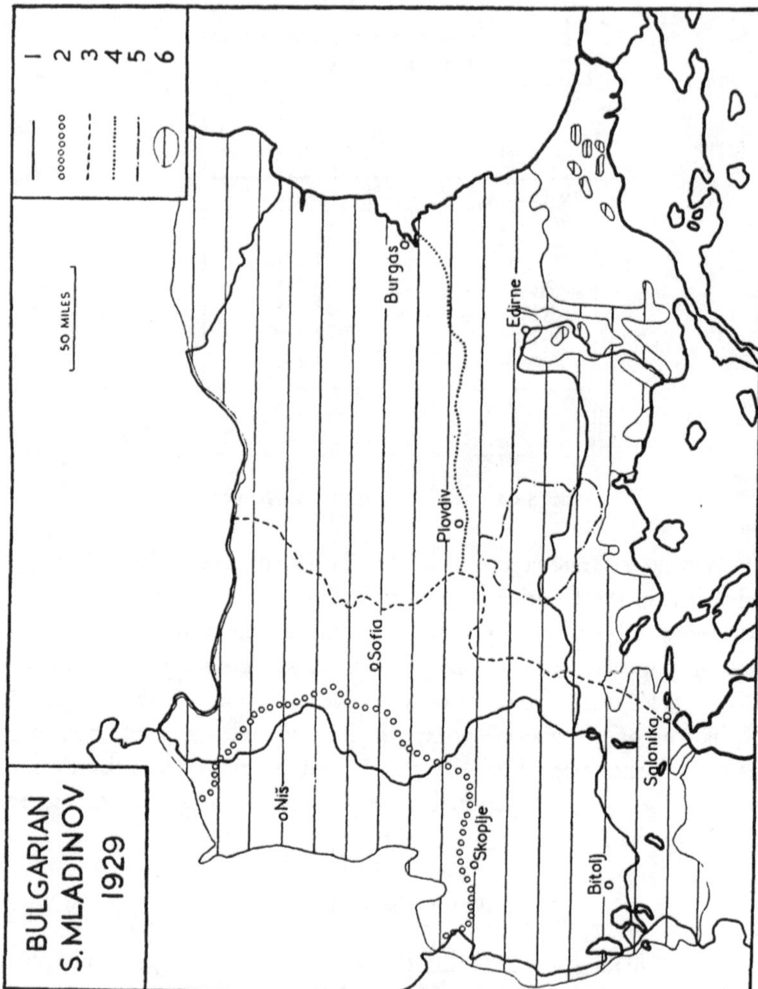
BULGARIAN
S. MLADINOV
1929
50 MILES
1
2
3
4
5
6
Burgas
Edirne
Plovdiv
Sofia
Niš
Skoplje
Bitolj
Salonika

language (Fig. 69). It was based on the work of T. D. Florinski, A. Iširkov, L. Niederle and J. Ivanov (whose maps have been referred to in earlier chapters) and on a map of the dialects of the Bulgarian language compiled by B. Konev.

It is not clear from the text whether Mladinov had considered the evacuation of Bulgarians from Greek Macedonia between 1919 and 1923, when drawing up his boundaries. It was not improbable that he intended the inference to be made that the transfer of population was not of sufficient magnitude to affect the basic distribution of Bulgarian speakers in Greek territory. The limits he attributed to the Bulgarian language were similar to those given by A. Iširkov and J. Ivanov. He included Vidin, Niš, Vranje, Tetovo, Prizren, Debar, Ohrid, Kastoría, Edhessa, Sérrai and Dráma within the Bulgarian sphere. His classification of Bulgarian dialects made an interesting comparison with those on the linguistic maps prepared by the Serbians. According to Mladinov the 'Timok-Prizren' region, including the districts of Niš, Leskovac, Vranje and Kumanovo, comprised part of the *West-Bulgarian* dialectal province. He referred to it as the *New-Bulgarian transitional zone*. A. Leskien, the Serbian philologist, had referred to this zone, in 1914, as of Serb speech where some Bulgarian linguistic traits were common.[1] A. Belić had also classified the Prizren dialect and the Timok dialect as Serb (Fig. 43). Pertinent also were the views expressed in a French map published in *Géographie Universelle*, Tome VII, 1934. It was presumably drawn up by Y. Chataigneau (Fig. 70). On this map the Prizren region was included within the *Sto* linguistic province, which forms the basis of the Serb written language. In the Timok region the *Sop* dialect was distinguished from both Serb and *West-Bulgarian*.

According to Mladinov, Macedonia proper formed part of the

[1] *Grammatik der Serbo-Kroatischen Sprache* (Heidelberg, 1914).

---

FIG. 69. THE DISTRIBUTION OF THE BULGARIAN LANGUAGE

The references in the key are as follows: **1.** Political boundaries (no date given); **2.** Eastern limit of the New-Bulgarian transitional dialect; **3.** Boundary between the East-Bulgarian and West-Bulgarian dialects; **4.** Northern limit of the archaic South-east Bulgarian dialect; **5.** Limits of the Central-Rodopi dialect; **6.** Limits and distribution of the Bulgarian language.

*West-Bulgarian* linguistic province. Leskien had classified *Macedo-Slav* as a Serb dialect. Belić had recognized two divisions here, Serbo-Macedonian and Bulgaro-Macedonian. Chataigneau distinguished *Macedo-Slav* as a separate dialect on its own merits.

The limit of the *East-Bulgarian* dialect given by Mladinov was identical with that expressed on Niederle's map (Fig. 38). The more

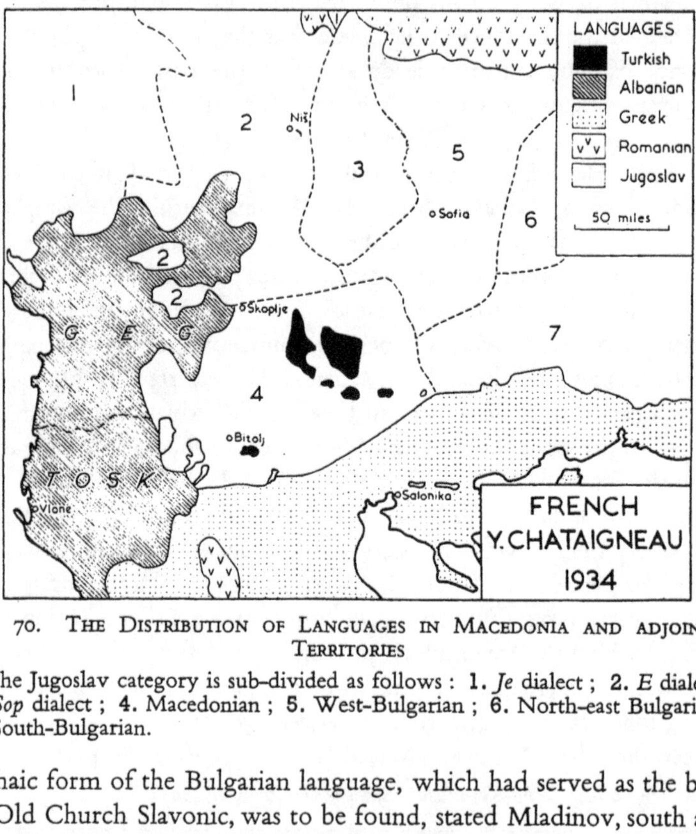


FIG. 70. THE DISTRIBUTION OF LANGUAGES IN MACEDONIA AND ADJOINING TERRITORIES

The Jugoslav category is sub-divided as follows : **1.** *Je* dialect ; **2.** *E* dialect ; **3.** *Sop* dialect ; **4.** Macedonian ; **5.** West-Bulgarian ; **6.** North-east Bulgarian ; **7.** South-Bulgarian.

archaic form of the Bulgarian language, which had served as the basis of Old Church Slavonic, was to be found, stated Mladinov, south and east of the Burgas-Plovdiv-Salonika line ; west and north of this line occurred the *New-Bulgarian* dialect, the development of which had been influenced by the Albanian, Romanian and Turkish languages. It is interesting to note that A. Belić in his map had dispensed altogether with the *West-Bulgarian* dialect by insisting that it was a transitional dialect between Serb and Bulgarian. He had virtually limited pure Bulgarian speech to the region designated by Niederle and Mladinov as *East-Bulgarian*.

The character of Mladinov's map pointed to the fact that the Bulgarians had not forgotten any of their claims on Greek and Jugoslav territory. He did not state categorically that Bulgaria had territorial claims on Greece and Jugoslavia to the extent of the limits prescribed on his map, or that language and national sentiment were necessarily coincident. But the map was published at a time when Jugoslavia was experiencing difficulty with the *Macedo-Slav* minority. Patently, the map was meant to express the need for boundary revision.

## V. Mikov's Map, 1936

Another Bulgarian map, compiled by V. Mikov and R. Vasilev, was published in 1936 (Fig. 71). It was based largely on Bulgarian sources and on older ethnographic maps such as those of A. Boué, G. Lejean, K. Sax, and J. G. von Hahn. The map contained a very useful list of Bulgarian contributions towards the ethnography of the Balkans, some of which have been already referred to, e.g. the work of V. Kănčev, A. Iširkov, J. Ivanov and D. M. Brancov. V. Mikov himself had published a number of monographs on Bulgarian ethnography of the eighteenth and nineteenth centuries.[1] A number of Russian works on the Balkans were also cited.

It is not proposed to deal with this map in detail since its distributions did not depart greatly from those given by J. Ivanov and A. Iširkov (cf. Figs. 47 & 48). Its most remarkable and significant feature was that it failed to incorporate any changes in the ethnography of Macedonia as a result of the population transfers. The Bulgarians, in fact, clung to the idea of *traditional Bulgarian territory* as it was in 1912-13 and refused to acknowledge the validity of any changes which had taken place since the Balkan Wars. This is characteristic of all Bulgarian maps.

[1] Н. Миковъ, Населението на Турция и България презъ XVIII и XIXв (София, 1915-1935). Naselsnieto na Turčiya i B'lgariya prez XVIII i XIX v (Sofia, 1915-35).

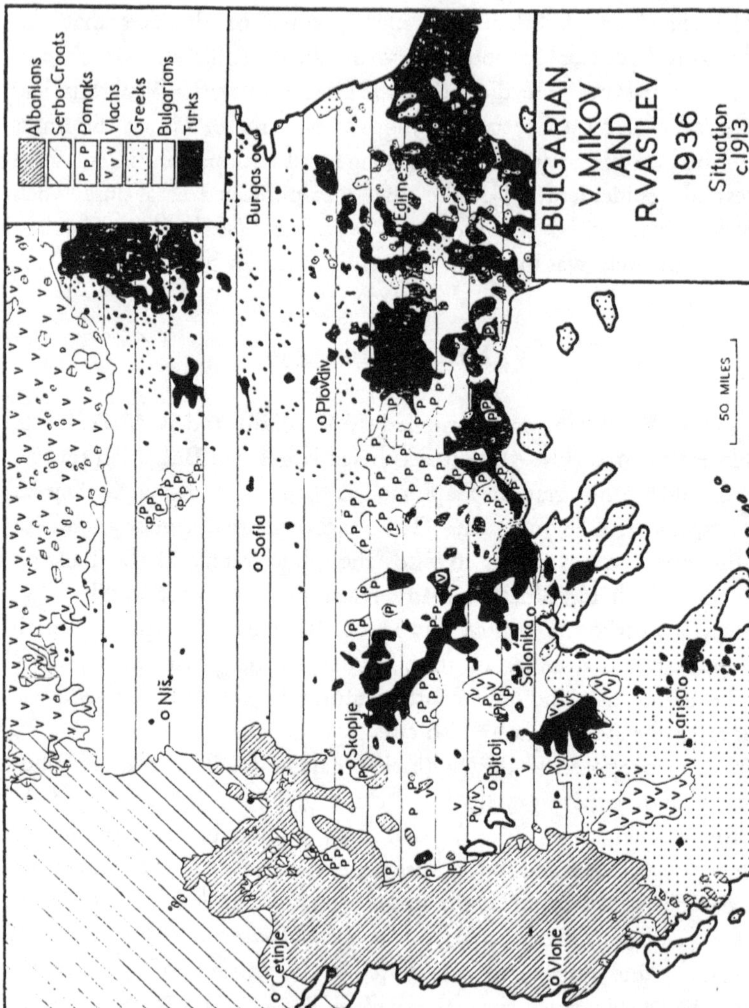
Albanians
Serbo-Croats
Pomaks
Vlachs
Greeks
Bulgarians
Turks
BULGARIAN
V. MIKOV
AND
R. VASILEV
1936
Situation
c.1913
50 MILES
Burgas
Edirne
Plovdiv
Sofia
Niš
Skoplje
Salonika
Bitolj
Larisa
Cetinje
Vlonë

FIG. 71

## Some Maps in British and Foreign Atlases, 1928–38

### *J. Halkin's Atlas*

An ethnographic map in the atlas of J. Halkin, one of the most popular of Belgian atlases, portrayed the ethnography of the Balkans as late as 1928 exactly according to J. Cvijić's map of 1918 and made no modifications of any kind.

### *British Atlases*

The two most popular general atlases produced in Great Britain in the inter-war years were the *Oxford Advanced Atlas* edited by John Bartholomew, and the *University Atlas* edited by G. Philip and H. C. Darby. The Oxford Atlas contained in the fourth and revised edition, 1931, an ethnographic map of Europe. It indicated large Turkish enclaves in the Konariote district of south-western Macedonia and in other parts of Greek Macedonia. The Greeks were given a scattered distribution in Greek Macedonia. No allowances apparently had been made for the exchange of Greeks and Turks in 1922. The Bulgarians were represented only in Bulgarian Macedonia, the remaining Slavs were referred to as Macedonian (south of Skoplje, to the Gulf of Salonika) and Serb (in northern Macedonia). No Vlachs were shown except in the Píndhos, and Albanians were indicated in Old Serbia but only as a minority in Ipiros.

In the *University Atlas*, a map showing the distribution of languages in eastern and southern Europe was included in the edition of 1937 and subsequent editions. Allowance had been made for the Graeco-Turkish exchanges and the Turks were represented as only a very small minority in the Vardar valley and in Dhitikí Thráki. The Greeks were given a linguistic frontier almost coincident with the political boundary in the north. Serbs, Bulgarians and *Macedo-Slavs* were shown in Macedonia. The Serbs were depicted as far south as Veles; Bulgarians were indicated in the Strumica salient of Serbian Macedonia. Bulgarian Pomaks were marked in the northern frontier zone of Dhitikí Thráki. The Macedonians were given limited territory in the southern half of Serbian Macedonia and in Greece, immediately south of the Jugoslav-Greek boundary. The *Macedo-Slavs* were portrayed in such a way as to indicate affinities with the Serbs in the north and the Greeks in the south. Vlachs were marked as in the G.S.G.S. map of 1918, and Albanians in Old Serbia according to the Jugoslav census-map of 1924. Albanians were not given a very liberal distribution in Ipiros, however, in spite of the fact that the map purported to show linguistic divisions.

Both British maps were influenced by Cvijić and his school of thought and both reflected the Jugoslav and Greek points of view rather than the Bulgarian and Albanian. The modifications in the ethnographic distributions of Turks and Bulgarians were not taken into account in the map in the Oxford Atlas until a later edition.

### *The Denoyer-Geppert Atlas*

American maps, in contradistinction to British and French maps, were not influenced so greatly by the Serbian ideas. A map of "The Peoples of Europe" published as one of the *Denoyer-Geppert Social Science Maps* in 1938, showed very small Turkish enclaves in the Vardar valley and Dhitikí Thráki. It placed the Greek ethnographic frontier north of the Graeco-Albanian political boundary but south of the Graeco-Jugoslav and Graeco-Bulgarian boundaries. No Serbs at all were shown in Macedonia or the Niš-Leskovac region but they were given territory in northern Albania. The *Macedo-Slavs* were ignored and Bulgarians were given a wide distribution in the Niš-Leskovac region, in Macedonia, and in northern Greece. Vlachs were ignored except in the Píndhos, and Albanians were given a very limited distribution indeed ; their frontier fell short of the political boundary both in the north and in the south, and only enclaves of Albanians were shown in Old Serbia.

### *Conclusion*

There were many other maps to be found in the numerous atlases appearing in the inter-war years in various countries of the world, but the examples considered are sufficient to demonstrate the singular lack of unanimity prevailing in Europe and America concerning the ethnographic distributions in Macedonia. Some confusion was occasioned by the failure to take into consideration the effects of the

Graeco-Turkish war on the ethnic composition of the population of Greek Macedonia. The chief deficiency, however, of such maps arose from the assumption that one map would suffice to incorporate all the ideas held by different authorities at different times about these distributions. The diversity incorporated in these examples emphasizes the fact that an ethnographic map out of its historical context and with its origins unacknowledged serves no useful purpose and only adds to the difficulty of interpretation.

## CHAPTER XII

# THE WAR YEARS AND AFTER, 1939-1950

### THE GREEK CENSUS, 1940

AN official Greek view of the ethnographic structure of Greek Macedonia immediately before the invasion by the Axis Powers is incorporated in the Greek census of 1940, the results of which are summarized below:

*Nationalities and Languages in Greek Macedonia*

| | 1928 | 1940 | |
|---|---|---|---|
| | *Mother-Tongue* | *Mother-Tongue* | *Nationality* |
| Greek .. .. | 1,165,553 | 1,487,571 | 1,602,381[1] |
| Slav .. .. .. | 80,789 | 84,751[2] | 74,751 |
| Jewish (Espagiol) .. | 59,146 | 48,874[3] | 48,874 |
| Armenian .. .. | 11,859 | 8,571[4] | 8,519 |
| Vlach .. .. .. | 13,475 | 26,750 | 6,750 |
| Gypsy .. .. | 3,387 | 5,046 | 5,045 |
| Russian[5] .. .. | 1,913 | 2,894 | 2,894 |
| Albanian .. .. | 1,119 | 5,945 | 1,445 |
| Turkish .. .. | 71,960 | 80,310 | — |
| Others .. .. | 3,276 | 8,419 | 8,571 |
| *Totals* .. | 1,412,477 | 1,759,130 | 1,759,130 |

Based on the Greek census of October 16, 1940; these 1940 figures are given in C. Christides, *Le Camouflage Macédonien* (Athènes, 1949).

[1] This figure includes 80,310 Turkish speakers, 20,000 Vlach speakers, 10,000 Slav speakers, 4,500 Albanian speakers.

[2] Includes Christian and Pomak dialects both of which, in the Greek view, approximate closer to Bulgarian than to any other language.

[3] In 1949 the number had fallen to 2,000 as a result of Nazi persecution during the occupation.

[4] The greater number of whom migrated to the Armenian Soviet after 1946.

[5] Mostly White Russians.

The returns of the 1940 census make interesting comparisons with those of 1928—comparisons which do not inspire confidence in Greek census methods. For example, the total number of Vlach speakers returned in 1940 was twice that of 1928 while the number of Albanians was more than five times as great. The fact that these figures are quoted without reference to the bilingual problem suggests an over-simplification of the language situation because many persons in Greek Macedonia in 1940 spoke both Slav and Greek languages equally fluently. The Greek claim, on the basis of these figures, that they composed over ninety per cent of the pre-war population of Greek Macedonia cannot be regarded as substantiated. On the other hand there would seem no reason to doubt the validity of their general thesis, that in none of the Macedonian eparchies (counties) did the Slavs constitute a majority.

## German Maps

A feature of the history of south-eastern Europe between the years 1933 and 1940 was the steady increase there of German influence. Those Danubian and Balkan states which were drawn within the orbit of the Reich had to pay the price of sustaining their economy by forfeiting their national sovereignty. The German Government inevitably extended their interest to the political structure of south-eastern Europe, and worked for control of Romania, Greece, Jugoslavia and Bulgaria. These states not only commanded the great through-routes to the Mediterranean, the Black sea and beyond, but, in addition, their agricultural and mineral resources were useful to a Germany preparing for war. There was a further consideration prompting German interest in this theatre—the revival of the traditional expansionist policy of Russia in eastern Europe. A reflection of German interest in the Balkan peninsula was the preparation in 1940 of detailed ethnographic maps for the use of the German General Staff and the German Foreign Office. The compilation of these maps fell to the lot of specially trained geographers recruited for state service.

*The German Staff Map of 1940*

A map depicting the ethnography of the Danubian lands and the Balkan peninsula was produced in 1940 by the division of the German General Staff responsible for war maps and survey. The method used was a straightforward representation of majority populations ; no refinements were introduced which might interfere with the map's simplicity. No references were given on the map about the sources used for its compilation but it must be supposed that the Division of Maps was well supplied with information relevant to the ethnographic situation in Macedonia, especially as there was a possibility of the German army being committed in this region.

The distributions given on this map showed some concentrations of Turks in the Maritsa valley of Greece and in the Bulgarian Rodopi mountains but none at all in Jugoslavia. Greek Macedonia and Dhitikí Thráki were shown as fairly solidly Greek, and Greek exclaves indicated in Albania around Gjinokastër and in the Semen valley. The Slavs of Serbian Macedonia were classified as Bulgarian with the exception of some small districts in the north where the population was shown as mixed Serbo-Bulgarian. The region around Pirot in Serbia was also marked as mixed Serbo-Bulgarian. The northern frontier zone of Greek Macedonia in the neighbourhood of Kastoría and Flórina was marked as Bulgarian. Small groups of Vlachs were depicted in the Píndhos and in the vicinity of Kastoría and Véroia. The Albanians were given a distribution which was in accord with their political boundary in the south and south-east but which, in the north, stretched over the boundary into Old Serbia as fas as Mitrovica.

*Manfred Straka's Maps, 1940*

The Geographical Division of the German General Staff also produced a number of maps of Jugoslavia compiled by Manfred Straka of the South-Eastern German Institute at Graz. He was principally concerned in experimenting with method. His main source of information appears to have been the Jugoslav censuses of 1921 and 1931, plus modifications of his own arising out of the failure of the Jugoslav figures to distinguish *Macedo-Slavs*.

On one of his maps he distinguished *kernel* districts, the ethnic character of which was demonstrably homogeneous. Where any mixed population occurred the map was left blank. The *kernel* district of the Serbs was sited, according to his interpretation, well to the north of Vranje, the *kernel* district of the *Macedo-Slavs*, in that part of Macedonia south of Skoplje. The Niš-Leskovac area, the Novi Pazar corridor and much of northern Serbian Macedonia were left blank, these areas apparently being transitionary between Bulgarian, Albanian, *Macedo-Slav* and Serb areas of characterization. Such an interpretation might be said to favour the partition of Macedonia.

On another of his maps he showed the ethnic affinities of 'majority' districts, i.e. small minorities were ignored and whole regions of Jugoslavia distinguished according to the character of the majority of their inhabitants. In this scheme Serbian Macedonia emerged as *Macedo-Slav* and Old Serbia as Albanian. The Niš-Leskovac area was still shown as mixed Serbo-Bulgarian and likewise the Novi Pazar corridor as mixed Serbo-Albanian. It is worth noting that few non-Bulgarian authorities, indeed not always the Bulgarians themselves, had insisted on claiming the Niš region as Bulgarian since 1878.

*The German Foreign Office Map, 1940*

More interesting, perhaps, than any of the German maps mentioned so far was that prepared by the Geographical Division of the Foreign Office late in 1940 (Fig. 72). It may not have been printed until 1941, on the eve of the German invasion of Jugoslavia. It dealt with south-eastern Europe and was described not as a map but as a cartogram, a significant distinction. It was based on the latest available local official statistics, the Jugoslav census returns of 1921 and 1931, the Bulgarian returns of 1926, the Albanian returns of 1918 and the Greek returns of 1928.

A unique and ingenious method of depicting distributions was adopted in this map. They were shown by the administrative units used in the various census returns. Each unit was divided into oblique strips of varying width between one and four millimetres. Each strip was sub-divided according to the strengths of the ethnographic elements in each district. Thus it may be seen from Fig. 73 that a

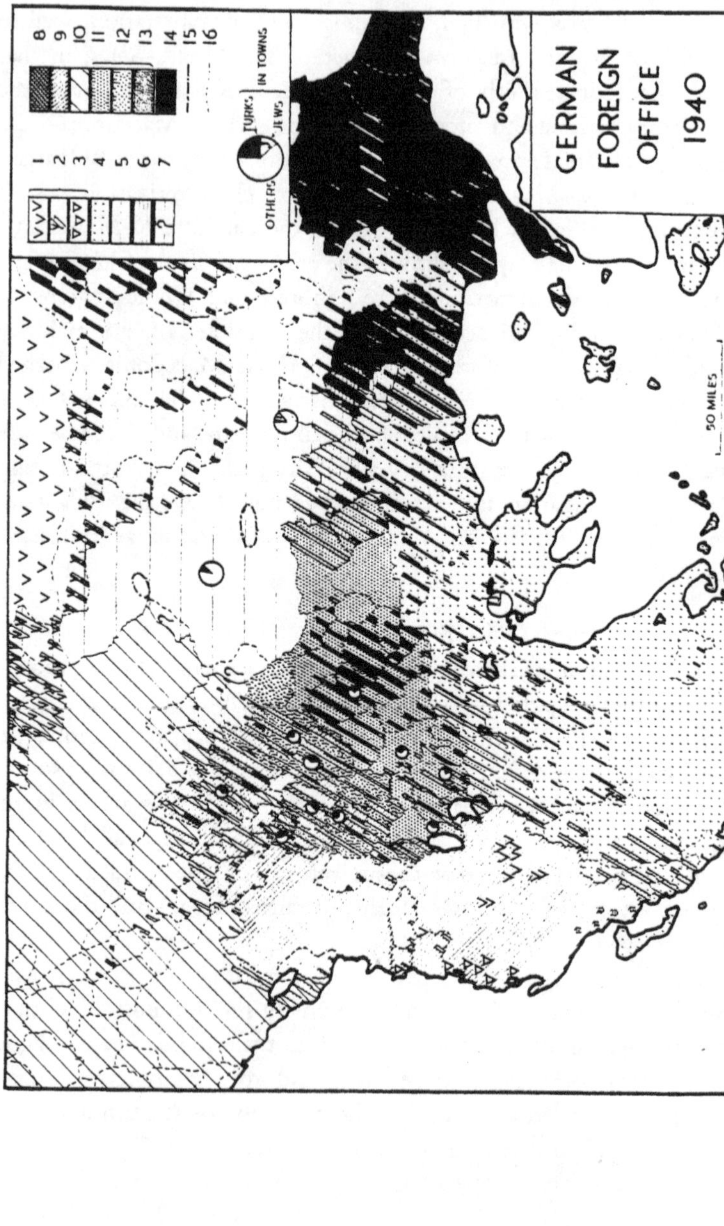
GERMAN
FOREIGN
OFFICE
1940
1
2
3
4
5
6
7
8
9
10
11
12
13
14
15
16
TURKS
JEWS
OTHERS
IN TOWNS
50 MILES

millimetre strip completely shaded represented 25 per cent of the total population of any particular district and this strip could be sub-divided to show any smaller percentage, e.g., one fifth of such a strip would be 5 per cent of the population and so on. Using this method it was possible to show the proportional strengths of the various elements, district by district. The emphasis given to any element thus corresponded closely to its *proportional numerical strength.* Of course, the exact location of minorities within each district could not be shown using this method but the overall picture obtained was precise and clear. All minority populations were represented according to their relative numerical importance and at the same time their general spatial distribution could be inferred but not density of settlement. Hitherto a series of maps had been necessary to show such facts, e.g., the Jugoslav census-maps of 1924.

*The Turks.* On this map the Turkish territory in Bulgaria was portrayed in the north-east and in the Rodopi mountains. Dhitikí Thráki was also marked as predominantly Turkish. Scattered minorities of Turks were shown throughout southern Jugoslavia, around Stip, Radovište and in the plain of Kosovo. The Jugoslav towns of Skoplje, Gostivar, Prizren, Pristina and Tetovo all had large Turkish populations, sometimes indeed rising to over half of the urban population. Included with the Turks were the *Gagauzi* or Christian Turks, most of whom were to be found in the Dobrudja.

*The Greeks.* Few Greeks were marked outside the boundaries of Greece, the exception being some villages in the vicinity of

---

FIG. 72

The references in the key are as follows : **1 & 2. Romanians and Vlachs** including summer settlements of the Vlachs in southern Albanian ; **3. Vlachs** (winter settlements) ; **4. Greeks** ; **5. Bulgarians** ; **6. Bulgarian Pomaks** ; **7. Bulgarians** (not counted in Jugoslav Census returns) ; **8. Gypsies** ; **9. Albanians** ; **10. Serbo-Croats** (including Bosnians) ; **11. Macedonians** with Bulgarian affinities ; **12. Macedonians** of doubtful affinity (disputed between Bulgarians and Serbs) ; **13. Macedonians** with Serbian affinities ; **14. Turks** including Gagauzi and Tatars ; **15.** Exclaves of Macedonian dialect in Bulgaria ; **16.** Administrative boundaries used as units for the construction of the cartogram.

*Note.* In the original, more details of the ethnographic structure of towns are given.

Gjinokastër and scattered exclaves in eastern Thrace. Nor did the method adopted show the Greeks in quite the strength in Macedonia which one had been led to expect from previous maps such as the League of Nations' map. In Dhitikí Thráki the Greeks were depicted virtually as a minority.

*The Slavs.* An elaborate classification of the Slavs was resorted to for this map, as a means of dealing with the problem of the Macedonian Slavs. The Serbs, distinguished from the Croats and Bosnians, were shown extending only as far south as Vranje, the Goljak and the Prokletije mountains. The Niš region was acknowledged to be Serb. The Novi Pazar corridor was shown as Bosnian. South of the Sar mountains the Slav population was classified as *Macedo-Slav*. Here the map departed from official Jugoslav statistics on the grounds, presumably, that they did not give a reliable picture of the situation. The *Macedo-Slavs* were not, however, treated as one group but divided into three categories.

(1) *Macedo-Slavs proper* were marked in Jugoslavia south of a line from Lake Ohrid through the Golešnica mountains and the plain of Ovče to Kratovo. They were also shown in south-western Bulgaria in the districts of Gorna Dzhumaya, Razlog, Melnik, Nevrokop and Petrich, and in northern Greece as minorities, in the districts of Flórina, Kastoría, Kilkís, Pélla, etc.

(2) *Serb Macedonians* were restricted to a few small exclaves in Old Serbia. This group was meant to represent *Macedo-Slavs* whose Serbian affinities were freely acknowledged even by the Bulgarians.

(3) *Disputed Macedo-Slavs* were shown in the neighbourhood of Skoplje, Kumanovo, Tetovo, Kičevo, Gostivar and Debar. The affinity of these Slavs was in dispute between Serbians and Bulgarians. In addition to the three main groups of *Macedo-Slavs* distinguished above, a reference was made to Bulgarians with *Macedo-Slav* cultural characteristics. The inference here was that the *Macedo-Slav* culture was akin to that found in many parts of Bulgaria.

The Bulgarians themselves were grouped into three divisions.

(1) *The Bulgarians proper* were all those Slavs confined within the political boundaries of Bulgaria, excepting those to be found in Bulgarian Macedonia (Pirin).

(2) *The Bosiligrad and Caribrod Bulgarians* were marked in those territories which were transferred from Bulgaria to Serbia in 1919.

(3) *The Pomaks* or Moslem Bulgarians were given a wide distribution in the Rodopi mountains and in parts of Dhitikí Thráki.

Finally, for the sake of comparison, and also to emphasize the

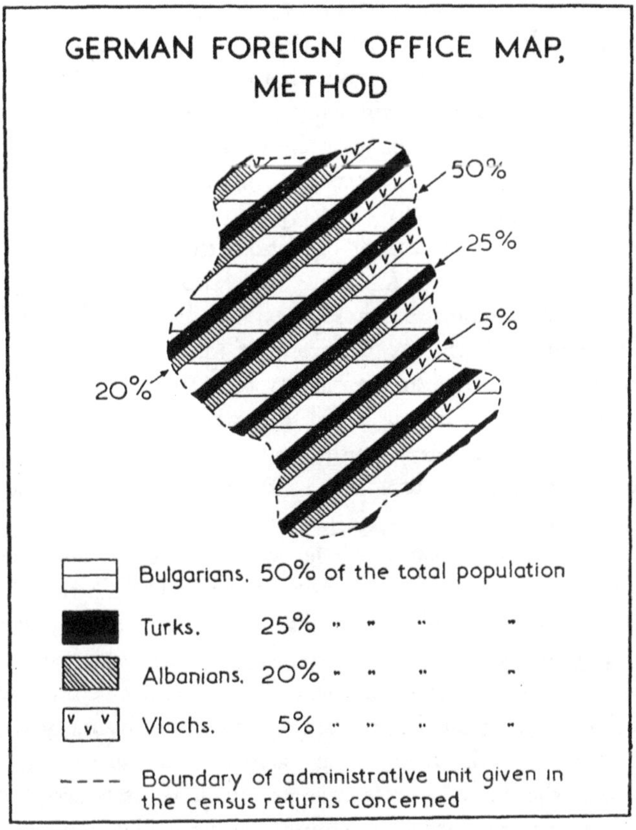


FIG. 73

difficulties attending the interpretation of Macedonian ethnography, the limits of the Bulgarians as marked by A. Išírkov, and of the Serbs and *Macedo-Slavs* as marked by J. Cvijić, were also included on the map.

*The Vlachs.* Only small minorities of Vlachs were recorded on the map, the most significant being in the vicinity of Kónitsa and Métsovon in Greece and Kruševo in Jugoslavia. For Albania, a slightly different form of depiction was used to show Vlachs. *Summer* settlements in the Korcë district were distinguished from *winter* settlements along the coastal and river plains. On some maps, notably Romanian, the practice had been to show territory occupied in either summer or winter as Vlach. This explains, perhaps, the extensive Vlach territory to be found on A. D. Atanasiu's map of 1919. The German map-compilers, aware of the stringency of Greek figures regarding the Vlachs, also incorporated on to their map the distribution of Vlachs in Greece according to the views of Th. Capidan. This distribution accorded with that given by G. Weigand (*cf.* Figs. 28 & 62).

*The Albanians.* With minor exceptions the whole of Albania was shown as Albanian. The Albanians were also marked in Old Serbia and eastern Macedonia in those areas attributed to them by the Jugoslav returns of 1921. A sub-group of Albanians was distinguished also in northern Greece in the vicinity of Ioánnina. Their depiction was not based on Greek figures and they were described as 'Albanians openly ignored in the Greek census returns.' Albanians acknowledged in the Greek census returns were indicated along the coast of the Kérkira channel.

*Jews, Gypsies and Tatars.* These peoples were all distinguished as separate groups. The Jews formed a particularly strong element in the Macedonian towns, Salonika and Skoplje being outstanding in this respect. Gypsies and Tatars were for the most part found in extra-Macedonian territory, in northern Bulgaria and the Dobrudja.

*Conclusion.* A great deal of care and thought obviously went into the preparation of this map. It appeared to be a dispassionate attempt to present the facts by reference to available statistics, but these statistics were not regarded as infallible and modifications were made when they appeared to be in error. Cartographically, the methods of depiction used marked a technical advance in the mapping of ethnographic statistics. The map provided an interesting contrast with the General Staff Map discussed above both with regard to distributions and method. In particular, the treatment of *Macedo-Slavs* was unique. The distribution of this group, also, differed from any other interpretations although it bore some likeness to that of A. Meillet. From the evidence given in the text accompanying the map, the affinity of the *Macedo-Slavs* was Bulgarian rather than Serb. This map must have been used in connection with the decisions made by the German Foreign Office concerning the political re-organization of the peninsula, and it would seem to have prepared the way for the partition of Macedonia which took place after the Bulgaro-German occupation.

*W. Krallert's Map of Jugoslavia, 1941*

Even more elaborate than the German Foreign Office map was that prepared by the German geographer, W. Krallert, with the aid of a staff of other German geographers after the occupation had taken place ; this map was confined to Jugoslavia. It was on a relatively large scale, 1 : 200,000, and no less than forty sheets were required to cover Jugoslavia. It was based on the official Jugoslav censuses of 1921 and 1931, many returns of which had hitherto not been published. Since the returns of ' mother-tongue ' did not differentiate between the various branches of the Southern Slavs, religious statistics and ' declarations of population made locally ' were used. The latter figures apparently distinguished between *Macedo-Slavs*, Bulgarians and Serbs ; the former statistics enabled the Moslem Slavs to be distinguished as a separate group, including presumably Moslem Serbs, Pomaks and Bosnians. Their grouping together was an innovation from which the inference was to be made that in the German view religious affinity was stronger than linguistic in this instance.

The method of depiction was also novel. Circles proportionate in size to the number of inhabitants in each ethnic group were plotted for each inhabited locality in Jugoslavia, a device facilitated by the use of a large-scale base-map. Thus it was possible to ascertain at a glance the strength and location of any specified group in any part of Jugoslavia. Uninhabited localities functioning as divides between different communities could also easily be discerned. The revelation of inter-mixture and overlap as a feature in the ethnic structure was a further advantage attached to this method. The validity of criteria and classifications adopted, and the debatable accuracy of statistics were, of course, open to criticism, but it may be said of this map that it was the most ambitious survey attempted since the Exarchate survey in 1901.

*The Turks.* It showed the Turks in small communities scattered over Old Serbia and Jugoslav Macedonia, with concentrations more particularly in the Radovište district and along the Stara river. Many towns were marked as having a strong Turkish element, particularly Tetovo, Gostivar, Prizren and Skoplje.

*The Slavs.* Only insignificant groups of Serbs were shown in Macedonia, and even in Old Serbia they were marked in strength only in the Mitrovica and Pristina districts. The greater part of the population of the region focusing on Pirot, Niš and Leskovac was, however, depicted as Serb; the population of the Caribrod and Bosiligrad districts was marked as Bulgarian. The Slav population of Macedonia was, on this map, shown almost entirely as *Macedo-Slav*. The 'Moslem Slavs' were shown only as relatively small communities mostly situated in north-western Macedonia and in the Novi Pazar corridor.

*The Vlachs.* Only a very few Vlach communities were registered on this map. The most important of these occurred in the vicinity of Kruševo and Bitolj.

*The Albanians.* From even a casual inspection of the map, the Albanians emerged as substantially the most important element in the population of Old Serbia and western Jugoslav Macedonia. In Old Serbia, Peć, Istok, Srbica and Podujevo were included in Albanian territory, leaving the Bogičevica mountains and the Mokra hills as fairly effective divides between territory populated by Serbs on the north, and by Albanians on the south side. The line of towns, Kumanovo, Skoplje, Tetovo, Gostivar, Kičevo and Struga, marked roughly the frontier between the Albanians and the *Macedo-Slavs*. Thus the Crni Drim, the upper Vardar valley, the Sar mountains, the Crna hills, the plains of Kosovo and Metohia were depicted, for the most part, as Albanian. Ulcinj and Gusinje in Montenegro, also, were shown as Albanian. It is worth noting that all these areas were incorporated into Albania consequent upon the re-organization of the political boundaries of Jugoslavia in 1941.

*The Political Re-organization of the Peninsula, 1941*

The problem confronting the Axis Powers, after the military occupation of Jugoslavia and Greece had been effected, concerned the restoration of political equilibrium in the peninsula. Considerations of ethnography played some part in the Axis attempt to solve their problem, for concessions were made to those dissentient minorities within Jugoslavia and Greece in such a way that Bulgarians, Croats, and Albanians benefited at the expense of Serbs and Greeks (Fig. 74). The Axis Powers thus resorted to the policy of *dividere et imperare*, so favoured by the Turks in their ascendency and not altogether ignored by the Allied and Associated Powers in 1919.

*The Partition of Jugoslav Macedonia and its Consequences.* Croatia was established as an independent state which embraced a very large area of the former Jugoslavia, including Bosnia and Hercegovina. Serbia, with considerably less territory than it had boasted in 1913, was re-established as a separate state. Montenegro, likewise, re-appeared but in a mutilated form. Old Serbia was handed over to Italian-controlled Albania together with part of western Macedonia. The greater part of Jugoslav Macedonia together with the districts of Vranje and Pirot was consigned to Bulgaria.[1]

The Bulgarian Government found itself, for a short time, administering territory which had for so long figured in Bulgarian revisionist claims. This was the bait with which German diplomats had tempted their Bulgarian allies but that they had been ready enough to swallow it was only too evident from the vast output of contemporary Bulgarian revisionary literature—G. P. Ghénov's *Actes et traités internationaux concernant la Bulgarie avec des notices explicatives et une carte de la Bulgarie et des pays voisins* (Sofia, 1941) ; A. Hajek's *Bulgariens Befreiung und staatliche Entwicklung unter seinem ersten Fürsten* (München, 1939) ; H. Bartek's many articles in *Zeitschrift für Geopolitik* ; Dimitar Jaranov's voluminous work on the Macedonians, provide sufficient

[1] See. (1) H. Grothe, "Die mazedonische Frage und die Lösung durch die Achsenmächte." *Archiv Wanderungswesen und Auslandskunde*, Bd. XIII (1943) ; (2) J. F. Gellert, *Politische Geographie von Mazedonien* (unpublished manuscript, 1945).

indication of Bulgarian sentiment at this time.[1] It is apparent that not only did the change of boundaries satisfy the Axis Powers but the mass of Bulgarians also. Even the Bulgarian Communist Party seems to have supported the 'liberation' of Macedonia.[2] It was with a crusading spirit that the Bulgarian Army undertook its mission of occupation. But events were to prove that a long period of separation had produced new generations of Macedonians and the passing of the years 'in exile' had perceptibly weakened the pro-Bulgarian element.

The Bulgarian occupation, short as it was, had two surprising and important consequences. First, it reduced Serbian influence and eliminated the so-called 'pure Serbian' element in the population. The present government of Jugoslavia has officially recognized these facts, but they are also supported from other sources. According to M. Kulischer, some 120,000 Serbs were forced to emigrate from Macedonia and resettle in Serbia, almost all the Serbs who had entered Macedonia since 1913.[3] T. B. Schlechtman, however, suggests that this figure was merely a target and that, in fact, only 43,000 Serbs moved out, according to a count of ex-Serb refugees coming from Macedonia which was made in Serbia in 1943.[4] Even so the number was high, and it certainly included all the ruling classes. Some 12,000 Jews, mostly from the towns, in particular Skoplje, were deported to Poland according to Schlechtman. The Jewish and the 'pure Serbian' population of Macedonia was reduced to negligible proportions by these measures. In this manner the way was prepared for the acknowledgement of the Sar mountains and the Crna hills as the southern limits of Serbia in 1945 (Fig. 90). Some Serbs have since returned to Macedonia but not in sufficient numbers to have much effect.

The second consequence, and the more unexpected, was the ultimate reaction of the Macedonians themselves to the Bulgarian occupation. According to first-hand accounts, gathered from the Macedonians themselves, the older element amongst the Macedonian population welcomed the Bulgarians with open arms and even the younger

[1] D. Jaranov, now in Bulgaria, is a Macedonian by birth.

[2] "Macedonia", *The World Today*, R.I.I.A. (April, 1949).

[3] *The Displacement of Population in Europe*, International Labour Office (Montreal, 1943).

[4] *European Population Transfers*, R.I.I.A. (London, 1946).

element believed that independence for Macedonia was in sight. No one was sorry to see the Serbians go. Bulgarians were the liberators and deserved recognition as such.

But this attitude of mind, especially amongst the younger element, did not last very long for it soon became apparent that the autonomy desired by the Macedonian intelligentsia was as far off as ever. Macedonia was incorporated into the Bulgarian war machine; conscription was introduced; requisitioning of food, transport and buildings followed. The Bulgarians had fallen victim to their own propaganda in believing and acting as if Macedonia was a part of Bulgaria. They insisted too closely on the Bulgarian character of the *Macedo-Slavs*. Bulgarian schools were established. Bulgarian became the official language, and Bulgarian officials stepped into the shoes of their departed Serbian fraternity. According to the accounts of the Macedonians themselves, the Bulgarians welcomed as liberators were no more welcome as masters in 1941, than the Serbians had been before them. Whereas the union of Macedonia and Bulgaria might have been possible in 1913, the progress made by the Macedonian nationalist movement, of which I.M.R.O. was one manifestation, was sufficient to make Bulgarian advances repugnant to the Macedonian intelligentsia in the years 1941-44. And even the Macedonian kulaks were not happy with the trend of events. "We have experienced enough of both the Serbians and the Bulgarians. Happy was the day when we were rid of both." Such a sentiment was expressed by a kulak speaking of his war experiences in 1949, and these sentiments appear to have been widespread, at least in northern Macedonia.

Opposition to the Bulgarian occupation crystallized in the Macedonian Communist Party in 1941, and was stimulated by Tito's successes further north. A new Macedonian Regional Committee was formed in August 1941 with the object of liberating Macedonia 'from the forces of fascism'. The movement, directed by Tito, not only offered armed resistance to the Bulgarian Army but indulged in underground political and cultural activity. Macedonia was represented at the Jugoslav Anti-Fascist National Liberation Council (A.V.N.O.J.) by Dimitar Vlahov, leader of *Obedinena*, the Macedonian

Communist group. By August, 1944, Jugoslav Macedonia was already established as one of the six Republics and the newspaper *Nova Makedonija* was being widely circulated. All territory of Jugoslavia south of the Sar mountains and the Crna hills was acknowledged by the Serbians to be Macedonian (Fig. 77). The policy of 'serbization' was completely abandoned.

By these measures the Jugoslav Communist Party succeeded in keeping Macedonia within the old framework of the state and the Bulgarian aims for the continued occupation of Macedonian territory were completely frustrated. Indeed as early as 1944 the Fatherland Front in Bulgaria was compelled to acknowledge not only the secession of Jugoslav Macedonia but the possibility of Pirin Macedonia going the same way. Bulgarian cultivation of I.M.R.O. had produced a harvest which was in danger of being garnered by her old rival. Paradoxically, the Bulgarian occupation of Macedonia had succeeded in a few short years in more or less destroying the prospects of any immediate union of Macedonia and Bulgaria.[1]

The *Macedo-Slavs*, whatever might have been their origins, had arrived at a stage in their national development when identification with either the Serbs or Bulgarians was no longer possible in theory or in practice.

*The Partition of Northern Greece and its Consequences.* Because of its vital sea frontage much of northern Greece remained under military occupation, and any plans which the Axis Powers might have had for its future administration were not carried out. But the Bulgarians were allowed to annex Dhitikí Thráki and to garrison Greek eastern Macedonia (Fig. 74). Whilst they were in possession, the Bulgarians tried all within their means to consolidate their position by increasing the Bulgarian element in the population and reducing the Greek. According to E. M. Kulischer, some 80,000 Greeks fled the province before the Bulgarian armies, and another 25,000 were forcibly deported after the occupation. Some 80,000 Bulgarian settlers succeeded them from Bulgaria and elsewhere.

[1] The conflict between the Jugoslav and Bulgarian communist parties over Macedonia is dealt with by E. Barker, *Macedonia: Its place in Balkan Power Politics*, R.I.I.A. (London, 1950).

T. B. Schlechtman's estimates are higher. He quotes the *Bulletin of International News* as stating that of 700,000 Greeks living in the Aegean provinces in 1938, only 375,000 were left by the end of 1943. He puts the number of Bulgarians entering the provinces during these years at 122,000. He did not give separate figures for Dhitikí Thráki. During the confused fighting at the time of the Greek Revolution, and the guerilla activity which followed, the Greek element was reduced still further. Hence it was not surprising that Greek Government forces experienced great difficulty in regaining control of Dhitikí Thráki and eastern Macedonia. During the Peace Conference of 1946, Bulgaria flatly refused to withdraw the claim to Dhitikí Thráki, an attitude which received considerable support from the U.S.S.R.

In the remainder of Greek Macedonia, including Salonika, the Germans remained in occupation, and Bulgarian influence was not so strong. Had the head of the Gulf of Salonika not been so vitally important from the strategic angle, most of this part of Macedonia must also have gone to Bulgaria to round off the new boundary.

The Germano-Bulgarian occupation revealed anti-Greek tendencies among the Slav-speaking elements of northern Greece. Even Greek estimates put the figure of collaborators as high as 65 per cent of the 'Slavophones.'[1] *Slavianski Narodnii Osvoboditelen Front* (S.N.O.F.), the Slav Communist Party in northern Greece, declared for the secession of Macedonia from Greece and its incorporation into the incipient Macedonian republic. When the Communists intensified their partisan warfare in this territory, the trickle of Greek refugees fleeing southwards grew into a stream, so that even before 1945 a steady decline in the Greek population of German-occupied Macedonia had set in. During the period of ideological conflict in the Balkan peninsula the decline continued, and resulted in the depopulation of much of Greek Macedonia between 1945 and 1949. Greek official estimates of the number of refugees from the northern provinces quoted figures approaching a million. Whilst no accurate statistics are available, there is little doubt that very few Greeks remained, and the total eclipse of Hellenism established by the colossal colonization of the years between

[1] A. A. Pallis, *op. cit.*

Fig. 74. Boundary Changes, 1941–43

The references in the key are as follows: **1.** Boundaries in 1938; **2.** Boundary modifications, 1941–43; **3.** Extension of Albanian boundaries contemplated in 1941; **4.** Parts of Serbia and Greece under Bulgarian occupation; **5.** Albanian gains; **6.** Bulgarian gains; **7.** Italian gains.

1923 and 1926 was the result. In this way one of the primary requisites of the Slav communist plan for a Greater Macedonia, the 'liquidation' of the Greek element in the population, was temporarily achieved, but only at a fantastic cost. What the permanent consequences of the depopulation will be depends on a number of factors, but, with the success of the Greek armies against the communists in 1949, resettlement of refugees has already begun afresh.

One other aspect of the Axis occupation of northern Greece remains to be considered. No changes were effected in the Graeco-Albanian boundary but changes were contemplated. The Italian Foreign Office pressed strongly for the extension of the Albanian boundary southwards to the Gulf of Arta, so that the whole of Ipiros would have been included within Albanian territory (Fig. 74). According to Greek allegations, the Albanian-speaking Moslems or *Chams* of Greek Ipiros co-operated with the Italians during the recent war, presumably with the idea of enlisting Italian support for boundary revision.[1] The emphasis given by German maps to Albanian distributions in Ipiros was a factor in favour of revision. The delicate strategic situation, however, precluded any boundary modification during hostilities. Here, too, the effect of the occupation once again militated against the security of Greek settlement and guerilla warfare played its part in forcing Greek settlers to trek south.

## British Maps

At the onset of the war in 1939 there was a revival of interest in Britain in the affairs of the Balkan peninsula, largely on account of its strategic importance. The word 'revival' is justifiably used because between 1933 and 1939 British influence and interest in the Balkans had reached a lamentably low level. The very great influence exerted by Great Britain and France on the destinies of Balkan nations immediately after the end of the 1914-18 war had gradually declined, at first slowly and then with gathering momentum after 1933, when it was almost completely superseded by German and Italian influence, more especially the former. This development needs no elaboration

[1] *Report by the Commission of Investigation concerning Greek frontier incidents to the Security Council* (May, 1947).

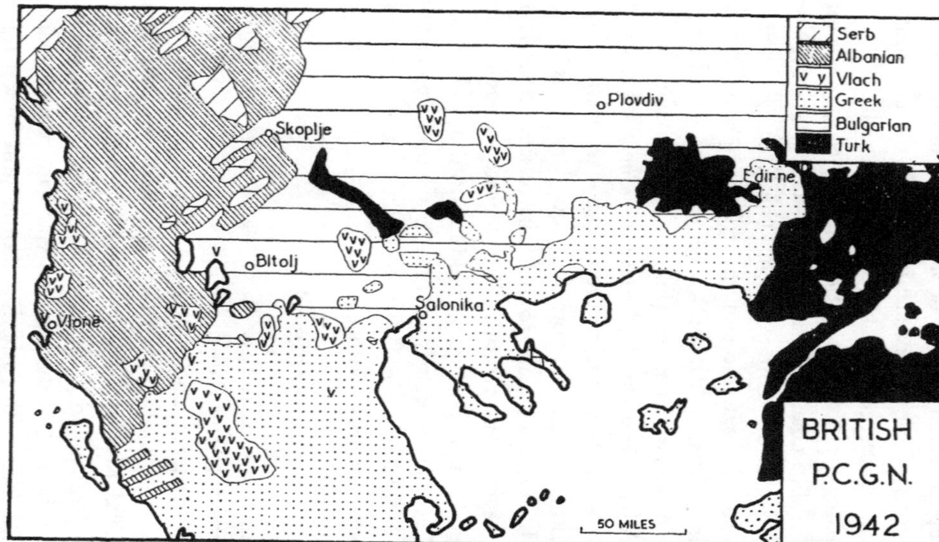


FIG. 75. THE LANGUAGES OF GREECE AND NEIGHBOURING COUNTRIES

In the original key, an area of 'Greek and Albanian' language is separately distinguished. It has been shown above diagrammatically, and the description dispensed with in the key.

here; it is apparent not only in diplomatic records but also in trade figures. The decline of British political and economic influence was accompanied by a dearth of literature published in the British Isles on Balkan affairs and by a general neglect of Balkan geographical studies. The study of Balkan languages—Serbo-Croat, Albanian, Bulgarian and modern Greek—was practically ignored in all British universities. The close relationship between the West and the Balkan world which had been promised in the years from 1918 to 1921 was never fulfilled. It was not surprising therefore that the year 1939 witnessed British intelligence concerning the Balkan peninsula at its lowest ebb. After 1939 British relations with Jugoslavia and Greece grew closer and some of the deficiencies in our knowledge of their affairs were remedied ; but the intimate understanding of the problems of the political geography of the peninsula, which could only have been gained by a long-term programme of intelligent interest in Balkan affairs, was never achieved.

In the realm of ethnography no large-scale detailed maps were compiled during the war, although the Geographical Section of the General Staff produced an ethnographic map of Central Europe which covered part of Jugoslavia. During the war and the crucial period of settlement that followed only a few sketch-maps were published interpreting the ethnographic situation in Macedonia.

### *The P.C.G.N. Map of 1942*

One of these sketch-maps was used to illustrate the *Gazetteer of Greece* issued by the Permanent Committee on Geographical Names for British Official Use in 1942. It showed the distribution of the languages of Greece and the neighbouring countries (Fig. 75). The method used was a simple shading of linguistic zones, the limits of which were interesting in so much as they differed considerably from those given on many British maps since 1915.

*The Turks.* Turkish-speaking people were, according to this map, to be found throughout eastern Thrace, in the Maritsa valley and in the Petrich district of Bulgaria, and in the Veles-Stip districts of Jugoslavia. No Turks were shown in Dhitikí Thráki and no reference was made to the Pomaks.

*The Greeks.* Greek speakers were given a widespread distribution in Dhitikí Thráki but in Macedonia the Greek linguistic frontier was placed far to the south of the political boundary. It ran from the Graeco-Bulgarian boundary east of the Mesta river through Sérrai, Dráma, Salonika, Véroia and Kastoría to the Albanian boundary. North of this line only exclaves of Greek language were marked, of which the Lake Dojran area was the most important, together with districts in the Struma valley and an odd group near Nevrokop in Bulgaria. The absence of Turks indicated that the transfers of populations had been considered, otherwise the distribution of Greeks in Macedonia was reminiscent of some pre-1919 maps. The inference to be made from the map was that the resettlement of Greek refugees between 1919 and 1926 had not materially altered the distribution of Greek language in Macedonia. It is improbable that the map compilers were influenced in their distributions by consideration of the Greek refugee movement which began in 1941 because Dhitikí Thráki, where this movement was greatest, was shown as almost purely Greek on the map.

*The Slavs.* The Slavs of Macedonia, on this map, were depicted as Bulgarian-speaking. They were shown not only in northern Greek Macedonia and Jugoslav Macedonia but also in the Niš frontier zone. This distribution was very similar to that claimed by the Bulgarians themselves (*cf.* Figs. 47, 48 & 71). It marked a radical departure from what had almost become the traditional British view—the depiction of Slav Macedonia as partly Serb and partly *Macedo-Slav*. On the P.C.G.N. map no Serbs at all were shown in Macedonia and only a few exclaves in Old Serbia.

*The Vlachs.* The distribution of the Vlach speakers would appear to have been taken from the map of A. J. B. Wace and M. S. Thompson. They, it might be remembered, had been very liberal in their estimates of the number of Vlachs in the peninsula (see p. 221) ; also, their distribution referred to the situation in 1914 which was very different from that in 1942 when undoubtedly many groups of Vlachs had lost their identity due first, to a changed economy, and second, to the process of 'hellenization' to which they had been subjected.

*The Albanians.* Albanian speakers were credited with considerable territory on this map. Northern Ipiros was marked as Albanian, as also was most of western Macedonia and Old Serbia. Albanian exclaves were also shown in Greek Ipiros. This distribution, also, differed considerably from that of many British maps which had appeared since 1915, especially with regard to the situation in Northern Ipiros.

*N.I.D. Maps, 1944*

During the war years a new series of Geographical Handbooks were produced by the Intelligence Division of the Admiralty. The policy of producing such works of reference had been inaugurated in 1915 by the Naval Intelligence Division (N.I.D.). The new handbooks on Jugoslavia and Greece contained many useful sketch-maps depicting Macedonian ethnography. The maps of A. Iširkov, 1915, of J. Cvijić, 1913, of the G.S.G.S., 1918, were produced on similar base-maps in such a way that the conflict of opinion expressed in the maps was immediately apparent. This kind of approach to the problems of Macedonian ethnographic distributions was, as far as it went, a much sounder approach than any hitherto used in British surveys.

The Greek Handbook in addition to these maps also contained a sketch-map showing the distribution of the languages of Greece and the adjoining territories. It was a typical 'compromise' map deriving its data from a number of sources, the Greek Gazetteer map mentioned above, the G.S.G.S. map of 1918 and the League of Nations' map of 1926. By an intelligent manipulation of the distributions given on these maps new limits for the various linguistic groups were produced, presumably the idea being to use the Greek Gazetteer map and the League of Nations' report of transfers of population, to bring the G.S.G.S. map of 1918 up to date.

The Turkish language was, according to this map, predominant in the Rodopi mountains of Bulgaria, in the Veles-Stip area of Jugoslavia and in parts of Dhitikí Thráki. The Greek linguistic frontier exceeded the Graeco-Albanian political boundary in Albania but fell short of the Graeco-Jugoslav boundary. The language of the Slavs of Macedonia was classified as *Macedo-Slav* and it was shown in Jugoslav and Greek Macedonia but not in Bulgaria (Fig. 84). The Bulgarian language was marked as extending into Jugoslav Macedonia in the Strumica district. The Serb language was excluded from Macedonia

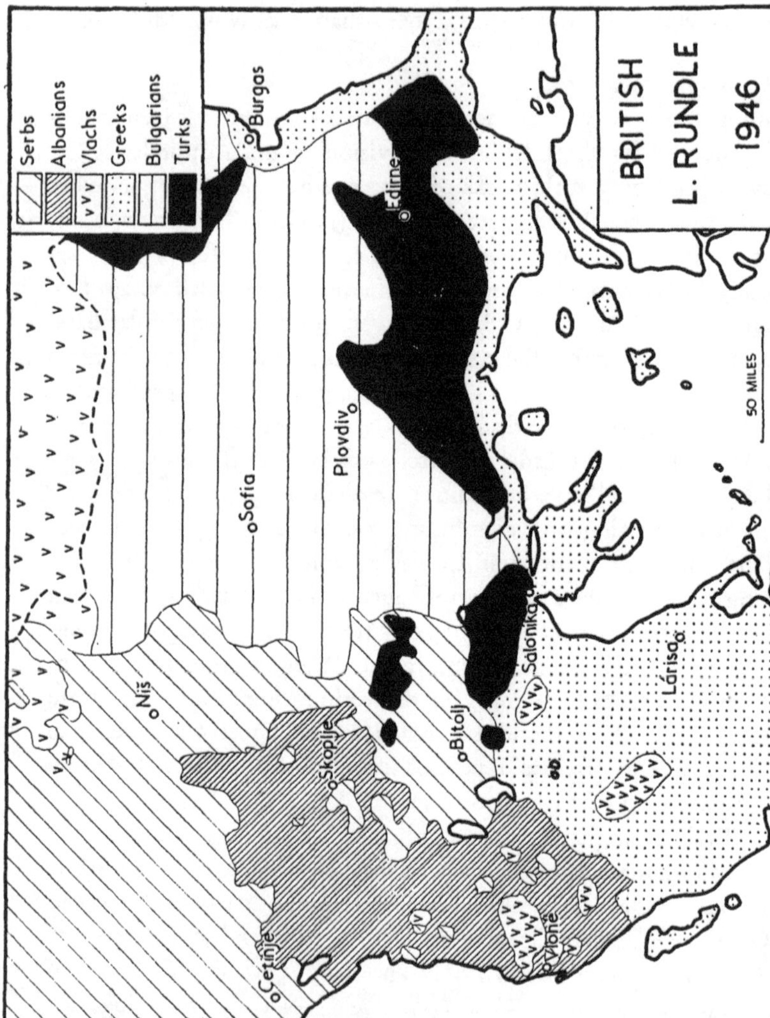


FIG. 76

In the original, Tatar is distinguished from Turk, and the Vlach category is referred to as Romanian. *Note.* The above map is a composite and is based on two of L. Rundle's sketches.

proper except in the extreme north. Although this purported to be a linguistic map, Pomaks were separately distinguished and some were marked in Dhitikí Thráki. The distribution of the Vlachs was taken from the G.S.G.S. map of 1918. Albanian language was given a widespread distribution in Old Serbia but not so liberal a distribution in western Macedonia or Ipiros as that given in the Greek Gazetteer.

*Some Other British Maps, 1945-6*

In *The Nationalities of Europe* which appeared in 1945, H. M. Chadwick, professor of Anglo-Saxon in the University of Cambridge, dealt at some length with the political significance of linguistic groupings in Europe. In a synthesis of the order he attempted, a detailed examination of Balkan ethnography was not to be expected. Nevertheless, it is possible to gain an indication of the author's ideas on Balkan linguistic distributions from a small sketch-map incorporated in the book. On this, the frontiers of Serbo-Croat, Albanian, Greek and Bulgarian speech roughly coincide with the political boundaries of 1938, a point of view not at all in accordance with the ideas expressed on any other ethnographic maps, and one from which fallacious deductions as to the ethnic validity of Balkan political boundaries might result.

The difficulties facing students of philology in tracing linguistic distributions in Europe was remarked upon by Stanley Rundle in his *Language as a Social and Political Factor in Europe*, which appeared in 1946. He expressed some concern on the failure of geographers to produce reliable linguistic maps. Two of his sketch-maps depicting Balkan linguistic distributions did indeed reflect his difficulties. These two maps are depicted in a composite form in Fig. 76. It can be seen that no allowance had been made for the transfers of population in the case of the distributions in Greece. In Jugoslavia the Serbian view of their distribution was accepted with modifications in favour of the Bulgarians in the frontier zone. There are many other obvious defects in Rundle's maps. They are defects of which Rundle himself was probably well aware, a faithful reflection of the difficulties to be encountered in any attempt to delimit ethnographic frontiers in Macedonia.

## THE JUGOSLAV CENSUS, 1948

As a result of the negotiations between the Serbo-Croat and the Macedonian communist parties, the Republic of Macedonia came into existence in 1945, and a provisional inter-republican boundary with Serbia was established. After the new constitution of Jugoslavia had been drawn up in 1946, the inter-republican boundaries were fixed in detail. The provisional Serbo-Macedonian boundary was modified slightly, and in its final form it was adjusted to the watershed of the Sar mountains from whence it ran south, leaving the Kačanik gorge in Kosmet, then north-west to Mount Gopan, then east across the Morava and Pčinja valleys to traverse the difficult country formed by the Siroka and Derman mountains, and finally to Mount Patarica and the Bulgarian border.

Except in the Morava valley, where the boundary-makers made modifications (*cf.* Figs. 77 & 90), the final boundary ran through territory of relatively high altitude and low population density. Obviously, considerations of economic orientation and accessibility played some part in the determination of its ultimate direction. It was claimed that the boundary was an ethnographic divide insomuch as it separated Serbs and *Macedo-Slavs*. But at the same time it ignored Albanian distributions in Macedonia, so that a considerable number of Albanians south of the Sar mountains was incorporated into the new republic. The new boundary coincided closely with that advocated for an independent Macedonia by D. Jaranov in 1930, and with that expressed in the I.M.R.O. map of 1934 (Fig. 58).

From the first, all the Slavs in the new republic were referred to by the Jugoslav Government as Macedonians rather than as Bulgarians. At the instigation of the Jugoslav Ministry of Public Education, a Macedonian alphabet, an orthography and a grammar were created on the basis of the vernacular. Problems of terminology were considerable but with the aid of technical terms borrowed from various sources, the new language had become the medium even of University instruction by 1947. The establishment of the new language was a concession to the Macedonian national feeling. But it might also be regarded as an attempt by the Jugoslav Government to crystallize such differences as did exist between Macedonian and Bulgarian vernaculars before those differences vanished altogether (Bulgarian had made considerable headway as the language of instruction in schools between 1941 and 1943).

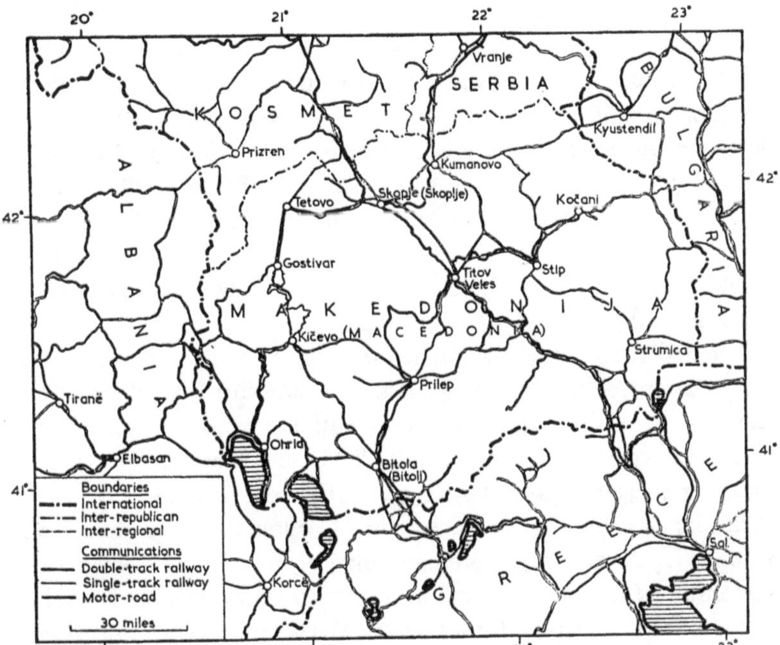


FIG. 77. THE REPUBLIC OF MACEDONIA

Based on *Federativna Narodna Republika Jugoslavije*, Razmer, 1: 750,000. Reprodukovano u Geografskom instituto Jugoslovenske Armije (Privremeno idzanje 1947 Godine).

The Bulgarian attitude to the new written language was one of tolerance in the first instance. Between 1945 and 1947 both Bulgaria and Jugoslavia were working together for the creation of a Greater Macedonia which, the Bulgarians believed, would be one of a trio of Slav states in a future Balkan federation.[1] Tolerance turned to

[1] The Bulgarian attitude is summarized in L. Mojsov, Бугарската Работничка Партија (Комунисти) и Македонското Национално Прашање (Скопје, 1948). Bugarskata Rabotnička Partije (Komunisti) i Makedonskoto Nacionalno Prašnje (Skopje, 1948).

alarm when the Jugoslavs began to send their Macedonian emissaries to Pirin Macedonia to spread the newly-established written language, and to prepare for a union of Pirin Macedonia with the new Jugoslav Republic. Of the 250,000 Macedonians calculated by I.M.R.O. to be in Pirin Macedonia, there is little doubt that 150,000 had become Bulgarians by 1945, both in speech and in national sentiment, and the secession of Pirin Macedonia for the sake of the few thousand Macedonians still hoping for independence, was a high price for the Bulgarian Government to stake in the gamble for a Greater Macedonia.[1]

Had the plans for federation proceeded as the Bulgarians had hoped, the gamble may have been worth risking, but the Jugoslav Government adopted the view that each of the six new Jugoslav republics should have equal representation in the federal scheme, with Bulgaria as the seventh. Instead of a joint Macedo-Bulgarian leadership of a Balkan federation, the Bulgarians were faced with the prospect of being but one of seven federal states, and thus virtually relegated to a minor role in Balkan affairs, and their enthusiasm for the Macedonian adventure correspondingly declined. The Cominform dispute, whatever its cause, gave the Bulgarian Government the opportunity of denouncing the Jugoslav policy in Macedonia as expansionist, and of renewing their own revisionary policy in this theatre. The natural outcome of the split between the two Slav states, was the abandonment of their joint plan for a Greater Macedonia. The split between the two Slav states undoubtedly led to a decline in guerilla activity in Greek Macedonia which enabled the pacification of the province to be achieved in 1949. With the *rapprochement* between the Jugoslav and Greek Governments in May, 1950, the pre-war balance of power in the Macedonian theatre was temporarily re-established—a fact which makes for the stability of the present political boundaries. But any change in the relations between the Jugoslav and Bulgarian Governments could well upset the equilibrium once again.

[1] But D. Mitrev, in a recent book on Pirin Macedonia written from the Macedonian standpoint ; Пиринока Македонија во Борба за Национално Ослободување (Скопје, 1950) ; Pirinska Makedonija vo Borba za Nacionalno Osloboduvanje (Skopje, 1950), gives the following figures of the ethnic composition of its population : Macedonian, 226,700 ; Turks 6,000 ; others 2,300.

Meanwhile, the Jugoslavs carried out a detailed census of the population of the Republic of Macedonia in 1948, the results of which are given in the table below :

*Nationalities in Macedonia,* 1948

| *Nationality* | *Totals* |
|---|---|
| Orthodox Serbs .. .. .. .. | 29,335 |
| Orthodox Croats .. .. .. .. | 2,680 |
| Slovenes .. .. .. .. .. | 777 |
| Macedonians .. .. .. .. .. | 788,889 |
| Montenegrins .. .. .. .. .. | 2,329 |
| Moslem Serbs .. .. .. .. .. | 417 |
| Moslem Croats .. .. .. .. .. | 24 |
| Moslem, undetermined .. .. .. | 1,565 |
| Bulgarians .. .. .. .. .. | 890 |
| Other Slavs .. .. .. .. .. | 1,331 |
| Albanians .. .. .. .. .. | 197,433 |
| Vlachs .. .. .. .. .. .. | 9,508 |
| Turks .. .. .. .. .. .. | 95,987 |
| Greeks .. .. .. .. .. .. | 1,013 |
| Gypsies .. .. .. .. .. .. | 19,500 |
| Other nationalities .. .. .. .. | 1,308 |
| *Grand Total* .. .. | 1,152,986 |

Data based on the census of March 15, 1948, and published with the permission of the Jugoslav Federal Institute of Statistics.

According to these figures, the Serbs formed only 2.5 per cent of the population of Macedonia in 1948. The Macedonians formed 68.4 per cent. The largest minority was that of the Albanians—17.1 per cent. The only other minority of any size was that of the Turks—8.3 per cent. Gypsies were distinguished as a separate nationality. Both language and religion were taken into account in the classification of nationalities.

## CHAPTER XIII

# SUMMARY AND CONCLUSIONS

THIS summary takes the form of a graphical portrayal of the many ideas held at different times on the distributions of the main ethnic groups—Turkish, Greek, Serb, Bulgarian, *Macedo-Slav*, Vlach and Albanian. The Pomaks have also been considered. These distributions have already been discussed in previous chapters but here they are presented in a rather different fashion. The base-map selected covers

FIG. 78. SUMMARY BASE-MAP

The political boundaries are those of 1950. In the maps which follow the towns given above are marked but not named.

Macedonia proper and incorporates the modern political boundaries (Fig. 78). The positions of several towns have been marked in order to facilitate comparison. The maps are selected and arranged in such a way that they tell their own story of the conflict in Macedonia from its origins down to the present-day.

### *The Causes of Diversity*

The striking diversity of opinion on ethnographic distributions in Macedonia which has existed since the first decades of the nineteenth century has been demonstrated by the method of depiction adopted in Figs. 79-87. The reasons for this diversity are many, and they have already been discussed for individual maps. It is useful, however, to attempt a brief recapitulation in which the causes of diversity are examined in increasing order of importance.

(1) *Misrepresentation of Facts.* Even scientists of the highest personal integrity were guilty of the practice of misrepresentation, excusing themselves on the ground that the end justified the means. In some cases, notably that of J. Cvijić, an unmerited, perhaps unconscious rationalization of false distributions was prompted by the irresistible spirit of patriotism of the period (Fig. 82).

(2) *Ignorance of the Ethnographic Situation.* Some of the diversity was, of course, due primarily to ignorance of Balkan ethnography. The wild speculations of F. A. O'Etzel, W. Müller and G. Kombst fall into this category. Their maps were not based on first-hand knowledge of the peninsula ; they were, in some part, the product of fertile and lively imaginations, the possession of which has continued to be part of the equipment of some authors of ethnographic maps right down to the present day. But it would not be fair to level this charge against the majority of map-compilers. Such men as A. Boué, G. Lejean and G. Weigand, to name but three, spent many years in the field and indeed were probably better informed of local conditions than many of those geographers today who rely on statistical information for the production of their maps. Lack of statistical data has not always been a serious handicap when it could be replaced by systematic field-work.

(3) *The Passage of Time.* From the moment when P. G. Safařik drew his famous map until the time when a team of German geographers constructed their cartogram in 1940, all kinds of fresh information concerning the ethnographic situation in Macedonia was assembled. It was not always accurate but it was of sufficient moment to modify previous distributions. Even whilst the Turks remained in control of Macedonia, investigations sponsored by opposing political 'parties' were carried out, including surveys of schools and churches, and local

counts of population, in addition to a fair amount of historical and philological research. Whilst it was not possible always to evaluate the results of such work, discriminating scholars could glean a little of the truth here and there. Opinions expressed by travellers and studies by consular officials all contributed to the store of intelligence gradually made available to map-compilers as the century matured. After the partition of Macedonia in 1913, official information provided by the three governments of Serbia, Greece and Bulgaria began to influence map-compilers in their delimitations of territory. More intelligence was provided concerning movement of population, and later, official censuses of the population made possible the production of maps of an entirely new character. The report of the League of Nations' enquiry into the mass-movements of population between 1912 and 1926 supplied data which produced decided modifications in the distributions of Greek Macedonia.

(4) *Methods of Depiction.* Even given the same data, map-compilers often produced two or more sets of distributions by using different methods of depiction. Much depended on the base-map used. The greater the scale, the easier it was to indicate small minorities. Distributions shown by counties, for example, concealed scattered groups such as the Vlachs. Nor was a mixture of ethnographic elements easy to represent. In the portrayal of majority populations by administrative areas, cases occurred in which a minority amounting to 49 per cent remained totally obscured. The vast majority of maps in flat colours ignored population density, whilst the use of symbols showed numerical strengths but not detailed distributions. Some maps indicated comparative strengths of ethnic groups but neither their numerical strength nor their distribution, e.g. the League of Nations' Map, 1926.

(5) *Criteria.* But all these explanations are not sufficient to account for the irreconcilable divergence of opinion expressed in many maps. Even given exactly the same information and similar methods, two or more sharply contrasting ethnographic maps were often produced by different map-compilers. Indeed a comparison even of contemporary maps reveals that scarcely two distributions may be found to agree. The fundamental reason for this conflict lay in the

Fig. 79. Twelve Views on the Distribution of the Turks, 1730–1944

The following groups have been variously regarded as 'Turks' at one time or another during the period under review: Turkish speakers, members of the *Islami Millet*, Moslem Vlachs, Moslem Bulgarians (Pomaks), Moslem Albanians, Moslem Albanians of 'Chamuria' (Chams), Moslem Albanians of Old Serbia (Arnauts), Moslem Serbs (Bosnians), Moslem Jews (Mammins or Domnes), Moslem Greeks, Moslem Gypsies, Christian Turks (Gagauzi), Moslem Slav refugees ('Mohadjers'), Arabs, Negroes, Levantines, Circassians, Persians, Turkomans, Yüruks, Tatars, Turko-Tartars, Huzuls, Patzinaks, Kumans, Konariotes, Vardariotes, Uzes, Seljuk and Osmanli Turks. This list is not exhaustive. In some cases, for example maps 1 and 2, mere proximity of territory to Constantinople seems to have been sufficient reason for its population to have been regarded as Turkish. The conflict of opinion on what constitutes a 'Turk' accounts in some measure for the variety of distributions given in these twelve maps.

The series of maps as a whole indicate a steady diminution of the Turkish ethnographic distribution, or what was conceived to be the Turkish distribution, between 1730 and 1944. This diminution may be directly related to the decline of Turkish political power in the Balkans during that period. With the development of nationalism, groups of Moslems which were previously associated with the Osmanli Turks, found, or were credited with, new allegiances. When the Turks were defeated in 1913, actual migration of Osmanli Turks began to take place, back to Thrace and Anatolia. This movement was reinforced after the Graeco-Turkish War of 1922 and the Osmanli element was further reduced.

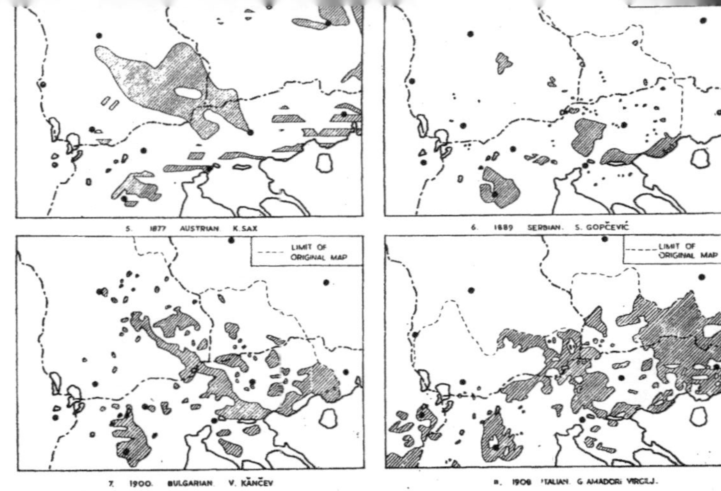


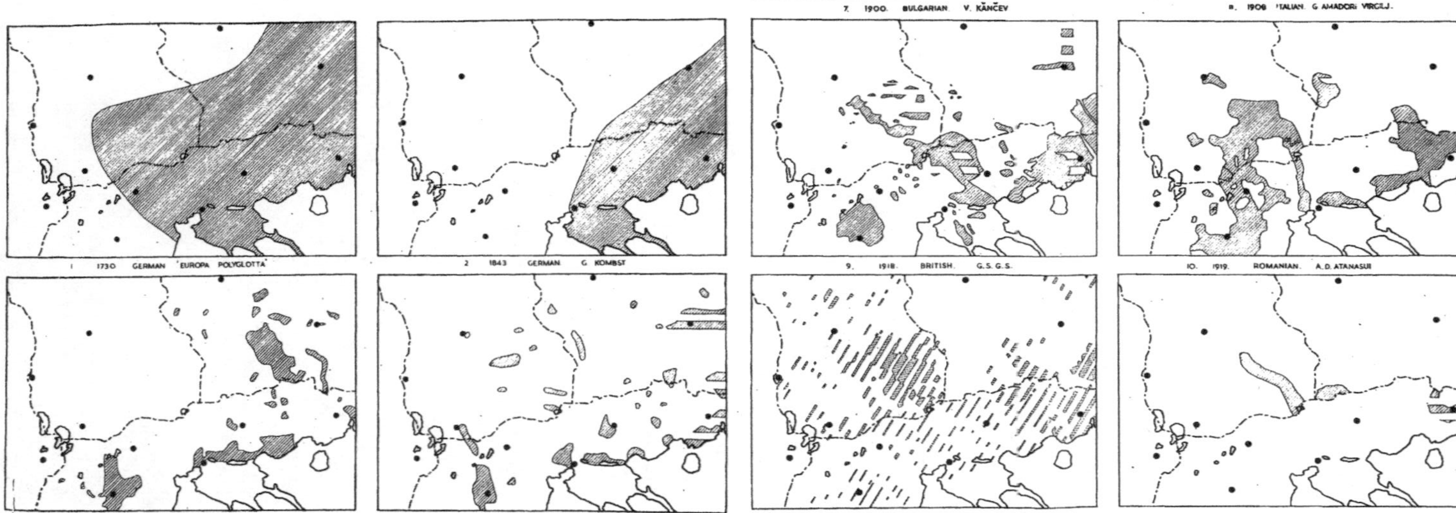

Fig. 80. Twelve Non-Greek Views on the Distribution of the Greeks, 1847–1944

The following groups have been variously regarded as Greeks at different times: members of the *Rumi Millet*, Greek speakers, Hellenes, Pelasgo-Greeks, Moslem Greeks (sometimes called Greek Pomaks), Suliotes, Greek Orthodox Albanians, Albanians under Greek influence, Albanian-speaking Greeks, Tosks, Graeco-Albanians, 'Hellenized' Vlachs, Greek Orthodox Vlachs, Graeco-Vlachs, Bulgarian Patriarchists, Graeco-Bulgarians, Bulgarophones (sometimes called Slavophones), Bulgarian-speaking Greeks, Graeco-Macedonians, Macédo-Slavs under Greek influence, Turks of Greek Orthodox faith and Greek Jews. The list is not exhaustive but the lack of agreement on what constitutes a 'Greek' helps to explain the variety of distributions depicted in this series of twelve maps. Interpretations emphasizing language tended to exclude the Greeks from Ipiros, and before 1923, from the Macedonian interior. Emphasis on religion (Map 5), or culture (Map 6), tended to give them a much wider distribution. The re-settlement of Greek refugees between 1923 and 1926 appreciably modified the ethnographic structure of Greek Macedonia but even so disagreement still existed concerning the Greek distribution (*cf.* Maps 11 and 12). Considering the series as a whole, ethnographic consolidation appears to have followed upon political consolidation. See also Fig. 88 for some pro-Greek views of Greek ethnographic distributions.

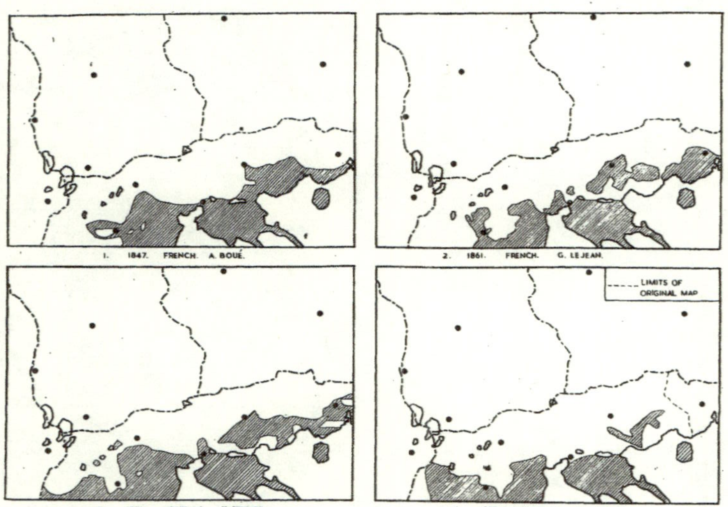

FIG. 81. TWELVE BRITISH VIEWS ON THE DISTRIBUTION OF THE SERBS, 1867–1945

The following groups have variously been regarded as Serbs at different times during the period under review: Serb speakers, Roman Catholic Serbs, Moslem Serbs (Bosnians), Montenegrins, Rascians, Clementi, Illyrians, Torlak speakers (Sop dialect), 'Serbized' Bulgarians, Slav partisans of Serb propaganda, Serbo-Macedonians, *Macedo-Slavs*, Albanians of Serb descent, Albanian Serbs, Arnauts (Albanian-speaking Moslems of Serb descent), 'Serbized' Albanians and 'Serbized' Vlachs. This list, although not exhaustive, indicates a lack of agreement on what constitutes a 'Serb', and explains in part the variety of Serb distributions given in the accompanying maps. Seen through British eyes, the Serb distribution apparently expanded southwards between 1867 and 1945. The acceptance of the idea that the *Macedo-Slavs* were partly or wholly Serb explains some of the changes in British distributions. Also, elements in the Macedonian population which had at one time been regarded as Albanian, Bulgarian, Greek, Romanian or Turkish, came to be regarded as Serb, particularly after 1915 (after Turkey and Bulgaria had entered the war of 1914–18). Changes in ideas do not wholly account for changes in distributions because some Serb migration also took place, particularly after 1913, from Serbia to Macedonia. A further factor promoting change was the partial success of Serbian propaganda in winning 'converts' to the Serb cause.

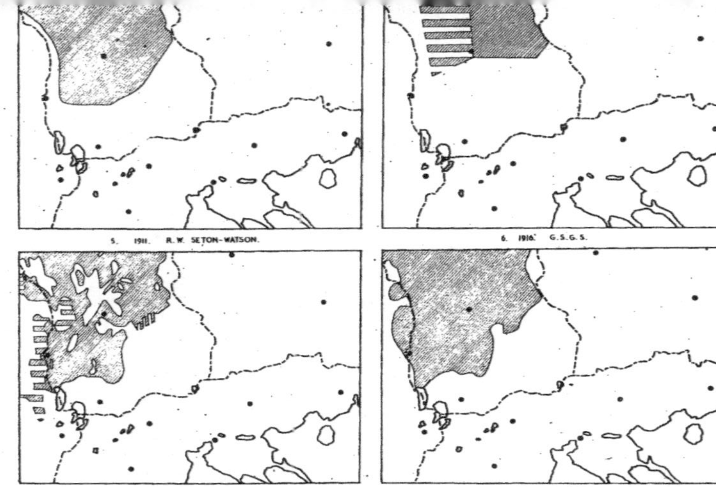

5. 1911. R. W. SETON-WATSON.

6. 1916. G.S.G.S.

7. 1917. R. W. SETON-WATSON.

8. 1917. E. STANFORD.

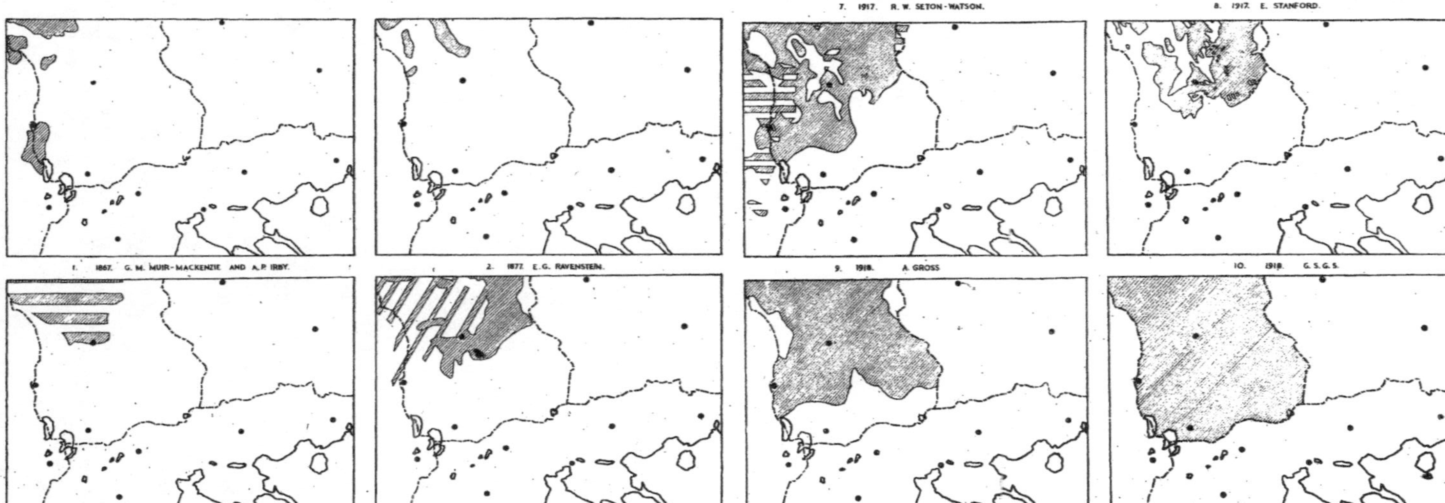

1. 1867. G. M. MUIR-MACKENZIE AND A. P. IRBY.

2. 1877. E. G. RAVENSTEIN.

9. 1918. A. GROSS

10. 1918. G.S.G.S.

3. 1906. H. N. BRAILSFORD.

4. 1909. A. STEAD.

11. 1919. E. STANFORD

12. 1945. H. M. CHADWICK

FIG. 82. EIGHT SERBIAN VIEWS ON THE DISTRIBUTION OF THE SERBS AND MACEDO-SLAVS, 1853–1924

Serbian interest in Macedonia in the mid-nineteenth century was negligible, but before the end of the century the Serbians were claiming Macedonia as a sphere of influence and were engaged in the 'ethnographic reclamation' of the Serb element in the Macedonian population. In the twentieth century attempts to identify Serbs and Macedonians were abandoned in favour of the *Macedo-Slav* theory which postulated *Macedo-Slavs* as 'incipient Serbs'. The success of this interpretation enabled Serbian claims to be pressed as far as the Aegean Coast, and in 1918, on Bulgaria itself. By 1924, *Macedo-Slavs* in Jugoslavia were regarded as having become Serbs. But in 1945, the Serbian claims were once again abandoned. The population of Macedonia was officially declared Macedonian and credited with a culture sufficiently different from that of the Serbs to justify separate political development (see also Figs. 77 & 89).

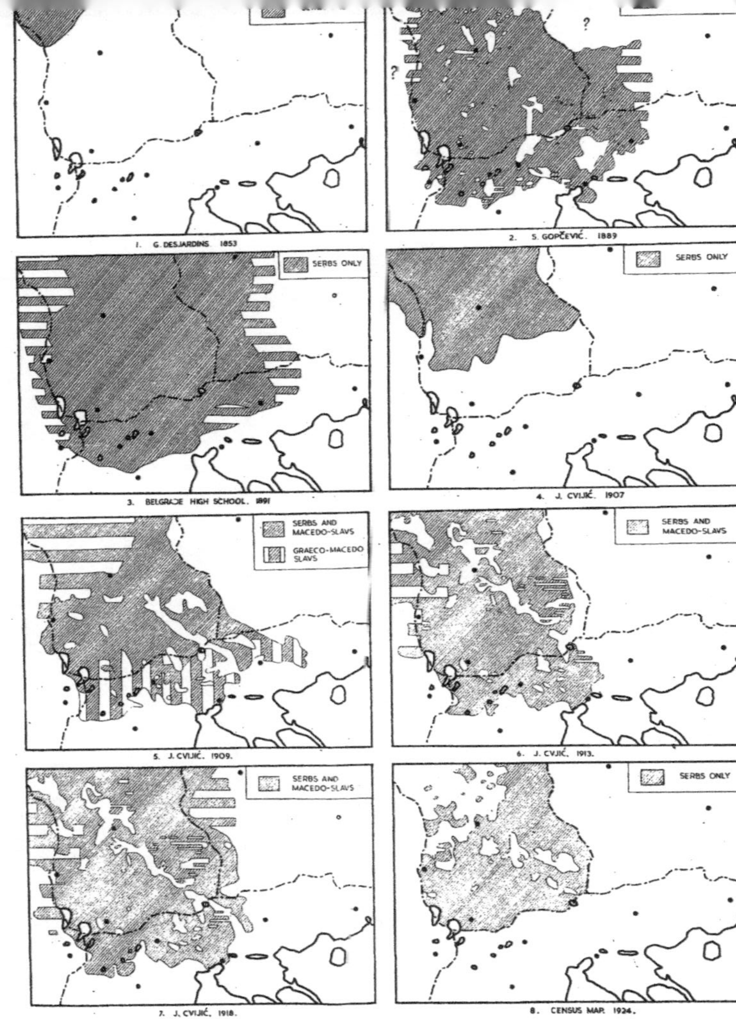

FIG. 83. TWELVE VIEWS ON THE DISTRIBUTION OF THE BULGARIANS, 1847–1940

The following groups have variously been regarded as Bulgarian at different times: Bulgarian speakers, members of the *Bulgari Millet*, Greek Orthodox Exarchists, Greek Orthodox Bulgarian Patriarchists (sometimes called Bulgarophones or Graeco-Bulgarians), Pomaks (Moslem Bulgarians), 'Serbized' Bulgarians (of the Niš district), Bulgarian Macedonians, Torlak speakers (Sop dialect), and the Bulgarians of Bosiligrad and Caribrod. This list is not exhaustive but it accounts in part for the variety of distributions incorporated in the accompanying maps. The most striking feature of the series as a whole is the remarkable recession, in the eyes of the world, of the Bulgarian ethnographic frontier between 1847 and 1940. With changes of fashion in ethnographic ideas, Serb, Macedo-Slav, Greek and Turkish elements each gained at the expense of the Bulgarian. The greatest loss from the Bulgarian point of view has been the Macedonian Slav element which before 1850 was unanimously acknowledged to be Bulgarian. The development of the idea that the *Macedo-Slavs* comprised an independent ethnographic group resulted not only in the exclusion of the Bulgarian distribution from Greek and Jugoslav Macedonia but also from south-western Bulgaria. Indeed, by 1948, the possibility of *Macedo-Slav* irredentism in Pirin Macedonia was a factor in the decision of the Bulgarian Government to ban Jugoslav Macedonians from entering Pirin Macedonia (see also Fig. 90).

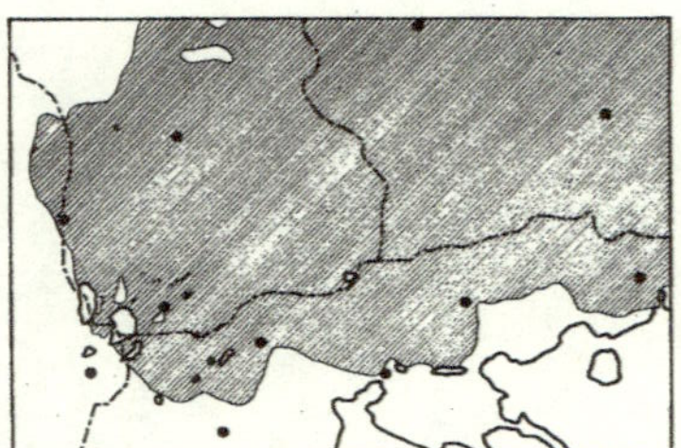

1. 1847. FRENCH. A. BOUÉ.

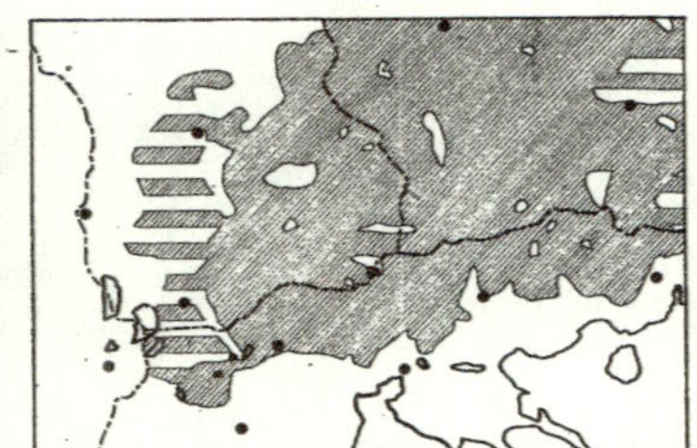

2. 1876. GERMAN. H. KEIPERT.

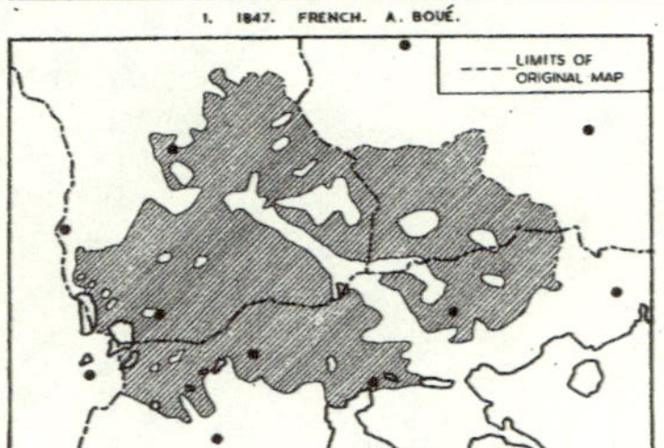


3. 1899. AUSTRIAN. F. MEINHARD.

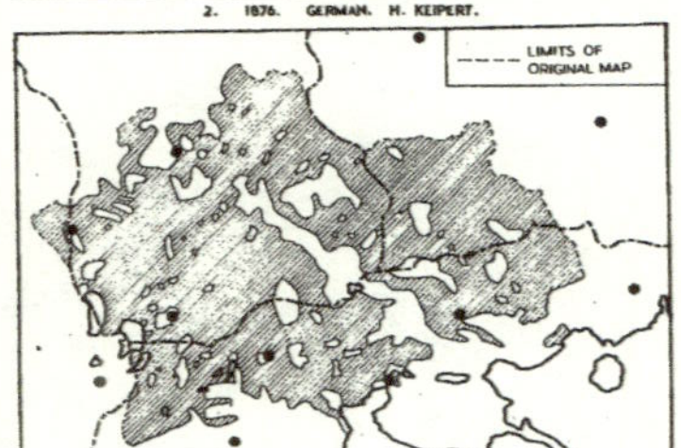


4. 1900. BULGARIAN. V. KĂNČEV.

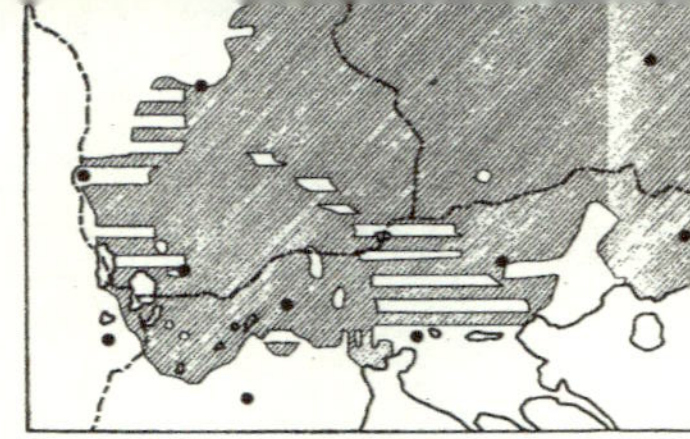

5. 1906. BRITISH. H. N. BRAILSFORD.

6. 1915. BRITISH. NEVILL FORBES.

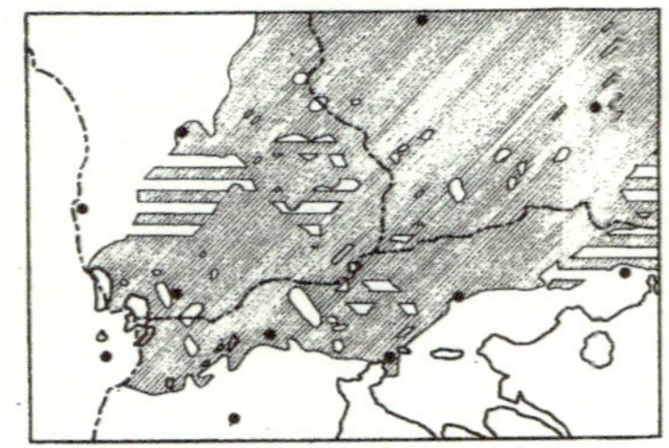

7. 1916. ITALIAN. A. DARDANO.

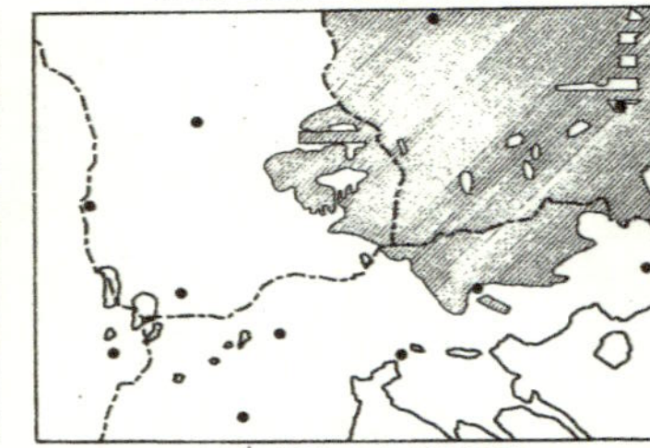

8. 1917. BRITISH. R. W. SETON-WATSON.

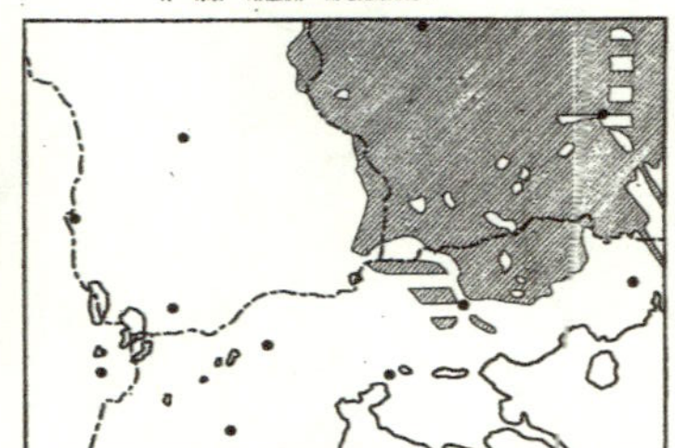

9. 1918. BRITISH. G.S.G.S.

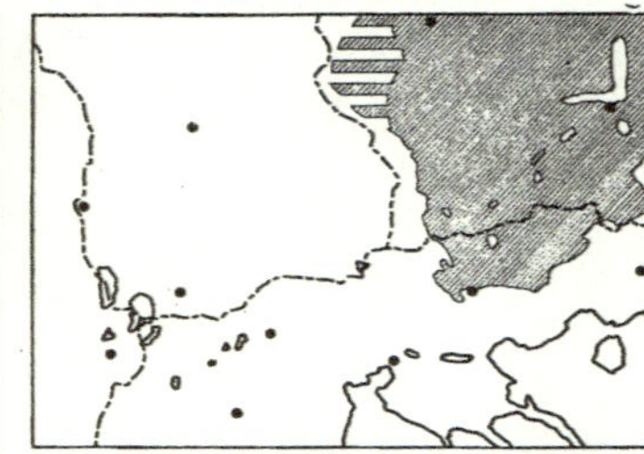

10. 1918. SERBIAN. J. CVIJIĆ.

11. 1928. FRENCH. A. MEILLET.

12. 1940. GERMAN. FOREIGN OFFICE.

FIG. 84. TWELVE VIEWS ON THE DISTRIBUTION OF THE MACEDO-SLAVS, 1909–1944

The *Macedo-Slavs* only began to figure on ethnographic maps about the beginning of the present century. The following groups have been variously regarded as *Macedo-Slav* since 1903: Persons speaking the Macedo-Slav dialect (no agreement exists on the distribution of the dialect, but the development of a standardized written language sponsored by the present Jugoslav Government is doing much to differentiate it both from Serb and Bulgarian); *Kulturlos* Slavs (i.e. Slavs without any specific national inclination); 'Slavized' Vlachs; Greek *Macedo-Slavs*; Bulgarian *Macedo-Slavs*; Serb *Macedo-Slavs*; 'Albanianized' *Macedo-Slavs* and Moslem *Macedo-Slavs* (sometimes called *Macedo-Slav* Pomaks). The variety of interpretations of *Macedo-Slav* distributions given in the accompanying maps indicates lack of precision in the definition of the group. In J. Cvijić's later maps, Serb *Macedo-Slavs*, for example, are portrayed as Serbs. The British maps tended to portray Bulgarian *Macedo-Slavs* as Bulgarians, and so on. Whether the *Macedo-Slavs* would ever have attained an independent ethnographic status had Bulgaria gained control of the territory in 1878 is a matter for conjecture, but Serbian propaganda certainly played a major part in severing their connection with the Bulgarians. The appearance of the *Macedo-Slavs* on ethnographic maps invariably reduced Bulgarian distributions in the Macedonian region. Whatever their origins, however, there is no doubt today that a strong feeling of political consciousness exists amongst them.

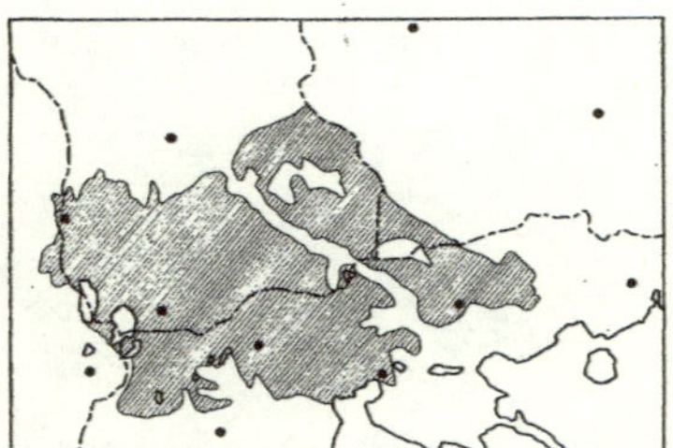

1. 1909. SERBIAN. J. CVIJIĆ

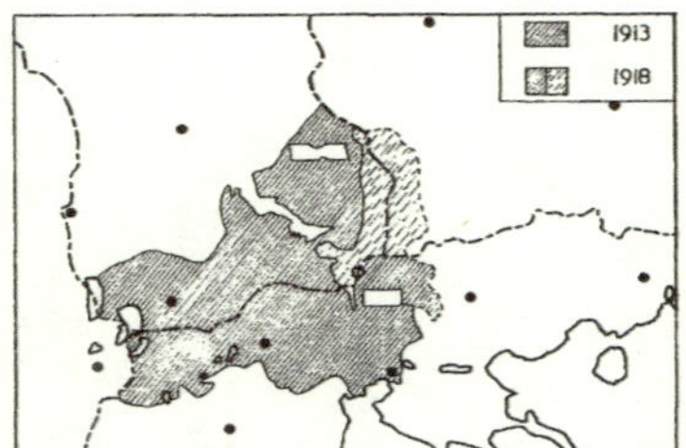


2. 1913 & 1918. SERBIAN. J. CVIJIĆ

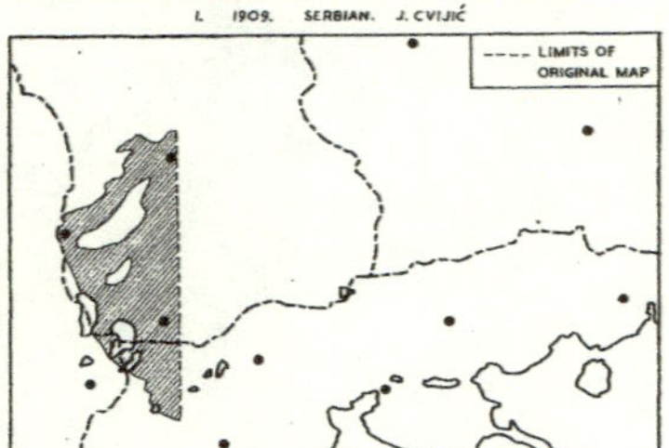


3. 1915. FRENCH. MINISTRY OF WAR

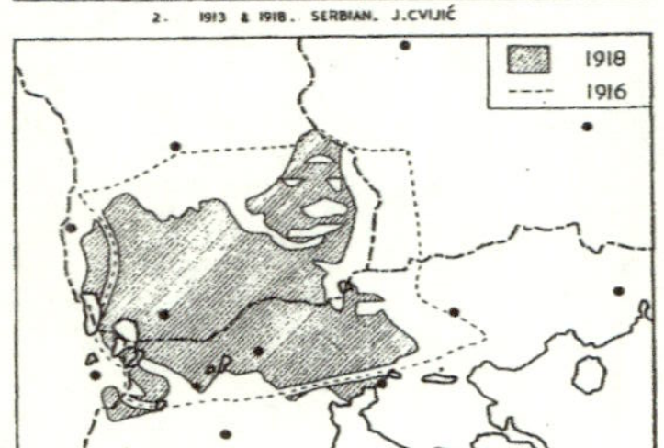


4. 1916 & 1918. BRITISH. G.S.G.S.

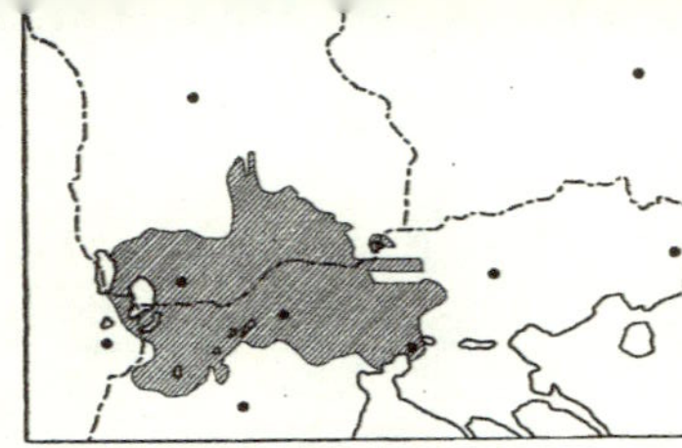

5. 1917. BRITISH. R.W. SETON-WATSON

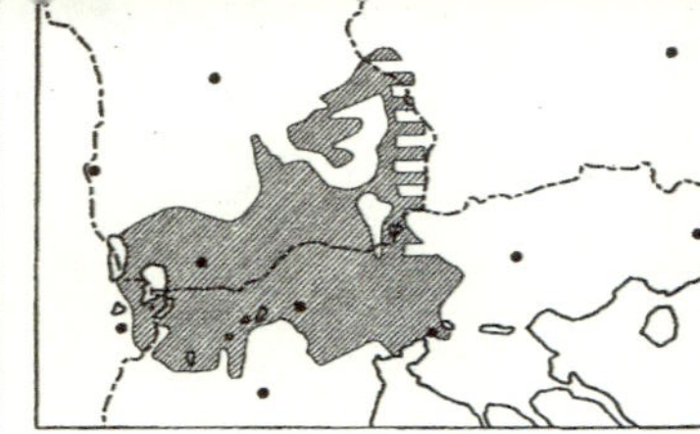

6. 1918 BRITISH. A. GROSS

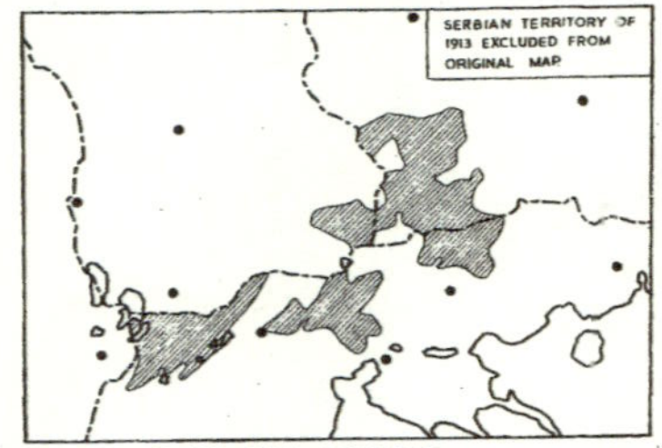


7 1918 GREEK G. SOTERIADES

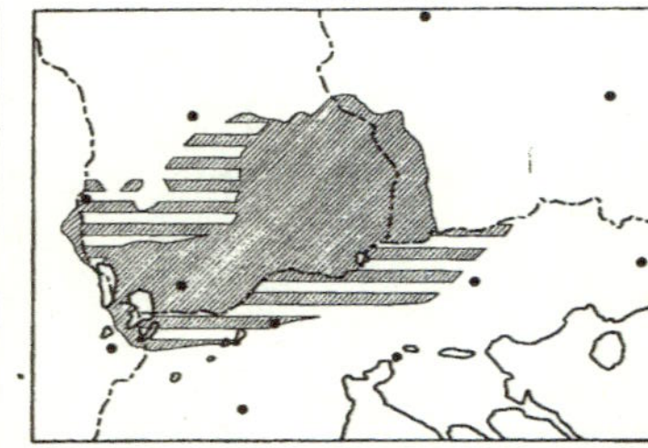

8. 1928. GERMAN. A. HABERLANDT

9. 1934 FRENCH. A. MEILLET

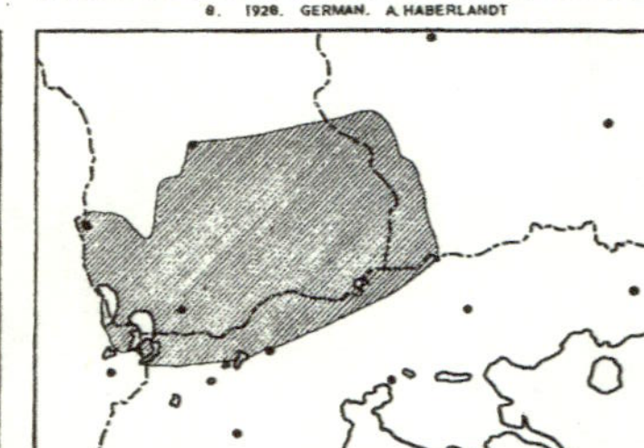

10. 1934. FRENCH. Y. CHATAIGNEAU

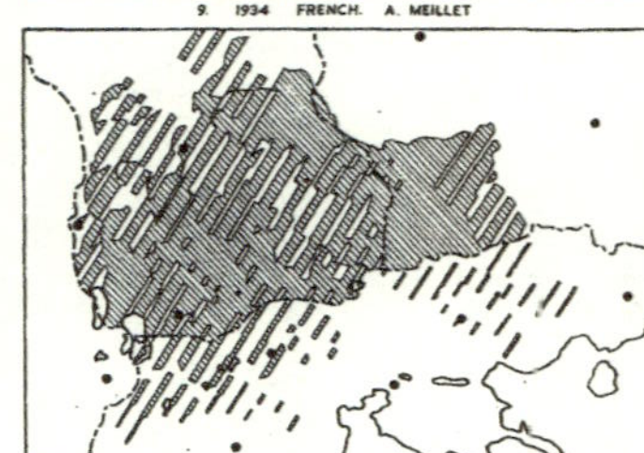

11. 1940 GERMAN. FOREIGN OFFICE

12. 1944 BRITISH. N.I.D. HANDBOOK

FIG. 85. EIGHT VIEWS ON THE DISTRIBUTION OF THE POMAKS, 1877–1940

Most authorities have been agreed in the past that the Pomaks are Bulgarian speakers who have embraced, or who have been forced to adopt, the Moslem religion. J. Ivanov, however, included Greek-speaking Moslems as Pomaks (Map 6). J. Cvijić excluded Moslem *Macedo-Slavs* from the Pomak group (Map 5). Pomaks have been variously regarded as Turks, as Bulgarians, or occasionally, as a distinctive group owing allegiance to neither Turks nor Bulgarians. The series of accompanying maps all show Pomaks in the Rodopi mountain district, but exhibit little agreement about the distribution of scattered Pomak exclaves. The concentration of Pomaks in Greek Eastern Macedonia which appears on the pre-1918 maps was reduced as a result of the Graeco-Turkish population exchange (V. Mikov's map does not acknowledge the validity of the exchange and refers to the situation in 1913). The Pomaks of Dhikití Thrakí (Western Thrace) were not affected by the exchange (Map 8).

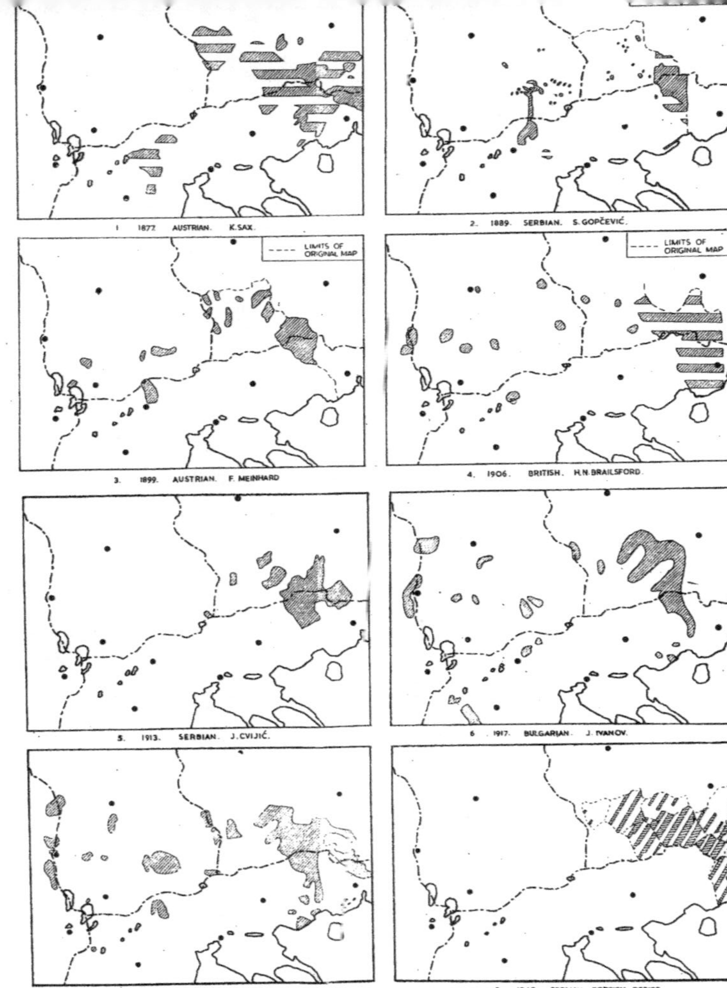

Fig. 86. Twelve Views on the Distribution of the Vlachs, 1821–1940

The following groups have variously been regarded as Vlach (sometimes referred to as Kutzo-Vlachs, Zinzares, Aromunes, Little Wallachians, Little Romanians, etc.). All persons in Macedonia, Albania and Greece speaking the Vlach dialect (the dialect is not readily recognizable as its original Latin quality has been greatly modified by borrowings from Greek, Albanian and the various Slav dialects, as well as from Turkish); 'Hellenized' Vlachs; 'Romanized' Vlachs; 'Slavized' Vlachs; 'Albanianized' Vlachs; Pindus Vlachs; Macedo-Vlachs; Frasheri (Vlachs of parts of southern Albania); Blachi (an historical reference); Boui or Boviens; Brouzi; Dacians, Dassarets or Massarets (the last six groups are historico-tribal groups); Perhebians (another tribal reference); Vlacho-Patzinaks (Megalo-Vlachs or Meglenites); Serbo-Vlachs and Bulgaro-Vlachs. The list is not exhaustive, but it serves to demonstrate the variety of opinions held about the Vlach group which explains in part the variety of distributions given in the accompanying series of maps. Views later than 1877 emphasized the scattered nature of their distribution, even if no unanimity existed on the precise location and size of enclaves. The migratory habit of the Vlachs adds to the difficulty of mapping their distribution. Since 1919, the tendency has been for the various Vlach elements to be absorbed by the Albanian, Greek, Jugoslav and Bulgarian groups (Map 12).

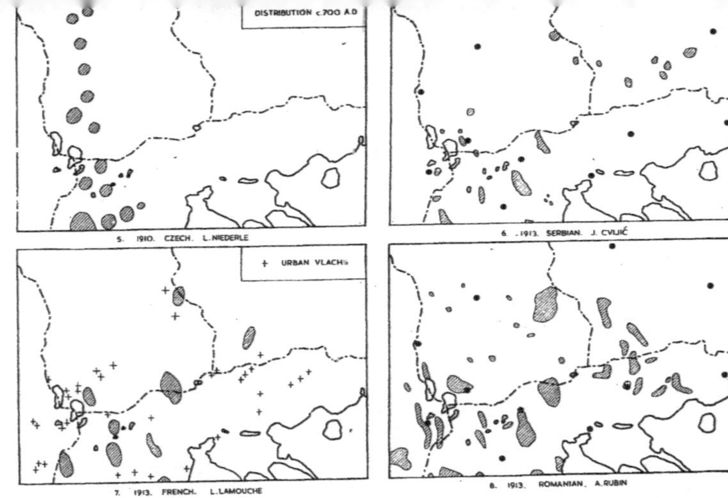


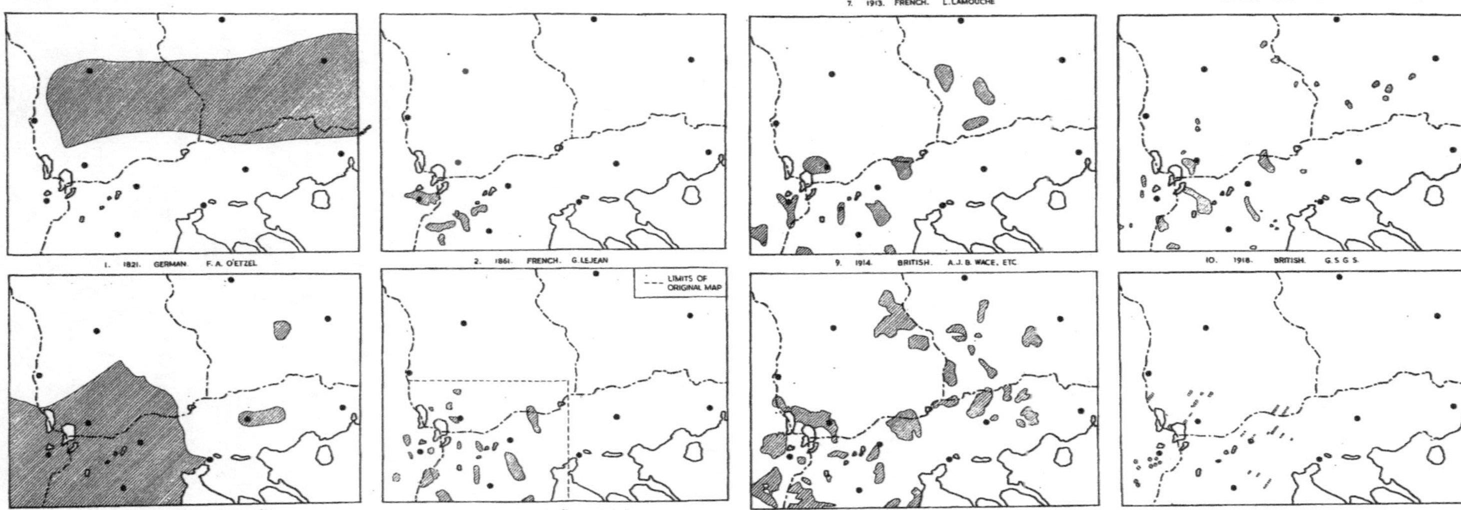

FIG. 87. TWELVE VIEWS ON THE DISTRIBUTION OF THE ALBANIANS, 1847–1942

The Albanians are sometimes called Shqiptars (Skipetars) and have also been described as Illyrians (L. Niederle), Thraco-Illyrians and Thracians. The following groups have been variously regarded as Albanian since 1847: Albanian speakers, Moslem Albanians, Arnauts (sometimes called Moslem Albanians of Serb descent), Christian Albanians of Serb descent, 'Serbized' Albanians, Gegs, Tosks, Suliotes, Mirdites, etc., Albanian-Serbs, Albanian-Greeks (Greek Orthodox Albanians), Chams (Albanian Moslems of 'Chamuria'), Frasheri (Albanian-Vlachs). There has thus been a lack of agreement on what constitutes an 'Albanian' which explains, in part, the variety of interpretations given in the accompanying maps. Before 1921, Serbian maps minimized the importance of the Albanian element in Old Serbia and western Macedonia on the grounds that Albanian speakers in these districts were of Serb descent (Maps 3, 4 and 6). Some British maps adopted the Serbian interpretation (Map 9). In the Serbian Census of 1921, the Albanian 'mother tongue' was conceded to be a test of nationality and the Albanian distribution in Jugoslavia was accordingly extended (Map 12). Italian maps have always sought to emphasize the Albanian element in Old Serbia and Macedonia (Maps 5 and 6).

Migration must also be taken into account in explaining the variety of distributions. The Macedonian geographer, J. F. Trifunoski, for example, has recently maintained that a considerable movement of Albanians took place into western Macedonia during the early and mid-nineteenth century and may well account for the differences of distribution shown in Maps 1 and 2.*

* Unpublished manuscript (Skoplje, 1948).

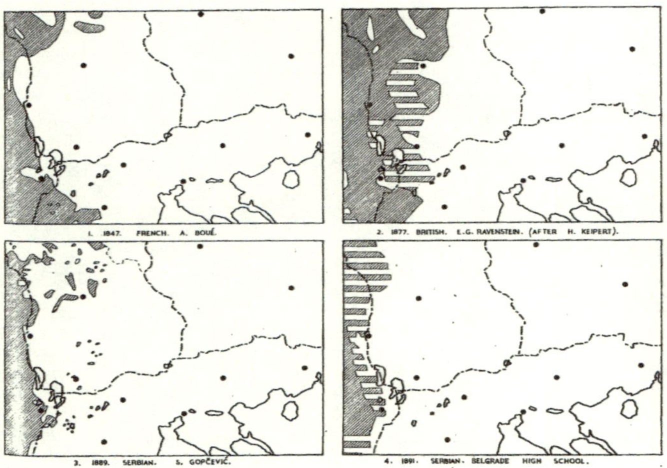


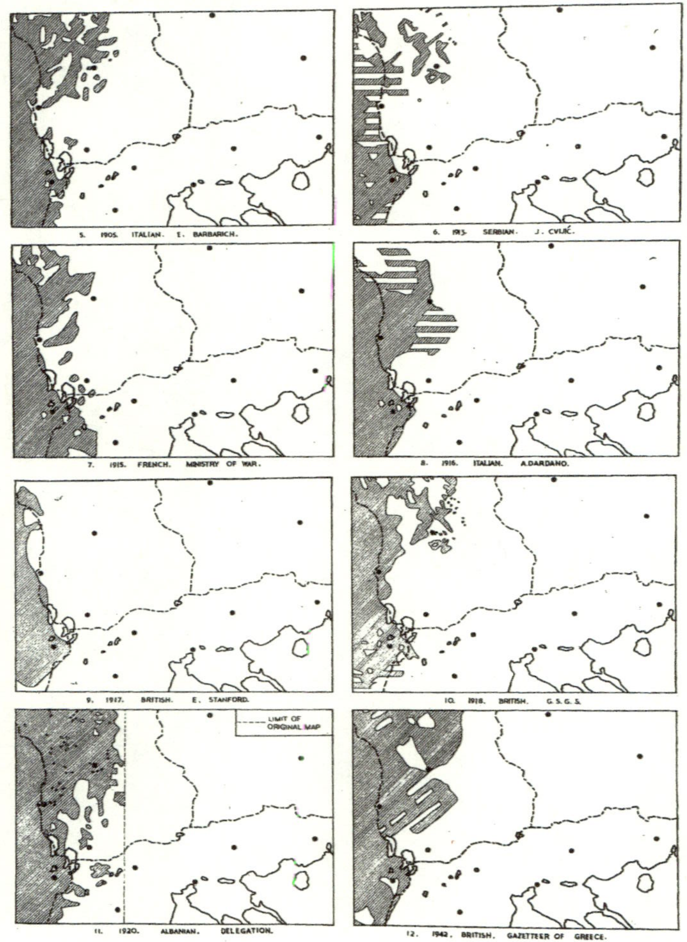

selection of criteria. Numerous data existed upon which the population of Macedonia might be indefinitely subdivided into some hundreds of different groups. The diversity of the maps reflected the lack of any commonly acceptable or durable criterion of ethnic affinity.

In many maps, language was the principal, indeed, sometimes the only, criterion used. To quote Dudley Kirk:

> Of the several components of ethnic nationality language is unquestionably regarded as the most important single element. In view of the pre-eminent importance of language in the transmission of cultural heritage it is understandable why language should have become the chief rallying point of nationality in Europe. Language is probably the greatest invention of mankind and the one that has made possible most of the cultural achievements which distinguish human beings from other creatures. In the modern world, language is the most important of cultural traditions.[1]

The measurement of linguistic distributions, however, proved extremely difficult in the case of Macedonia. Official census material was by no means reliable, and in conducting their own researches champions of the linguistic criterion found themselves confounded by a wealth of archaic forms preserved in Macedonian vernaculars and by the almost complete absence of written language-records of a date later than Old Church Slavonic. The problem as to whether Macedonian dialects were closer to Serb or Bulgarian was complicated by the introduction of Serbian and Bulgarian written languages into Macedonian schools and by the absence of any sharply defined linguistic divides of a kind which are usually associated with long-established political boundaries. The problem of differentiating between various Slav dialects or languages was more acute than that of differentiating between Slav, Greek, Albanian, Turkish, Ladino and Romanian respectively, but even these languages were closely inter-related and bilingualism had also to be considered (see Appendix A, p. 328).

Even if philologists could have agreed amongst themselves on linguistic groupings, language itself was not the only test of ethnic

[1] Dudley Kirk, *op. cit.*

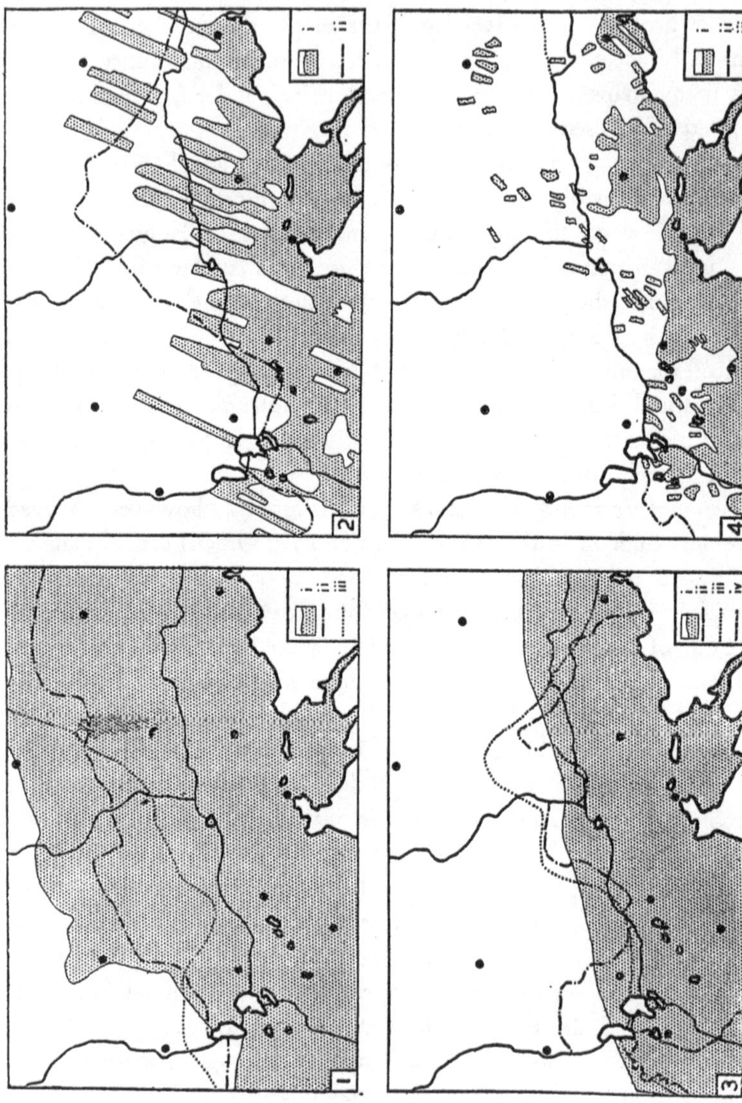

affinity to be found in Macedonia. Religion was equally admissible. Turks and Greeks, indeed, claimed it as the supreme criterion. Nor did language and religion exhaust the list of possibilities. Other loyalties were distinguished by discriminating observers which, they claimed, bound together people of different language and religion; similarity of economy, historical associations, common customs and observances, long-established political allegiances, like material cultures, pseudo-racial ties, even place of birth and 'political nationality' had their advocates.[1]

The plurality of criteria led many scholars to abandon the employment of single tests of affinity and to indulge in a combination of two or more criteria when drawing up their classifications. The most common combination was language for Bulgaria, Jugoslavia and Northern Albania and religion for Greece and Turkey. Hence the endless variety of contrasting ethnographic mosaics which resulted.

[1] See Dudley Kirk, *op. cit.* (Ch. X) for a useful discussion of the criteria of nationality.

---

FIG. 88. GREEK ASPIRATIONS IN MACEDONIA AND ADJOINING TERRITORIES, 1877-1946

The references in the key are as follows:

**Map 1. i.** Greek ethnographic distribution according to E. Stanford's Greek-inspired map of 1877; **ii.** Northern limit of the Empire of Constantinople in the 14th century (based on Spruner-Menke, *Historischer Handatlas*, 1878); **iii.** Northern boundary of a Greater Hellenic State proposed by H. Kiepert in his Ethnocratic Map of 1878.

**Map 2. i.** Greek ethnographic distribution according to A. Synvet's map of 1877; **ii.** British plan for the northern boundary of Greece in 1878 (based on Wyld's map of Turkey-in-Europe, 1878).

**Map 3. i.** Greek ethnographic distribution according to C. Nicolaides (1914); **ii.** Extent of Greek schools in 1899 (based on the maps of R. von Mach, 1900); **iii.** Territory occupied by Greece during the Balkan Wars (based on C. Nicolaides' map, 1914); **iv.** Greek demands in Ipiros in 1913 (based on the map "L'Epire du Nord" in *Histoire Diplomatique de la Grèce* (1926) by E. Driault).

**Map 4. i.** Greek ethnographic distribution according to G. Soteriades (1918); **ii.** Greek demands in Ipiros in 1919 (based on the Greek Memorandum to the Peace Conference, 1919); **iii.** Greek demands in Bulgaria (based on the Greek Memorandum of 1919).

*Note.* These claims in Ipiros and in the Rodopí of Bulgaria were reiterated in the *Memorandum du Gouvernement Hellenique sur l'Albanie et l'Epire du Nord* (April, 1946) and the *Memorandum du Gouvernement Hellenique sur la Frontière Greco-Bulgare* (July, 1946).

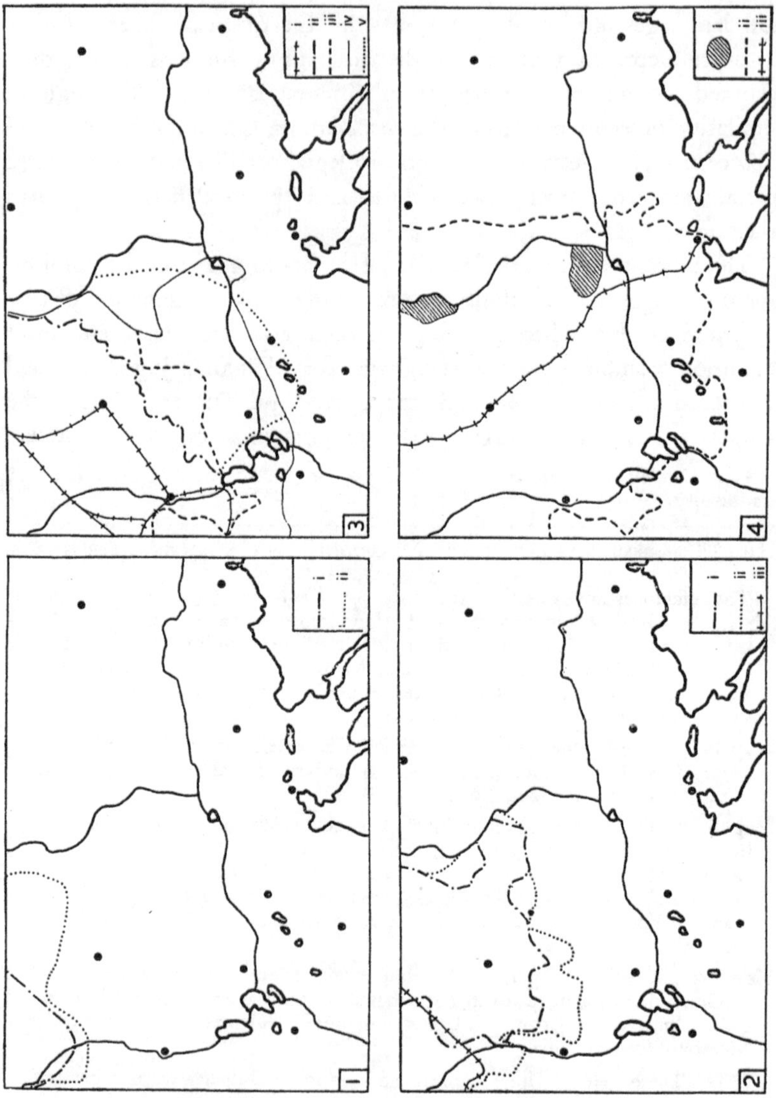

Each map became the vehicle of the ideas of its particular compiler—ideas which were partly the product of their age and environment and which differed accordingly. It is patent that even if the population of Macedonia had remained unchanged and static, the variety of ethnographic maps would still have remained comparatively rich.

(6) *The Grouping of Ethnographic Elements.* A further cause of diversity arose from the practice of combining certain sub-groups into major ethnographic groups, the inference being that the individuality of the sub-group was less than its affinity with the major group. The effect of this practice may be illustrated by reference to the Turks who were variously understood to be :

(*a*) Osmanli Turks only;
(*b*) Osmanlis plus such pre-Osmanli groups as *Vardariotes*, *Uzes* and Konariotes;
(*c*) Osmanlis and Turkomans;
(*d*) Osmanlis and Tatars;

---

FIG. 89. SERBIAN ASPIRATIONS IN MACEDONIA AND ADJOINING TERRITORIES, 1853-1918

The references in the key are as follows :

**Map 1.** **i.** Serbian ethnographic distribution according to G. Desjardins (1853) ; **ii.** ' Old Serbia '—representing Serbia's territorial claims in 1878 (based on Wyld's map of 1878).

**Map 2.** **i.** Extent of Serbian schools in 1899 (based on the maps of R. von Mach, 1900) ; **ii.** Limits of ' Old Serbia ' according to J. Cvijić (1907) ; **iii.** Rail-route to San Giovanni di Medua and Antivari proposed by J. Cvijić in 1909.

**Map 3.** **i.** Rail-routes from Serbia to the Adriatic ports in 1912-13, based on 1. J. Cvijić ("Der Zug nach Adria", *Petermann's Geog. Mittheilungen*, 1912) ; 2. Miss M. I. Newbigin (*Geographical Aspects of Balkan Problems*, 1915) and 3. a map in *The Geographical Journal*, 1913 ; **ii.** Southern limit of Serbian claims in Macedonia in 1912 (based on the Secret Annex of 1912) ; **iii.** Southern limit of the Serbs in 1913 according to J. Cvijić's map of 1913 ; **iv.** Territory occupied by the Serbs during the period of the Balkan Wars, 1912-13 (after ' Balkanicus ') ; **v.** Southern limit of Macedo-Slavs with Serb affinities according to J. Cvijić (1913).

**Map 4.** **i.** Bulgarian territory acquired by Serbia under the terms of the Treaty of Neuilly, 1919 (these acquisitions represented only part of Serbian claims on Bulgaria) ; **ii.** Extent of Serbs and Macedo-Slavs according to J. Cvijić's map of 1913 ; **iii.** The Morava-Vardar route from ' Old Serbia ' through Macedonia to Salonika.

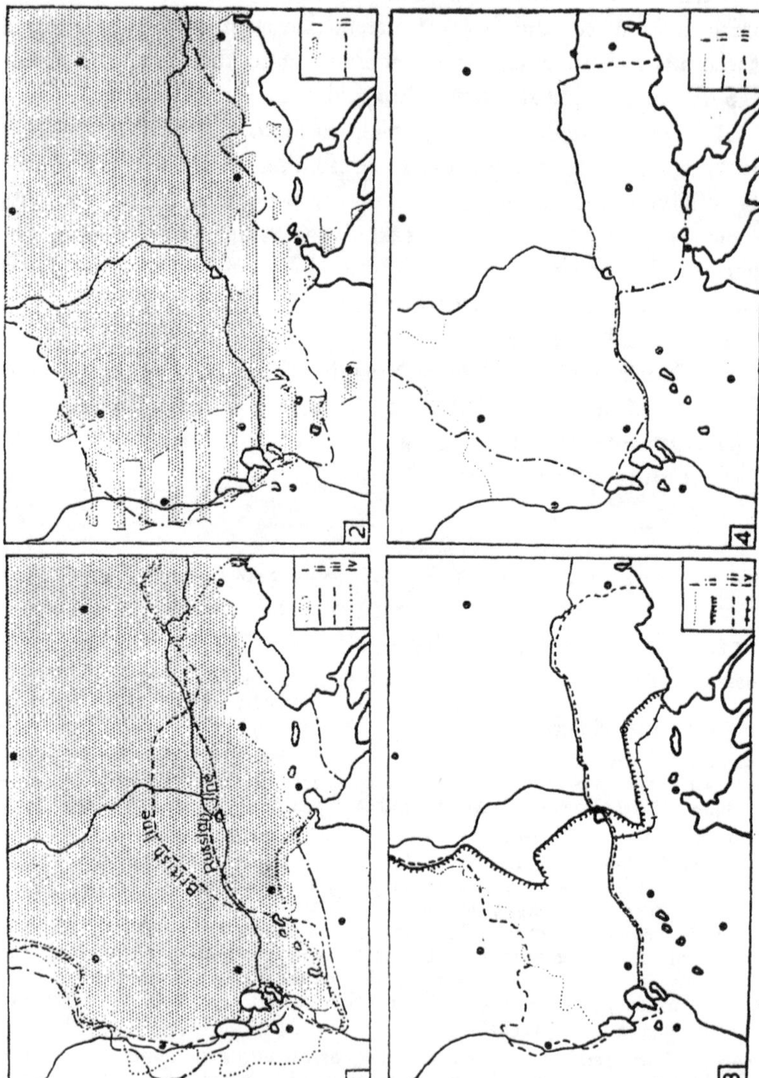
2
4
British line
Russian line
1
3

(*e*) Osmanlis, Tatars and Gagauzi (Christian Turks);
(*f*) Osmanlis and Gypsies;
(*g*) Osmanlis and Pomaks (Moslem Bulgarians);
(*h*) Osmanlis, Pomaks, Chams (Moslem Albanians of Chamuria), Arnauts (Moslem Albanians of Old Serbia), and Bosnians (Moslem Serbs).

This list does not exhaust the possible combinations but it is sufficient to demonstrate the way in which such combinations are likely to produce differing distributions.

*The Limitations of Ethnographic Maps*

Ethnographic maps are the product of an era in the history of south-eastern Europe which is not yet at an end, when scientific investigation into the nature of society was, and is, proceeding against a

---

FIG. 90. BULGARIAN ASPIRATIONS IN MACEDONIA AND ADJOINING TERRITORIES, 1877-1950

The references in the key are as follows:

**Map 1.** **i.** Bulgarian ethnographic distribution according to A. Boué (1847); **ii.** Extent of the Bulgarian Exarchate in 1876 (after J. Ivanov); **iii.** Bulgarian boundaries proposed at the Conference of Constantinople in 1876 (the British line was farther north than that put forward by the Russians); **iv.** Bulgarian boundaries proposed under the terms of the Treaty of San Stefano in 1878.

**Map 2.** **i.** Bulgarian ethnographic distribution according to N. Zaryanko (1890); **ii.** Extent of Bulgarian schools in 1899 (after the maps of R. von Mach, 1900).

**Map 3.** **i.** Northern limit of territory recognized by Serbia as Bulgarian in 1912 (after the Secret Annex of 1912); **ii.** Territory occupied by the Bulgarian army at the close of the Balkan Wars, April, 1913 (after 'Balkanicus'); **iii.** Revision of boundaries in Macedonia in favour of Bulgaria proposed by A. J. Toynbee in 1915 (*Nationality and the War*, Pl. iv); **iv.** Western limit of territory tentatively conceded by Greek diplomats to Bulgaria during negotiations for the delimitation of boundaries after the first Balkan War (after L. Dominian, *Frontiers of Language in Europe*, 1917).

**Map 4.** **i.** Provisional limits of the People's Republic of Macedonia according to *The Constitution of the Federal People's Republic of Yugoslavia*, 1946. This was the key republic in the scheme for Balkan Federation under discussion between Jugoslav and Bulgarian politicians, 1945-47. Upon the failure of the scheme, Bulgarian claims on the republic have been renewed (1949); **ii.** Territory occupied, and administered as part of Bulgaria, 1941-43; **iii.** Territorial claims of Bulgaria upon Greece put forward at the Peace Conference of 1946 (after *La Bulgarie et les Questions de la Paix*, Paris, 1946).

background of fervent national developments. The maps suffer from the inevitable reaction of these two processes; they are therefore symptomatic of a period of maladjustment between the community and the state. They reflect the evils and the merits of their age.

It is evident that the value of such maps is largely historical. That some political geographers give credence to their 'scientific validity' raises important issues of their value and their use and misuse in the field of political geography. Their application to the solution of problems of political boundaries has often had grave consequences in the past in the form of disrupted economies, in the creation of dissatisfied minorities and in the painful mass-movements of long-established populations; and even when exchanges of minorities are effected, memories remain as an incitement to policies of revision and revenge. In the case of Macedonia, the history of the last ten years has provided instances of the creation of ethno-political boundaries which have varied widely in direction according to the circumstances. The German ideas on the distribution of the Serbs in 1941 were very different from those entertained by Britain and France in 1921. The Serbian views of their own distribution in 1919 was altogether different from official ideas found in Communist Jugoslavia in 1945.

Ethnographic maps of Macedonia are inseparable from their historical context. No one map is ever likely to give a satisfactory picture of the ethnographic distributions to be found there, at least not until long years of political, economic and social consolidation have brought a measure of stability to the region. In the meantime no useful purpose is served by trying to produce a compromise map of the kind to be found in many political geographies and atlases.

### *The Value of the Maps as Source-material*

However ineffectual the various attempts to delimit Serb, Bulgarian, Greek and other distributions might have proved cartographically, the process has had far-reaching consequences for the political geography of the Balkan peninsula. Ethnographic maps and the ideas they expounded were not only influenced by, but exerted an imponderable influence upon, the history of the Balkan nations. Even although the ideas originally expressed may

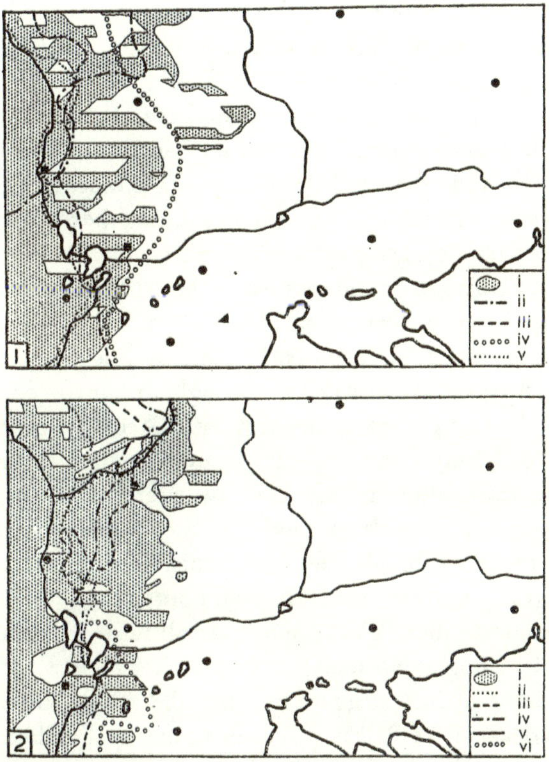


FIG. 91. ALBANIAN ASPIRATIONS IN MACEDONIA AND ADJOINING TERRITORIES, 1877–1950

The references in the key are as follows :

**Map 1.** **i.** Albanian ethnographic distribution according to K. Sax (1877) ; **ii.** Albanian boundaries proposed by H. Kiepert in his Ethnocratic Map of 1878 ; **iii.** Albanian boundary proposed by Austrian and Italian delegates at the London Conference of 1913 (after C. Woods' map in *The Geographical Journal*, 1918) ; **iv.** Albanian boundary proposed by the Provisional Albanian Government, 1913 (after C. Woods) ; **v.** Albanian boundary fixed by the Conference of London, 1913.

**Map 2.** **i.** Albanian ethnographic distribution according to A. D. Atanasiu (1919) ; **ii.** Revision of the Albanian boundary proposed by J. S. Barnes in 1918 ; **iii.** Boundary proposed by the Provisional Albanian Government of 1919 (after H. W. V. Temperley) ; **iv.** Albanian boundary in 1943 ; **v.** Boundaries of 'Kosmet', the Autonomist Albanian District of the Republic of Jugoslavia (claims to this District are now being pressed by Albanian politicians, 1949) ; **vi.** Proposed south-eastern boundary of Albania in 1943 (after a map in *L'Albania*, Reale Societa Geografica Italiana, Bologna, 1943).

have been divorced from reality, their power to suggest relationships, to stimulate aspirations and to mould policies, has been a very real force (Figs. 88–91).

For this reason, the value of these maps lies in the light which they throw on the Balkan political scene in the nineteenth century. They demonstrate the full extent and significance of the Macedonian problem and its impact upon inter-state relationships during the formative period of the modern political geography of the peninsula. They were instrumental in the shaping of boundaries and in the delimitation of spheres of influence. The Macedonian national movement which did not crystallize until the war years of 1940–45 was basically a delayed response to the suggestive ideas propagated through the maps compiled by J. Cvijić in the first decade of the century. Greek pretensions in the Balkans in 1922, Bulgarian revisionist policy since 1913, the awakening of Albanian nationalism in the last few years of the recent war, the partitions of Jugoslavia and Greece during the German occupation, the move for a Greater Macedonia, the strained relationships between Greece and her northern neighbours, 1945–50, the present Jugoslav-Bulgarian ideological dispute, are all related instances of the influence of ethnographic ideas upon practical politics. One might declare without undue exaggeration that the whole history of the Macedonian part of the Balkan peninsula in the last hundred years has been characterized by this fascinating interaction between ethnographic ideas and nationalist aspirations.

## Appendix A

## G. WEIGAND'S VIEWS, 1924

G. WEIGAND will be remembered for his map on the ethnography of south-western Macedonia compiled in 1895 (Fig. 28). During the War of 1914-18 he was commissioned by the German General Staff to enquire into the ethnography of Macedonia. Every facility was granted to him to carry out his task. He was given a staff of six Germans and a number of Bulgarians, but no Serbians served with him for obvious reasons—Serbia was the enemy. His enquiries did not extend to Greek Macedonia as the Germans and Bulgarians were not in occupation of that territory. He established his headquarters at Niš. By the year 1918, when the Germans and Bulgarians were forced to withdraw, he appears to have completed a survey of the ethnography of Serbian Macedonia, Old Serbia, northern Albania and the Niš-Leskovac region. The work was never officially published but Weigand summarized his ideas in a book published in 1924.[1] It contained no map, although a manuscript map was prepared which would seem to have been lost during the War of 1939-45.

Weigand's work, as might be expected, had a Bulgarian flavour but he substantiated his arguments by philological and historical data which made them difficult to refute. His ideas compelled attention if only on account of his enviable knowledge of Balkan languages and of the opportunity which fell to his lot of investigating the ethnographic situation with the aid of a trained staff. He attempted nothing less in 1925 than a precise reconstruction of the ethnography of Macedonia from the earliest times. He believed, rightly, that ethnographic distributions had no value without perspective. According to his ideas, five main periods could be distinguished in the evolution of the modern ethnography of the peninsula. Each period shed some light on contemporary ethnographic problems.

[1] *Ethnographie von Makedonien* (Leipzig, 1924).

*Weigand's Chart of Language Development in the Balkan Peninsula*

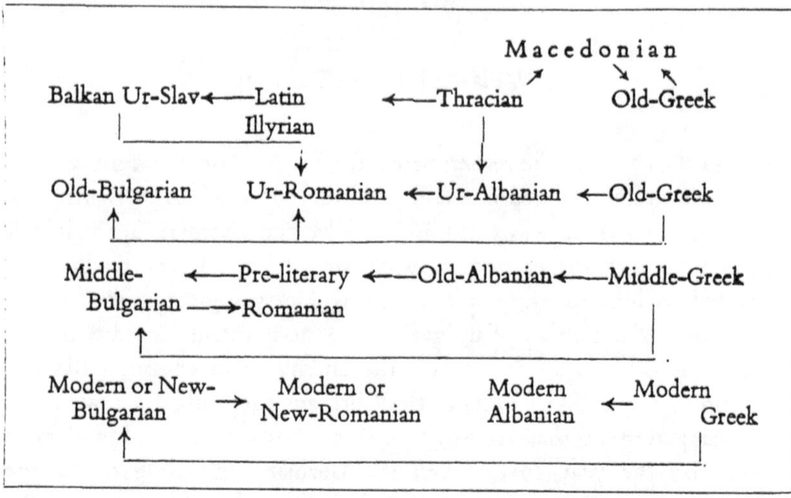


(1) *The Pre-Classical Era*

The south of the peninsula was originally inhabited by the Macedonians of unknown origin who were greatly influenced by Greek culture, but who had no written language and were an 'eng verwandtes Volk.' They occupied southern Macedonia proper and the Aegean littoral. In the north-west were found the Illyrians and in the central and northern parts of the peninsula, the Thracians. A thick wooded zone, the 'Bulgarwald', formed the boundary between the Illyrians and Thracians in the neighbourhood of the valley of Morava and Timok.[1] The significance of this reconstruction was that it prepared the way for Weigand's contention that the Morava-Timok region ultimately became as effective a divide between the Serbs and the Bulgarians as it had been between Illyrians and Thracians.

(2) *The Classical Period*

Greek and Roman influence penetrated into the peninsula from the

[1] Many early nineteenth-century atlases make a particular point of distinguishing the 'Bulgarian Forest' (e.g., *Atlas Universel, Historique et Géographique* Par A. Houzé, Paris, 1841).

south and west respectively during this period. Hellenism gained a hold on Macedonia and Thrace as far north as the Balkan crest-line but Latin influences prevailed in Illyria and Moesia.

(3) *The Völkerwanderung*

(*a*) From the fourth century onwards, the Goths, Vandals and Huns overran Moesia, Thrace and Macedonia, but they exerted only a minor influence on the ethnic composition of the contemporary population. The area north of the Balkan range remained in culture a mixture of Latin and Thracian elements, and that of the north-west of the peninsula, a mixture of Latin and Illyrian. Pure Thracian survived in the mountainous area south of Sofia and in the Rodopi mountains.

(*b*) During the sixth century, the Slavs pushing south down the Drava and Sava from Central Europe inundated Moesia, Macedonia and the western Balkans. The Thracian pastoralists, however, survived in the coastal districts. These early Slavs or 'Urslavs,' as Weigand called them, were unable to prevent the revival, later, of the Greek influence in the towns of Macedonia.

(*c*) Two new ethnic groups were born during the Slav invasions from the remnants of the Thracians in the Sofia-Niš-Skoplje region. They were :

(i) the Albanians, descendants of the pure Thracians, who were pushed westwards into the mountains of modern Albania during the migrations ;

(ii) the Wallachians, descendants of the latinized Thracians, who were dispersed in two main directions—northwards over the Danube via the Dobrudja, and southwards into Macedonia and Thessalía. Thus the paths of Albanians and Wallachians, both of the Thracian stem, diverged once and for all. It is interesting to note that Weigand derided the popular theory that the Albanians were of Illyrian stock or culture.

(*d*) The Bulgars were steppeland folk who gradually moved westwards from the land between the Volga and the Don and crossed the Danube about the seventh century. They allied themselves with the Slavs and eventually became completely slavized, bequeathing little more than their name to the Slav population. *Altbulgarisch* or old-Bulgarian became the language which prevailed over the whole of the eastern and southern part of the Balkans and it was used by Cyril and Methodius, the Greek missionaries, as the basis of Old Church Slavonic. Much influenced by Greek syntax, the language became the medium of church services all over the Balkan peninsula and was even used in Moravia, parts of Russia and Romania. From *Altbulgarisch* modern Bulgarian was ultimately derived.

(4) *The Mediaeval Period*

From the ninth century to the thirteenth century, a struggle ensued between the Bulgarians and the Byzantine Emperors for control of the southern part of the peninsula. Bulgarian influence in this area was successfully established by:

(i) the First Bulgarian Emperors, Simon (892-927), and Samuel (980-1014). Macedonia became the chief centre of Bulgarian power and capitals were founded at Prilep, Prespa and Ohrid. The Bulgarian Patriarchate of Ohrid was established and remained in being even when the Byzantine Emperors regained control of the *Egnation Way* between 1094 and 1230;

(ii) Asseniden (1210-1250), who established the Second Bulgarian Empire.

The Wallachians were also prominent in political affairs during the thirteenth century and Weigand noted the existence of the Empire of Great Wallachia based on Thessalía.

By the middle of the thirteenth century the Serbs, who hitherto had played little part in Balkan affairs began to extend their power farther south. Skoplje became the imperial residence of the Serbs under Kings Uros I and II (1243-1321). Under Stefan Dušan the Serbian Empire was extended into northern Macedonia and Thessalía (1331-55). Thus, during the course of the thirteenth and fourteenth centuries, Bulgarian and Byzantine influence in Macedonia was superseded by that of the Serbs.

The sequence of events was interrupted by the arrival of more nomadic raiders—the Patzinaks and Kumans (Huzuls). They have left evidence of their existence in the Meglen (Moglena or Karáj) region and in place-names such as Kumanovo in northern Macedonia. They, too, quickly became assimilated into the native population, some becoming Slavs and others Vlachs (see p. 116).

### (5) *The Turkish Period*

The final stage in the evolution of the ethnography of the Balkans according to Weigand took place under the Turks. The region south of a line from Prizren, through Niš and Vidin to the Danube, was profoundly affected. Here the *Ciflik* and *Wakuf* were firmly established and serfdom maintained in consequence. Turko-Arabic culture modified the language, habits of life, economy and mode of thought of the inhabitants of this area. Evidences of this were to be found in food and drink, clothes, household practices, animal husbandry and in military and administrative terminology.

### *Weigand's Conclusions*

On the evidence outlined above Weigand concluded that Macedonia and Bulgaria were more closely associated than Serbia and Macedonia, and that the common experiences of Macedonians and Bulgarians had endowed them with an ethnic homogeneity which found support in the evidence provided by philology. In his opinion the Niš-Leskovac-Skoplje-Prizren region had served as a mark between the Serbs and the Bulgars and therefore the greater part of Macedonia itself was not a transitionary region but an integral part of the Bulgarian province. The dialect of the Slavs of Macedonia was much closer to Bulgarian than to Serbian. As an illustration of the analogy between the Serb, Macedo-Slav and Bulgarian languages, Weigand compared the dialect of Bitolj with that of Sofia (Sop dialect) and Serbo-Croat. The comparison is given below. Note the Macedonian post-positive article *no*. From this extract it would appear that Weigand's case was justified.

| | | | | | |
|---|---|---|---|---|---|
| Serb. | mómče nékakvo | uće | u | bérbernicu | da se |
| Mac. | édno mómče | vléglo | vo | édna berbérnica | za da se |
| Blg. | ednó momčé | vlézlo | v | edná berbérnica | êa da se |
| Eng. | A youth | entered | a | barber's parlour | in order to be |

| | |
|---|---|
| Serb. | óbrije. Bérberin bio čóvek vragólast i šaljif, pa se |
| Mac. | ízbričit. Bérberot bil čóek majtápčija i šegádzia, i |
| Blg. | izbrѣ́sne. Berbérino bil čovék majtapčíja i šegadžia, i |
| Eng. | shaved. The barber was a facetious and droll fellow and |

| | |
|---|---|
| Serb. | chtéde málo nášaliti sa ónim mladíćem, te ga zápita: |
| Mac. | sákaše da náprait málku šéga so mómčeno, ta go zápita: |
| Blg. | sákaše da naprávi málko šegá s-momčéto, ta go zapita: |
| Eng. | wished to have a little joke with the youth, and asks him |

| | |
|---|---|
| Serb. | „ ódakle si ti mómče ?" ónaj mu réče, da je ródom |
| Mac. | „ otkáde si ti mómče ?" tóa mu réče, toi je róden |
| Blg. | „ otdéka si ti momčé ?" ono mu réče, oti je ródom |
| Eng. | " Where do you come from lad ?" The latter tells him that he is native |

| | |
|---|---|
| Serb. | iz Sárajeva. a bérberin nástavi : „ čúo sam, da ima |
| Mac. | vo Sarájevo. a bérberot pródolžit : „ sum čul, óti íma |
| Blg. | ot Sarájevo. a berbérino prodѣ́lži : „ čul sam, óti ima |
| Eng. | to Sarajevo. And the barber continues : " I have heard that there are |

| | |
|---|---|
| Serb. | tákovich junáka u Sárajevu, kóji se bríju bez |
| Mac. | tákvi júnaci vo Sarájevo, što se bričat bez |
| Blg. | takiva junáci v Sarájevo, koíto se brѣ́snat bez |
| Eng. | such heroes in Sarajevo, who shave themselves without |

| | |
|---|---|
| Serb. | vrúće vóde i sapúna, no sámo stúdenom vódom." |
| Mac. | tópla vóda i sápun, a sámo so studena vóda." |
| Blg. | tópla vóda i sapún, a sámo sѣ́ studéna vóda." |
| Eng. | warm water and soap, and only with cold water." |

| | |
|---|---|
| Serb. | óno mómče, da pokaže, káko je i óno júnak, |
| Mac. | mómčeno, da pókažit, óti i tóa je júnak, |
| Blg. | momčéto, da pokáže, óti i onó i junák, |
| Eng. | The lad, in order to show, that he also is a hero, . . . |

Appendix B

# BIBLIOGRAPHY

"Europa Polyglotta linguarum genealogiam exhibens una cum literis, scribendis modis omnium gentium." [*n.s.*, 165 × 200mm.]. Opera G. Henselii, delineata (Nuremberg, 1730).

"Völkerkarte von Europa." [*n.s.*, 540 × 420mm.]. Von Dr. F. A. O'Etzel, Technisches Bureau (Berlin, 1821).

"Carte de l'Europe, présentant le tableau des peuples qui l'habitent, classés par rapport à la dépendance relative des nations et à l'affiliation des langues." [*n.s.*, 390 × 275mm.]. Carte No. 7 [in] *Atlas Physique, Politique et Historique de l'Europe* par M. A. Denaix (Paris, 1829).

——— Nouvelle Edition (Paris, 1855).

"Planche Ethnologique de les Pays et les Peuples de l'Europe, de l'Asie antérieure et de la Berbérie dans leur Etat Actuel," 2ᵉ Edition. [*n.s.*, 635 × 485mm., part of] *Atlas Ethno-Géographique ou Länder und Völkerkarten* von Wilhelm Müller (Paris and Leipzig, 1842).

"Slovenský Zeměvid." [Scale,] 30 Iemĕp. mil [=49mm., 584 × 469mm.]. Od P. J. Safařika (Praze, 1842).

"Ethnographic Map of Europe." [*n.s.*, 595 × 470mm.]. By Dr. Gustaf Kombst [in] *The National Atlas of Historical, Commercial and Political Geography*, constructed by A. Keith Johnston (Edinburgh, 1843).

"[Conjectural] Ethnographic Map of Europe in the earliest times, illustrative of Dr. Prichard's 'Natural History of Man' and his researches into the physical history of mankind." [*n.s.*, 600 × 480mm., in] *Prichard's Ethnographical Maps* (London, 1843).

"Ethnographische Karte des Osmanischen Reichs, europäischen Theils und von Griechenland." Massstab, 1 : 3,800,000, [360 × 340mm.]. Dem Herausgeber des Physikal. Atlas von seinen ethnologischen Freünde Wien 3 März, 1847. A. Boué. [In] Heinrich Berghaus, *Physikalischer Atlas*, Zweiter Band, VIII, Tafel XIX (Gotha, 1848).

*Ibid.* "Ethnographische Karte von Europa." Massstab, 1 : 6,000,000, [4 shts. each 410 × 340mm.], auf F. v. Stülpnagel geogr. Zeichnung Zusammengestellt von H. Bgs. im März, 1845, im stich sich vollendet (Gotha, Nov., 1847).

Землѣ Коима Пребиваю Србльи, од Д. Давидовић. Београду, 1848.

"Zemle Koima Prebivayu Srbl'i", od D. Davidović (Beograd, 1848).

"Ethnographic Map of Europe, according to Dr. Gustaf Kombst." Scale of English Miles, 300 [=53mm., 600×49mm.,] by A. K. Johnston [in] *The Physical Atlas of Natural Phenomena* by Alexander Keith Johnston. A new and Enlarged Edition (Edinburgh, 1856).

*Ibid.* "Distribution of Man in Europe according to Language" [*n.s.*, 200×140mm.].

Србія и Землѣ у Коима се србскіи говори са некимъ Частима, Ноограничны и Земаля Уоравлно и на Цынкусу Резаио, оді Проф. Г. Дежардена у Београду, 1853.

"Srbiya i Zemle u Koima se Srbskiya Govori sa Nekim' Castima, Nograničny Zemalya uoravl'no i na Cynkusu Rezano," odi Prof. G. Dežardena [G. Desjardins] u (Beograd, 1853).

"Carte Ethnographique de la Turquie d'Europe et des états vassaux autonomes." Echelle, 1 : 2,500,000, [520×410mm.]. Par G. Lejean [in] "Ethnographie de la Turquie d'Europe," *Ergänzungsheft Nr. 4 zu Petermann's Mittheilungen* (Gotha, 1861).

"Croquis des westlichen Gebietes der Bulgarischen Morava." Massstab, Turkische Reitstunden, 5 [=90mm., 480×530mm.]. Von J. G. von Hahn und F. Zach, *Drukschriften der K. Akad, D. Wissensch. Philos. Histor.*, CIXI Bd. (Wien, 1861).

Этнографическая Карта Славянскихъ Народностей. Масштабъ, 1 : 4,200,000 [633×790mm., by] М. Ф. М. Моск. 1867.

"Etnograficheskaya Karta Slavyanskikl' Narodnostey." Masshtab', 1 : 4,200,000, [635×790mm.]. M. F. M.[irkovitch] (Moscau, 1867).

"[Ethnographic] Map of the South Slavonic Countries." Scale, English miles, 100 [=28mm., in] *Travels in the Slavonic Provinces of Turkey-in-Europe* by G. Muir Mackenzie and A. P. Irby (London, 1867).

"Mapa Slovanského světa ole Safařika, Czeorniga, Kozlena, Boeckha, Lejeana, Koeppena, i národopisné mapy Ruskée zr. 1867." [Scale,] 1 : 6,400,000. Od J. Erben (Praha, 1868).

"Carte Linguistique, Ethnographique et politique actuelle de l'Europe Orientale." [*n.s.*, 305×475mm.]. Par C. Delamarre (Paris, 1868).

"Die Ausdehnung der Slaven in der Türkei und den angrenzenden Gebieten." Massstab, 1 : 3,700,000, [400×330mm.]. Von A. Petermann [in] *Petermann's Mittheilungen*, Bd. XV, Tafel XXII (Gotha, 1869).

"Map of Europe showing distribution of the Celtic, Teutonic, Greco-Latin, Thraco-Illyrian and Sclavonic Families." [Scale,] English Miles, 600 [=50mm., 325×220mm., in] *An Ethnographic Atlas*, S. Bagster [pub.] (London, 1870).

" Carte des Races Européennes ou tableau des peuples Européens rangé par familles de langues." [*n.s.*, 660×470mm.]. Par F. L. Passard (Paris, 1873 ?).

" Ethnographische Ubersichtskarte des Europäischen Orients." Massstab, 1:3,000,000, [424×512mm.]. Zusammengestellt von H. Kiepert. Mit Erlauterndem Text. D. Reimer [pub.] (Berlin, im Mai, 1876).

" Wyld's Map of Turkey-in-Europe [with] Ethnographical Colouring." Scale, British Miles, 200 [=110mm., 480×550mm.] (London, 1876).

" T.[urkey] in Europe Showing Percentage of Muhammedans, [and] T.[urkey] in Europe Showing Nationalities according to H. Kiepert." [In] *The Geographical Magazine*, Markhams (London, October, 1876).

" Populations de la Turquie d'Europe." Echelle de 1 : 4,444,000, [280×190mm.]. Par Elisée Réclus [in] *Nouvelle Géographie Universelle*, Tom I, p. 133 (Paris, 1876).

" Die Neueste Eintheilung. Die Türkischen Gebiete und Die Confessionen in der Türkei." Im Massstabe von 1 : 2,500,000, [410×335mm., in] *Petermann's Mittheilungen*, XXII Bd., Tafel XIII (Gotha, 1876).

" Les Roumains du Sud. Carte Ethnographique de la Macédoine, Thessalie, Epire et l'Albanie " par Nic. Densusianu et F. Damé. [*n.s.*, 380×290mm., in] *L'Element Latin au Orient : Les Roumains du Sud*, etc. (Paris, Bucarest, 1877).

[Bacon's New] Ethnological Map of Europe, showing the distribution of races throughout the continent. Scale, English Miles, 300 [=45mm., 600×430mm.] (London, 1877).

" Carte Ethnographique et Oro-Hydrographique de la Turquie d'Europe, de la Serbie, du Montenegro, de la Grèce, de L'Archipel et de L'Ouest de L'Asie Mineure." Echelle, 1 : 1,300,000, [840×1120mm.]. Par F. Bianconi [in] *Ethnographie et Statistique de la Turquie d'Europe* (Paris, 1877).

" [Ethnographic] Map of the South Slavonic Countries." Scale of English Miles, 300 [=35mm., 325×250mm., in] *Travels in the Slavonic Provinces of Turkey*, by G. Muir Mackenzie and A. P. Irby, 5th Edition (London, 1877).

" Carte Ethnographique de la Turquie d'Europe et dénombrement de la Population Grecque de l'Empire Ottoman." Echelle, 1 : 3,400,000, [335×375mm.]. Par A. Synvet (Paris, 1877).

" An ethnological Map of European Turkey and Greece." Scale, miles, 100 [=48mm., 340×405mm.]. With introductory remarks on the distribution of races in the Illyrian Peninsula and statistical tables of Population. Edward Stanford [pub.] (London, 1877).

" Ethnographische Karte der europäischen Türkei und ihrer Dependenzen zu Anfang des Jahres 1877." [*n.s.*, 430×260mm.]. Von Karl Sax [in] *Mitteilungen geographischen Gesellschaft*, XXI Bd., Tafel III (Wien, 1878).

"Ethnographische Karte von Epirus, vorzüglich nach den Angaben von Aravandinos zusammengestellt." Massstab, 1 : 500,000 [450×440mm.]. Von H. Kiepert [in] *Zeitschrift der Gesellschaft für Erdkunde zu Berlin*, XIII Bd., Tafel V (Berlin, 1878).

"Tableau ethnocratique des pays du Sud-est de l'Europe." [*n.s.*, 205×260mm.]. Par H. Kiepert [in] *Notice explicative sur la Carte Ethnocratique des pays Hellenique, Slaves, Albanais et Roumains, etc.* (Berlin, 1878).

"Völkerkarte der Balkanhalbinsel." Massstab, 1 : 6,000,000, [170×230mm., in] *Richard Andree's Allegemeiner Handatlas* (Bielefeld und Leipzig, 1881).

"Politisch-Ethnographische Uebersichtskarte von Bulgarien, Ost-Rumelien und den benachbarten Balkanländern." Massstab, 1 : 8,000,000, [310×345mm.]. Geographisches Institut (Weimar, 1885).

"Srbija, opisdržave zemlje i naroda" [in] *Karia rasprostranjenja Srba* [by] Vladim Karić. (Beograd, 1887).

"Ethnographische Karte von Altserbien und Makedonien." Massstab, 1 : 300,000, [1000×1250mm.]. Von S. Gopčević [in] *Militärgeographisches Institut k. und k.* (Wien, 1889).

"Ethnographische Karte von Altserbien und Makedonian." Massstab, 1 : 750,000, [400×500mm.]. Von S. Gopčević [in] *Petermann's Mittheilungen*, XXXV Bd., Tafel IV (Gotha, 1889).

Karta Славянскихъ Народностей. Издaie В.В. Комароба, Составилъ Н. С. Зарянко Масштабѣ, 1: 1,200,000, [700×600mm.] (С. Петербургъ, 1890).

"Karta Slavyanskikh' Narodnostey." Izdanie V. V. Komarov, Sostavil' N. S. Zaryanko. Masshtab', 1: 1,200,000, [700×600mm.]. (St. Peterburg', 1890).

Етнографска Карта Српских Земаља. Carte Ethnographique Serbe. Размера, 1: 1,600,000, [570×510mm.]. Израда и Издаюе. Српске Великошколске Омладине (Београд, 1891).

"Etnografska Karta Srpskih Zemalja, Carte Ethnographique Serbe," Razmera, 1 : 1,600,000, [570×510mm.]. Izrada i Izdanje Srpske Velikoškolske Omladine (Beograd, 1891).

"Europa um 1880 [ethnographical]." Massstab, 1:15,000,000, [400×330mm.]. Ent. v. Dr. G. Gerland [in] *Berghaus Physikal. Atlas*, VII, Abteilung Völkerkunde, No. 67 (Gotha, 1892).

"Kleine Völker- und Sprachen-Karte von Mitteleuropa." Massstab, 1:6,000,000, [350×280mm.]. Gez. v. Dr. K. Peucker (Wien, 1893).

"Europa, Sprachen- und Völkerkarte (für das Jahr 1880)." Mittlerer Massstab, 1: 12,000,000, [480×380mm.]. Bearbeitet von E. Debes [in] *Neuer Handatlas über alle Teile der Erde* herausgegeben von E. Debes (Leipzig, 1895).

"Europa, Sprachen- und Völkerkarte (für das Jahr 1880)." Zweite vermehrte und verbesserte Auflage (1898).

——— Vierte [ . . . ] Auflage (1913).

"Ethnographische Karte der Südwestlichen Balkanhalbinsel." Massstab, 1 : 750,000, [465 × 325mm., in] *Die Aromunen, Ethnographische-Historische Untersuchungen,* von Gustaf Weigand, I Bd. (Leipzig, 1895).

"Populations de l'Europe Centrale." Echelle de 1 : 6,000,000, [224 × 255mm., in] *Atlas de Géographie Général* par le Colonel Niox, No. 31 bls. (Paris, 1899).

"Carte ethnographique et linguistique de la Macédoine." Echelle, 1: 1,006,000, [375 × 340mm.]. Supplément à l'ouvrage *La Macédoine* par le Dr. Cléanthes Nicolaidès (Berlin, 1899).

"Karte der Schulsphären der Türkischen Balkan-Halbinsel." Massstab, 1 : 3,700,000, [490 × 390mm.]. Entworfen und gezeichnet von Richard von Mach [in] *Petermann's Mittheilungen* (Gotha, 1899).

"Ethnographische Karte von Makedonien." Massstab, 1 : 1,500,000, [260 × 245mm.]. Von F. Meinhard in Sofia [in] *Deutsche Rundschau für Geographie und Statistik,* XXI, Heft X, p. 433 (Wien, 1899).

"Ethnographic Map of the Balkan Peninsula" [Scale,] 1 : 6,000,000, [175 × 230mm., in] *The Times Atlas,* New Edition (London, 1900).

"Carte ethnographique de la Macédoine d'après la carte du Docteur H. Kiepert." Echelle, kiloms. 60 [=55mm., 380 × 345mm.]. (Paris, 1900).

"Cinq Cartes Ethnographique de la Macédoine," [A. Boué 1847, H. Kiepert 1876, E. Stanford 1877, Serbian High School 1891, V. Kănčev 1900]. Par Paul Milioukoff (St. Petersburg, 1900).

"Cartes Ethnographiques des Vilayhets Salonique, Cossovo et Monastir." [Scale,] 1 : 250,000. Lith. par l'Institut Cartographie (Sofia, 1901).

——— [Scale,] 1: 200,000. Erhard [pub.] (Paris, 1905).

"Die Sprächlichen Interessensphären in und um Makedonien u. Altserbien." Massstab, 1 : 8,000,000, [130 × 95mm.. This is an inset map in] *Makedonien, Altserbien und Albanien, Kartographische, Historische und Statistische Beilagen zum Verständnisse Frage,* von K. Peucker (Wien, 1903).

"Carte Ethnographique Serbe avec les limites méridionales de la Vieille-Serbie et celles de la Serbie du Tsar Douchan." Ancienne Edition des Etudiants de l'Université de Belgrade revue et corrigée par Milan J. Antonovitch (Belgrade, 1903).

"Cartes des Populations Chrétiennes : Bulgare, Grecque et Koutzovlaque en Macédoine, 1904." Echelle, 1 : 1,500,000, [240 × 300mm., in] *La Macédoine et sa population chrétienne* par D. M. Brancoff (Paris, 1905).

*Ibid.* "Carte des Ecoles Chrétiennes de la Macédoine, 1902." Echelle, 1: 1,500,000, [240 × 300mm.].

"Carte des Ecoles Chrétiennes de la Macédoine. Echelle, 1 : 400,000, [1020×840mm., anonymous] (Paris, 1905).

"Carta Etnografica della regione albanese." Scala di 1 : 750,000, [710×320mm., in] *Albania* [by] Eug. Barbarich (Roma, 1905).

Политичко–Етнографска Скица Македоније и Старе Србије. 1: 4,000,000, [380×320 mm., in] Основе за Географију и Геологију Македоније и Старе Србије, написао J. Цвијић. Књигапрва (Веоград, 1906).

"Političko-Etnografska Skice Makedonije i Stare Srbije." 1 : 4,000,000, [380×320mm., in] *Osnove za Geografiju i Geologiju Makedonije i Stare Srbije*, Hapisao J. Cvijić. Knjigaprva (Beograd, 1906).

"Ethnographical Map of Macedonia." Scale, Miles, 52 [=1 inch, 120×160mm., in] *Macedonia : Its Races and their Future* by H. N. Brailsford (London, 1906).

"La Vieille Serbie." Echelle de 1 : 1,000,000, [120×115mm., in] *Remarques sur l'Ethnographie de la Macédoine* par J. Cvijić, 2e Edit. (Paris, 1907).

Этнографическая карта Славянства. Масштабъ, 1: 8,000,000, [960×1090 mm., in] Славяское Племя, на Т. Д. Флорински (Кіевъ, 1907).

"Etnograficheskaya Karta Slavyanstva." Masshtab, 1 : 8,000,000, [960×1090mm., in] *Slavyaskoe Plemya*, na T. D. Florinski (Kiev, 1907).

*Ibid.* Карта Разселен ія Спавянъ во второй половинѣ IX Столѣтія. Масштабъ, 1: 12,600,000, [385×320 mm.].

*Ibid.* "Karta Razselen iya Spavyan vo Vtoroy Polovine IX Stoletiya." Masshtab, 1 : 12,600,000, [385×320mm.].

"Vilayets de Jannina—Monastir—Salonique et Adrinople, Sanjak indépendant de Tchataldja et Partie Européenne du Vilayet de Constantinople. Carte indiquent la prépondérances des différentes nationalités sur les divers points des Vilayets." Echelle, 1 : 2,100,000, [410×250mm.]. Carta no. 8 [in] *La Questione Rumeliota* [by] Giovanni Amadore-Virgilj. Istituto Geografico de Agostini (Bitonto, 1908).

*Ibid.* "Vilayet d'Adrinople. Carte des écoles grècques et bulgares et des églises grècques." Scala di 1 : 800,000.

*Ibid.* "Vilayet de Salonique. Carte des écoles chrétiennes." Scala di 1 : 600,000.

*Ibid.* "Vilayet de Monastir. Carte des écoles chrétiennes." Scala di 1 : 600,000.

Етнографска карта на Славянството въ Европа. Мѣрка, 1: 5,000,000, [1000×830mm.]. На Т. Д. Флорински, В. Комаровъ идр. Издава Славгнска Бесѣда въ (София, 1908).

"Etnografska Karta na Slavyanstvoto v Europa." Merka, 1 : 5,000,000, [1000×830mm.]. Na T. D. Florinski, B. Komarov idr. Izdava Slavgnska Beseda v (Sofiya, 1908).

"Ethnographical Map of Servia." [*n.s.*, 330×320mm., in] *Servia by the Servians* compiled by Alfred Stead (London, 1909).

"Carte ethnographique de la nation Serbe." Echelle, 1:2,750,000, [350×320mm., in] *The Annexation of Bosnia and Herzegovina and the Serb Problem.* By Yovan Cvijić. (Translated from the Serb by I. L. Christitch, London, 1909).

"Balkánský Poloostrov, v. VII-VIII stol." [*n.s.*, 350×340mm., in] *Slovanské Starožitnosti* sepsal Lubor Niederle, Dílu II, Svazek 2, V. (Praze, 1910).

"Empire Ottoman. Carte Statistique des Cultes Chrétiens." Echelle de 1 : 600,000, [700×410mm.]. Dressée par R. Huber (Caire, 1910).

Етнографска Скица Старе Србије. Размер, 1:750,000, [400×330 mm.] од Д-ра. Ј. Цвијића. Основе за Географију и Геологију Македоније и Старе Србије, Књига Јрећа, Српска Краљевска Академија (Веоград, 1911).

"Etnografska Skice Stare Srbije." Razmer 1 : 750,000, [400×330mm.]. Od Dr. J. Cvijić. *Osnove za Geografiju i Geologiju Makedonije i Stare Srbije*, Knjiga Jreć, Srpska Kraljevska Akademija (Beograd, 1911).

"Carte Ethnographique de la Race Slave" [*n.s.*, 240×200mm., in] *La Race Slave* par L. Niederle (Paris, 1911).

"The Southern Slavs" [Ethnographical. *n.s.*, 180×180mm., in] *The Southern Slav Question and the Hapsburg Monarchy* by R. W. Seton-Watson (London, 1911).

"Carte géographique des communes Roumaines de Turquie." Dressée par Mr. V. Diamandi d'après Charles Peuker et l'état-major autrichien. Echelle, 1:1,125,000, [220×155mm., in] *Les Roumains de Macédoine* [by] Alexandu Rubin (Bucarest, 1913).

"Répartition des populations roumaines de la Macédoine et de l'Albanie." [Scale,] 1:1,600,000, [260×200mm.]. Par Léon Lamouche [in] *Questions Diplomatiques et Col.* XXXV, p. 420-21 (Paris, 1913).

"Ethnographische Karte der Balkanhalbinsel." Massstab, 1 : 1,500,000, [670×550mm.]. Von Dr. J. Cvijić [in] *Petermann's Mittheilungen*, März Heft (Gotha, 1913).

"Carte Ethnographique de la Macédoine, de l'Albanie, de la Vieille Serbie et de la Thessalie." [In] *Les Roumains, Koutso-valaques, les Populations Macédoniennes et la Crise Balkanique* par Const. Noé (Bucarest, 1913).

"Dialectes de la Macédoine d'après A. Belitch." [*n.s.*, 125×150mm., in] *Enquête Dans Les Balkans*, Dotation Carnegie (Paris, 1914).

*Ibid.* "Carte Ethnographique de la Macédoine, Point de Vue Bulgare." Echelle de 1 : 1,500, 000, [280 × 220mm.] D'après Vasil Kănčov[Kănčev].

*Ibid.* "Limites Ethnographiques de la Bulgarie avant la traité de San-Stephano." [*n.s.*, 85 × 120mm.].

"Coup d'oeil sur l'Ethnographie des Balkans." Echelle de 1 : 4,000,000, [200 × 100mm.]. Par Cte. Adrien D. Arlincourt [in] *L'Ethnographie*, N. Ser., No. 2 (Paris, 1914).

"Carte Ethnographique de la Péninsule Balkanique." [*n.s.*, 195 × 150mm., in] *Histoire de l'Empire Ottoman* par le Vte. de la Jonquière, vol. II (Paris, 1914).

"Karte von vergrösserten Griechenland, mit einer farbigen Völker- und Spracheneinteilung der von den Balkanstaaten neueroberten Gebiet." Massstab, 1 : 3,000,000, [310 × 310mm., in] *Griechenlands Anteil an den Balkan-Kriegen, 1912-1913* von Dr. K. Nicolaides (Berlin, 1914).

"The Distribution of Vlachs in the Balkan Lands." [Scale,] Miles, 100 [=78mm., 230 × 150mm., in] *The Nomads of the Balkans. An account of the life and customs of the Vlachs of the Northern Pindus* by A. J. B. Wace, etc. (London, 1914).

"The Balkans (ethnographical)." [*n.s.*, 240 × 150mm., in] *Nationality and the War* by A. J. Toynbee, Pt. IV (London, 1915).

"Carte de l'Extension Ethnique de la Nation Serbo-Croat-Slovène." Echelle, 1 : 1,500,000, [595 × 675mm.]. Par St. Stanoïévitch et D. J. Derocco (Belgrade and Nish, 1915).

"The Distribution of the Peoples of the [Balkan] Peninsula." [*n.s.*, 120 × 80mm., in] *Geographical Aspects of the Balkan Problems* by M. I. Newbigin (London, 1915).

"Part of Europe showing languages having political significance, based on Sheet No. 12c, Sept. 1911, Debes' Handatlas and other sources." Scale, 1 : 9,000,000 or 1 inch = 142 miles, [325 × 270mm.]. By Leon Dominian [in] *Bulletin of the American Geographical Society*, vol. XLVII, No. 6 (New York, 1915).

"Carte ethnographique de la Macédoine grecque, représentant la répartition ethnique au mois d'août 1915." Echelle de 1 : 200,000. Lithogr. Contogoni (Athens, 1915).

"Carte Ethnographique de la Turquie d'Europe avant les Guerres Balkaniques." Echelle kiloms., 100 [=47mm., 395 × 240mm.]. Empruntée à l'ouvrage de M. Amadori Virgilj "La Question Rumeliota" [in] *Revue Hebdomadaire* (Paris, 1915).

"Das Bulgarentum auf den Balkanhalbinsel im Jahre 1912." Massstab, 1 : 1,500,000, [605 × 410mm.]. Von Prof. D. A. Ischirkoff unter Mitwerkung von D. L. Miletitsch, Prof. Dr. B. Zoneff, Prof. J. Ivanoff, Prof. Dr. St. Romanski. [In] *Petermann's Mittheilungen*, Bd. LXI, Tafel XLIV (Gotha, 1915).

"The Balkan Peninsula, Ethnological." [Scale,] Miles, 150 [=1 inch, 150×170mm., in] *The Balkans, A History of Bulgaria, Serbia, Greece, Rumania and Turkey* by Nevill Forbes, Arnold J. Toynbee, D. Mitrany, D. G. Hogarth (Oxford, 1915).

"Carte ethnographique de la Macédoine du Sud, dressée à la veille des dernières guerres Balkaniques." [*n.s.*, 505×195mm.]. Par J. Ivanoff [in] *Revue Hebdomadaire* (Paris, 1915).

[Ethnographical] "Map of Jugoslav Territory." Scale, 1:1,500,000, [620×550mm.]. By Dr. Niko Zupanić, issued for the Jugoslav Committee (London, 1915).

"Ethnographical Map of Central and South Eastern Europe." Scale, 1:4,000,000, [490×715mm., issued by the] G.[eographical] S.[ection] G.[eneral] S.[taff], War Office (London, 1916).

"Länder- und Völkerkarte Europas." Massstab, 1:4,000,000, [1140×840mm.]. Von Professor Dr. Dietrich Schäfer. 3 Auflage. Abgeschlossen (Berlin im August, 1916).

"Carta Etnico-Linguistica dell' Oriente Europeo." [Scale,] 1:3,000,000, [in] *L'Europa, Etnico-Linguistica. Atlante descrittino in tre Carte speciali colorite con Testo dimonstratino.* L'Istituto Geografico de Agostini (Novara, 1916).

"Carte des langues parlées dans la Péninsule des Balkans." [Scale,] 500 kilomètres [=500 mm., 120×150 mm., in] *Les peuples des Balkans* [by] Eugène Pittard. (Paris, 1917 ? ).

Етнографска Карта на Македония. Мѣрка, 1:1,500,000, [290×190mm. in] Българитѣ бъ Македония омъ Й. Ибанобъ (София, 1917).

"Etnografska Karta na Makedonija." Merka, 1:1,500,000, [290×190mm. in] *B'lgarite b Makedonija* om J. Ivanov (Sofija, 1917).

"The Allies Peace Terms. Suggestions for the New Map of Europe on a national basis." [Scale,] statute miles, 150 [=30mm., 405×345mm.]. Supplement to the *National Review*, March issue (London, 1917).

"The Races of the Balkan Peninsula." [Scale,] English Miles, 150 [=65mm., 280×415mm., in] *The Rise of Nationality in the Balkans* by R. W. Seton Watson, etc. (London, 1917).

"A Sketch Map of the Linguistic Areas of Europe." Scale, 1:3,220,177 of nature, [935×740mm.]. E. Stanford (London, 1917).

"Sprachen und Religion in Europa und die Grenzen zwischen west und osteuropäischer Kultur." Massstab, 1:10,000,000, [600×470mm.]. Von Paul Langhans [in] *Petermann's Mittheilungen* (Gotha, 1917).

"Die Bulgaren in ihren historischen ethnographischen und politischen Grenzen." Atlas mit 40 Landkarten. Vorwort von D. Rizoff, etc., [In English, French, German and Russian, 300×300mm.]. (Berlin, 1917).

"Hellenism in the Near East. An ethnological Map." Scale, 1:2,155,000, [510×630mm.]. Compiled by Professor George Soteriades, Edward Stanford (London, 1918).

"Ethnographic Map of Europe." [*n.s.*, 240×165mm., in] *Stakes of War* by L. Stoddard, etc. (New York, 1918).

"Ethnographical Map [of part of Europe, showing Germany, Northern Italy, Poland and] South Eastern Europe." Scale, 1:1,500,000, [4 sh. each of 610×610mm.]. G.[eographical] S.[ection] G.[eneral] S.[taff], War Office (London, 1918).

"Carte Ethnographique de l'Europe." Echelle, 1 : 5,000,000, [1060×860mm., by] J. Gabrys, 1re Edition, librairie central de nationalités (Lausanne, 1918).

"Carte Ethnographique de la Péninsule des Balkans." Echelle de 1:3,000,000, [485×440mm.]. Carte hors texte à la fin du volume, *La Péninsule balkanique : Géographie Humaine* par Jovan Cvijić (Paris, 1918).

"Carte Ethnographique de la Péninsule des Balkans." Echelle de 1 : 3,000,000. Par J. Cvijić [reproduced in] *Annales de Géographie*, XXVII (Paris, 1918).

"Ethnographic Map of the Balkan Peninsula." Scale, 1 : 3,000,000. By Jovan Cvijić [reproduced in] *The Geographical Review*, vol. V (New York, 1918).

"The Daily Telegraph Language Map of Eastern Europe." [Scale,] English Miles, 100 [=72mm., 700×990mm.]. By Alexander Gross. *The Daily Telegraph Map No. 4. Peace Conference Map of Eastern Europe etc.*, Geographia (London, 1918).

"Races of Eastern Europe." [Scale,] Statute Miles, 50 [=35mm., 830×1310mm.]. By Alexander Gross, *Daily Telegraph Map No. 25*, Geographia, (London, 1918).

"Carte ethnographique de la Macédoine Orientale, représentant la répartition ethnique vers le commencement de 1913." Echelle de 1 : 750,000, [260×350mm., in] *La Région de Cavalla* par J. Ivanoff (Berne, 1918).

"Sketch Map of Albania [. . .] show[ing] the proportion of Albanians to other nationalities, etc." Scale, 1 : 2,000,000, [120×170mm., by] J. S. Barnes [in] *The Geographical Journal*, vol. LII (London, 1918).

"Zones of Civilisation of the Balkan Peninsula." Scale, 1 : 3,000,000, [495×405mm.]. By Jovan Cvijić [in] *The Geographical Review*, vol. V (New York, 1918).

"Yugo-Slavia, Provisional. Nationalities." Scale, 1:1,700,000, [500 × 450mm.]. Naval Staff I.[ntelligence] D.[ivision] (1919).

"The Yugo-Slavs." Scale, 13,700,000, [240 × 240mm., in]. *The Peoples of Austria-Hungary, I. Hungary, Atlas*, Plate XLII, Naval Staff Intelligence Division (1919).

"Limites Ethniques Bulgares au Sud-Ouest ; D'après les cartes parues après le Congrès de Berlin, 1878." Echelle de 1 : 1,500,000, [320 × 390mm., in] *Les Bulgares Devant Le Congrès De La Paix. Documents Historiques, Ethnographiques et Diplomatiques avec quatre cartes en couleurs* par J. Ivanoff, 2[m] Ed. (Berne, 1919).

"Völkerkarte der Balkanhalbinsel." Mittlerer Massstab von 1 : 1,500,000, [620 × 820mm.], von J. I. Kettler. *Carl Flemmings Völkerkarte Nr. 3* (Berlin, 1919).

"The New Map of Europe. Approximate Boundaries of Peoples." [Scale,] 400 miles [=30mm., 265 × 205mm.]. By H. A. Miller [in] *The Jewish Pogroms in Ukraine* by J. Batchinsky, etc. (Washington, 1919).

"Ethnographische Karte." Massstab, 1 : 7,500,000, bearbeitet von Alois Fischer [in] *Europa's Neue Staatengrenzen* (Wien, 1919).

"Les Roumains de Serbie, région du Timok et de la Morave, d'après l[es] carte[s] ethnographique[s] par G. Lejean 1861, Mackensie [*sic*] et Irby 1867, H. Kiepert 1869, T. D. Florinsky 1911, G. Valsan et G. Guiglea 1913." [*n.s.*, 190 × 160mm.]. Dess. par A. D. Atanasiu (Paris, 1919).

"Carte Ethnographique des Macédo-Roumains (Koutzo-valaques)." [*n.s.*, 290 × 230mm.]. Par A. D. Atanasiu (Paris, 1919).

"Map of the Races of Europe and adjoining portions of Asia and Africa." [Scale,] 500 Statute Miles [=80mm., 610 × 500mm.]. Drawn by Edwin A. Grosvenor and Gilbert Grosvenor. *National Geographic Society* (Washington, 1919).

"Limites ethniques bulgares au Sud-Ouest d'après les cartes antérieures à la querelle de races." Echelle de 1:1,500,000, [320 × 370mm., in] *Les Bulgares et leurs manifestations nationales* par J. Ivanoff (Berne, 1919).

*Ibid.* "Carte Ethnographique des Slaves du Sud en 1913." Echelle de 1 : 3,700,000, [250 × 200mm.].

"Yugoslavia, [showing] area of Serbo-Croatian, Albanian and Slovene Speech." Scale, English Miles, 50 [=1 inch, 335 × 265mm., in] *The Peace Conference Atlas*. Edward Stanford Ltd. (London, 1919).

"Carte Ethnographique de l'Epire du Nord en 1913." 1 : 200,000, [700 × 950mm.]. Grand Quartier Général De L'Armée Hellénique. Etat Major avant 2[me] Bureau. Section des Affaires Politiques (Athens, 1919).

"Die Völker Europas." Massstab, 1 : 3,000,000, [1770 × 1520mm., part of] *Physikalischer Wandatlas*, VII, Arbeitung, "Völker, Sprachen, Staatskunde" von H. Haacke (Gotha, 1919).

"Carte ethnographique de la Macédoine." Echelle de 1 : 1,500,000, [270 × 190mm., in] *La Question Macédonienne* par J. Ivanoff (Paris, 1920).

"Carte annexée à la 'Notice sur l'Albanie et le Montenegro', publiée par le Ministère de la Guerre français en 1915." Echelle, 1 : 1,500,000, [145 × 200mm., in] *La Question Albanaise*. Etudes Recueillies et publiées par les Soins de Démètre Beratti (Paris, *n.d.*).

"Carte Ethnographique de l'Albanie." Echelle, 1 : 1,000,000, [265 × 470mm.]. Délégation de la Colonie Albanaise de Turquie (London, 1920).

"Carte Ethnographique de l'Albanie." Echelle, 1 : 1,500,000, [200 × 240mm.]. Délégation de la Colonie Albanaise de Turquie (London, 1920).

"Distribution of nationalities in South-eastern Europe." [Scale,] 1 inch equals 95 miles. [200 × 280mm.]. C. S. Hammond and Co. (New York, *n.d.*).

"Europe, ethnographical." Scale, 1 : 15,000,000, [270 × 205mm., in] *The Times Atlas of the World* (London, 1920).

"Völkerkarte von Europa." Massstab, 1 : 3,200,000, [905 × 730mm.]. Von Ernst Schwabe. Dritte Auflage (Leipzig, 1921).

"Albania [showing] ethnographical lines." Scale, Miles, 40 [=28mm., 130 × 210mm., in] *A History of the Peace Conference of Paris* by H. W. V. Temperley, vol. IV (London, 1921).

*Ibid.* "Bulgaria (ethnographical)." Scale, Miles, 80 [=50mm., 255 × 195mm.].

"Ethnographical Map of Turkey in Europe, 1912." [Scale,] 80 km. [=30mm., 295 × 150mm., in] *Venizelos* by H. A. Gibbons, etc. (London, 1921).

"Veroispovesti U Kraljevini S.[rba] H.[rvata] S.[lovenaca] Po Opštinama. Prema popisu od 31 Januara, 1921 god. Confessions Dans Le Royaume S.H.S. Par Communes. D'après le récensement du 31 janvier, 1921. Catolici—Catholiques." [Scale,] 1 : 1,500,000, [500 × 460mm., in] *Prethodni Resultati Popisa Stanovništva U Kraljevini Srba, Hrvata I Slovenaca 31. Januara 1921 Godine. Résultats Préliminaires Du Récensement De La Population Dans Le Royaume des Serbes, Croates et Slovenes Du 31 Janvier, 1921.* Publié Par La Direction De La Statistique D'Etat [Belgrade]. (Sarajevo, 1924).

*Ibid.* "Pravoslavni—Orthodoxes" [same scale, etc.].

*Ibid.* "Muslimani, Evangelisti i Grko-Katolci—Musulmans, Evangélistes et Uniates" [same scale, etc.].

*Ibid.* "Confessions Dans Le Royaume S.H.S. Par Communes. Orthodoxes, Musulmans, Katholiques, Evangélistes, Uniates" [same scale, etc.].

*Ibid.* "Maternji Jezici U Kraljevini S.H.S. Po Opštinama—Langues Maternelles Dans Le Royaume S.H.S. Par Communes. D'Après le récensement du 31 janvier 1921. Arbanaški—Albanaise, Nemački—Allemande." [Scale,] 1 : 1,500,000, [500 × 460mm.].

*Ibid.* "Turski—Turque, Madžarski—Hongroise" [same scale, etc.].

*Ibid.* "Rumanski—Roumaine, Drugi Slovenski—Des Autres Slaves" [same scale, etc.].

*Ibid.* "Srpsko-Hrvatski i Slovenački—Serbo-Croate et Slovène" [same scale, etc.].

*Ibid.* "Langues Maternelles Dans le Royaume S.H.S. Par Communes, Serbo-Croate et Slovène, Turque, Allemande, Albanaise, Hongroise, Roumaine, Des Autres Slaves" [same scale, etc.]

"Kingdom of the Serbs, Croats and Slovenes. Religions by Communes. Census of January 31, 1921." Scale, 1 : 375,000, [200 × 190mm., in] *The Geographical Review*, vol. XV (New York, 1925).

*Ibid.* "Kingdom of the Serbs, Croats and Slovenes. Languages by Communes" [same scale, etc.].

"Races and Frontiers." Natural Scale, 1 : 5,500,000, [1100 × 820mm., in] W. and A. K. Johnston's *Effective Class Room Atlas of Europe and the Mediterranean* (Edinburgh, 1926).

"Die Völker Europas in sprachlicher Gliederung (Bevölkerungszahl 1925)." [Scale,] 700 km. [= 33mm., 270 × 180mm.]. Entworfen von Prof. Dr. Arthur Haberlandt. *Illustrierte Völkerkunde* herausgegeben von Dr. Georg Buschan, Zweiter Teil—"Europa und seine Randgebiete" von Dr. A. Byhan, Prof. Dr. A. Haberlandt, Prof. M. Haberlandt (Stuttgart, 1926).

"Ethnographical Map of Greek Macedonia showing the proportion of the different ethnographical elements in 1912 (before the Balkan Wars) and in 1926 (after the Settlement of Refugees)." [*n.s.*, 700 × 520mm., in] *Greek Refugee Settlement.* Translation, League of Nations (Geneva, 1926).

*Ibid.* "Map showing the settlement of rural and urban refugees in Greece (1923-1926)." Scale, 70 km. [= 93mm., 700 × 915mm.].

"Die volkische Zugehörigkeit der Dörfer im skopischen Feld zu seiten des Vardar in der letzten Zeit der turkischen Herrschaft." Massstab, 1 : 400,000 [260 × 240mm., in] *Makedonien, Landschaft- und Kulturbilder* von Dr. Leonhard Schultze-Jena (Jena, 1927).

"Karte der Völker Europas nach Sprache und Volksdichte." Massstab, 1 : 3,000,000, [1720 × 1340mm.], bearbeitet von A. Haberlandt. Freytag und Berndt (Wien, 1927).

"Kingdom of the Serbs, Croats and Slovenes. Languages." Scale, 1 : 3,750,000, [205×190mm., in] *The New World* by Isiah Bowman. 4th Edition (New York, etc., 1928).

*Ibid.* "Kingdom of the Serbs, Croats and Slovenes. Religions" [same scale, etc.].

*Ibid.* "Albanian Ethnography, Religions and Boundaries." Scale of miles, 20[=13mm., 65×145mm.].

*Ibid.* "The Vlachs, or pastoral nomads of Rumanian affiliation, in the central Balkan Lands. From the British General Staff ethnic map, 1918." [Scale,] miles, 50 [=12mm., 65×65mm.].

*Ibid.* "Greek territorial claims at the close of the World War. [With Greek] Ethnic distributions as of 1919." [*n.s.*, 115×75mm.].

*Ibid.* "Conflicting views on the ethnography of the Macedonian region by a Bulgarian, Ischirkoff, and a Serbian, Cvijić." Scale of Miles, 50 [=25mm., 120×100mm.].

"Linguistic regions of Europe together with religious boundary separating areas of the Greek Orthodox Church [ . . . ] from those of Roman Catholicism and Protestantism [ . . . ] and boundary separating [ . . . ] Greek and Cyrillic alphabets from areas [ . . . ] in which the Latin alphabet is used" [*n.s.*, 90×95mm., in] *The Geographical Basis of European History* by John Kirtland Wright (New York, 1928).

"Carte Linguistic de L'Europe." [*n.s.*, 385×300mm., in] *Les Langues dans L'Europe Nouvelle avec un Appendix de L. Tesnière sur la Statistique des Langues de L'Europe* par A. Meillet, 2e Edition (Paris, 1928).

"Péninsule Balkanique, Groupes Ethniques." Echelle de 1 : 12,000,000 [115×100mm.]. *Atlas Classique* par Joseph Halkin, Planche 28 (Namur, 1928).

"Jugoslavia" [by] E. A. Monfosco [in] *Le Minoranze Nazionali contemplate dagli Atti internazionali. II Paesi Danubiani, balcanici e Turchia*, 1 : 3,000,000, (Florence, 1929). [Reproduced in] E. Meynen, "Volks- und Sprachenkarten Mitteleuropas," *Deutsches Archiv für Landes- und Volksforschung*, II Jg., Heft 4 (Leipzig, 1938).

"Karte des Bulgarischen Sprachgebiets." Massstab, 60 km. [=30mm., 450×300mm., in] "Geschichte der Bulgarischen Sprache" von Stephan Mladinov (in the series, *Grundriss der Slavischen Philologie und Kulturgeschichte* herausgegeben von Reinhold Trautmann u. Max Vasmer (Berlin, 1929).

"[Ethnographical] Map of South-West Macedonia in 1923." [Scale,] 1 : 250,000, [280×310mm., accompanying] "The Measurement of Macedonian Man" [by] M. M. Hasluck etc., *Biometrika*, XXI (Cambridge, 1930).

"Distribution of Races in the Balkan Peninsula and Asia Minor in 1913." Scale, 1:10,000,000, [190×140mm., in] *Historical Atlas* by William R. Shepherd, 7th Edition (London, 1930).

"Europe, Ethnographical. Bonne's Projection." Scale, 1:15,000,000, [345×210mm., in] *The Oxford Advanced Atlas* [by] John Bartholomew, Fourth Edition—revised (Oxford, London, etc., 1931).

——— Seventh Edition (1942).

"Staaten und Völker in Europa, 1919." [77×82mm.]. Von Dr. Karl Haushofer [in] *Raumüberwindende Mächte* (Munich, 1934).

"Répartition des Parlers dans les Pays Balkaniques." Echelle de 1:10,000,000, [110×135mm.]. Par Y. Chataigneau, *Géographie Universelle*, VII, 2 (Paris, 1934).

Етнографска Карта на Българитѣ. Мѣрка, 1:1,000,000, [1000×850 mm.]. Презъ Последнитѣ Години преди освобждението. Съставилъ, В. Миковъ, Издава Найденъ, Р. Василевъ (София, 1936).

"Etnografska Karta na B'lgarite." Merka, 1:1,000,000, [1000×850mm.]. Prez Poslednite Godini Predi Osvobozhdenieto. S'stavil, V. Mikov, Izdava Nayden, R. Vasilev (Sofia, 1936).

"Macedonia, Boundary between Greek and Slav languages." Scale, 60 Miles [=24mm., 100×90mm., in] *Borderlands of Language in Europe* by Vaughan Cornish (London, 1936).

"Etnička Karte Kraljevine Jugoslavije. Carte Ethnographique du Royaume de Yougoslavie." Echelle, 1:2,500,000 (Beograd, *n.d.*).

"Linguistic Divisions and Political Frontiers of Eastern Europe." Scale, 1:10,000,000, [175×225mm., in] *Philip's Library Atlas.* Edited by George Philip and H. C. Darby. First Edition (London, 1937).

"Schematische Karte der Nationalitätenverteilung um 1910." Massstab, 1:10,000,000, [330×170mm., in] "Neugriechenland, Eine Landeskunde Ostmakedoniens und Westthraciens mit besonderer Berücksichtigung der Geomorphologie, Kolonistsiedlung und Wirtschaftsgeographie" von Joachim H. Schultze *Ergänzungsheft Nr. 223 zu Petermann's Mittheilungen* (Gotha, 1937).

"The Peoples of Europe based chiefly on language." Scale, 80 miles = 1 inch [720×1050mm.]. By Samuel B. Harding. *The Denoyer-Geppert Social Science Maps* (Chicago, 1938).

"Yugoslavia (nationalities)." [Scale,] 100 miles [=21mm., 100×95mm., in] *Atlas of Today and Tomorrow* [by] Alexander Radó (London, 1938).

"Geografski raspored Jugoslovena prema većinama u opštinama." 1:1,500,000, [by] G. Ružičić (Beograd, 1939).

"Languages [of Europe]." Scale, kilometers, 300[=15mm., 100×90mm., in] *Environment and Conflict in Europe, Eighteen Basic Maps*, American Geographical Society Special Publication No. 24 (New York, 1939).

"Jugoslawien: Volkische Gliederung Kerngebiete." Massstab, 1:1,500,000, [520×480mm.]. Entworfen von Dr. Manfred Straka, Südostdeutsches Institut, Graz. General Stab des Heeres (1940).

"Jugoslawien: Volkische Gliederung Mehrheitsgebiete unter besonderen Berücksichtung des Deutschtums." Massstab, 1:1,500,000, [520×480mm.]. Entworfen von Dr. Manfred Straka, etc. (1940).

"Die Völker des Donauraumes und der Balkanhalbinsel." Massstab, 1:3,000,000, [460×570mm.]. General Stab des Heeres. Abteilung für Kriegskarten u. Vermessungswesen (Berlin, 1940).

"Volkstums- und Sprachenkarte von Südosteuropa." Massstab, 1:1,500,000, [820×960mm.]. Bearbeiter: H. P. Kosack, H. Lembke, K. Masuch, H. Waldbauer, H. Wintz. Herausgegeben von dem Geographischen Dienst des Auswärtigen Amtes (Berlin, 1941).

"Volkstumskarte von Jugoslawien." 40 Blätter im Massstab 1:200,000, [each sheet 395×565mm.]. Unter Mitarbeit von E. Brunner, G. Ebel, G. Krallert, V. Paulsen und F. Volkmann bearbeitet von Wilfried Krallert auf grund der amtlichen Zählungsergebnisse über die Muttersprache von 1931, mit Trennung der Sudslawischen Völker mit Hilfe der Konfessionsstatistik und örtlicher Angaben (Wien, 1941).

"The Languages of Greece and neighbouring Countries." Scale, 1 inch = 90 miles. *A Gazetteer of Greece*, Permanent Committee for Geographical Names (London, 1942).

"Jugoslavia: Racial Groups." [*n.s.*, 410×350mm.]. G.[eneral] S.[taff] I.[ntelligence] Topographical. *Jugoslavia* Part III, (1942?).

"Albania. Distribuzione della popolazione." [Scale,] 50 kms. [=23mm., 140×220mm., in] *L'Albania* [by] B. Castiglioni, F. Milone, A. Sestini. Reale Societa Geografica Italiana (Bologna, 1943).

"The languages of Greece and adjoining areas." [Scale,] 200 km. [=26mm., 100×125mm., in] *Greece*, vol. I. Geographical Handbook Series. Naval Intelligence Division of the Admiralty (Cambridge, 1944).

"A Bulgarian view of the ethnography of Macedonia and surrounding areas, 1912." Based on A. Ischirkoff. [Scale,] 150 km. [=30mm., 170×85mm., in] *Jugoslavia*, vol. II. Geographical Handbook Series. Naval Intelligence Division (Cambridge, 1944).

*Ibid.* "A Serbian view of the ethnography of Macedonia and surrounding areas 1913." Based on J. Cvijić. [Same scale, 170×105mm.].

*Ibid.* " A British view of the ethnography of Macedonia and surrounding areas, 1918." Based on G.S.G.S. [map] No. 3703A. [Same scale, 170×90mm.].

*Ibid.* " Conflicting interpretations of ethnic distributions in Macedonia, 1912-20." Based on A. Ischirkoff, J. Cvijić and the G.S.G.S. map. [Scale,] 100 km. [=23mm., 100×90mm.].

*Ibid.* " The distribution of Slovene and Serbo-Croat dialects." Based on A. Leskien, A. Meillet and A. Vaillant. [Scale,] 100 miles [=20mm., 90×105mm.].

*Ibid.* [Various maps showing the distributions of religions and minorities according to the Jugoslav census of 1921].

" South-Western Balkans: The Distribution of Nationalities (1920)." Scale, 50 miles [=18 mm., 105×75mm., in] *The New Europe* [by] Walter Fitzgerald (London, 1945).

" Europe, Linguistic Boundaries." Scale, 500 English miles [=29mm., 145×90mm., in] *The Nationalities of Europe and the Growth of National Ideologies* by H. Munro Chadwick (Cambridge, 1945).

" Languages of Yugoslavia " [*n.s.*, 105×85mm., in] *Language as a Social and Political Factor in Europe* [by] Stanley Rundle (London, 1946).

*Ibid.* " Languages of Albania, Greece, Bulgaria and Turkey-in-Europe." [*n.s.*, 105×110mm.].

*Ibid.* " The languages of Europe." [*n.s.*, 120×100mm.].

" Macédoine ' Historique ' : Superficie et Population." [Scale,] 1:2,100,000, [180×135mm., in] *Le Camouflage Macédonien à la lumière des faits et des chiffres* [by] Ch. Christidès (Athènes, 1949).

*Ibid.* " Macédoine ' Geographique ' (point de vue Slave) : Superficie et Population." [Same scale and size.].

*Ibid.* " Principales Minorités en Macédoine Géographiques (1940) par arrondissements." [*n.s.*, 135×110mm.].

" Makedonija [showing] narodnostna meja Makedoncev na jugu in vzhodu." [Scale,] 100 km. [=22mm., 105×80mm., in] *Jugoslavia* [by] A. Melik (Ljubljana, 1949).

" Peoples of Europe." Scale 83 miles to the inch, [120×85mm.]. Edited by William M. Austin. *Webster-Knowlton-Hazen European History Maps* (Chicago, 1950).

# INDEX

www.ingramcontent.com/pod-product-compliance
Lightning Source LLC
LaVergne TN
LVHW101439200926
848483LV00004B/404